Recent Titles in This Series

144 **Bor-Luh Lin and William B. Johnson, Editors,** Banach spaces, 1993

143 **Marvin Knopp and Mark Sheingorn, Editors,** A tribute to Emil Grosswald: Number theory and related analysis, 1993

142 **Chung-Chun Yang and Sheng Gong, Editors,** Several complex variables in China, 1993

141 **A. Y. Cheer and C. P. van Dam, Editors,** Fluid dynamics in biology, 1993

140 **Eric L. Grinberg, Editor,** Geometric analysis, 1992

139 **Vinay Deodhar, Editor,** Kazhdan-Lusztig theory and related topics, 1992

138 **Donald St. P. Richards, Editor,** Hypergeometric functions on domains of positivity, Jack polynomials, and applications, 1992

137 **Alexander Nagel and Edgar Lee Stout, Editors,** The Madison symposium on complex analysis, 1992

136 **Ron Donagi, Editor,** Curves, Jacobians, and Abelian varieties, 1992

135 **Peter Walters, Editor,** Symbolic dynamics and its applications, 1992

134 **Murray Gerstenhaber and Jim Stasheff, Editors,** Deformation theory and quantum groups with applications to mathematical physics, 1992

133 **Alan Adolphson, Steven Sperber, and Marvin Tretkoff, Editors,** p-Adic methods in number theory and algebraic geometry, 1992

132 **Mark Gotay, Jerrold Marsden, and Vincent Moncrief, Editors,** Mathematical aspects of classical field theory, 1992

131 **L. A. Bokut′, Yu. L. Ershov, and A. I. Kostrikin, Editors,** Proceedings of the International Conference on Algebra Dedicated to the Memory of A. I. Mal′cev, Parts 1, 2, and 3, 1992

130 **L. Fuchs, K. R. Goodearl, J. T. Stafford, and C. Vinsonhaler, Editors,** Abelian groups and noncommutative rings, 1992

129 **John R. Graef and Jack K. Hale, Editors,** Oscillation and dynamics in delay equations, 1992

128 **Ridgley Lange and Shengwang Wang,** New approaches in spectral decomposition, 1992

127 **Vladimir Oliker and Andrejs Treibergs, Editors,** Geometry and nonlinear partial differential equations, 1992

126 **R. Keith Dennis, Claudio Pedrini, and Michael R. Stein, Editors,** Algebraic K-theory, commutative algebra, and algebraic geometry, 1992

125 **F. Thomas Bruss, Thomas S. Ferguson, and Stephen M. Samuels, Editors,** Strategies for sequential search and selection in real time, 1992

124 **Darrell Haile and James Osterburg, Editors,** Azumaya algebras, actions, and modules, 1992

123 **Steven L. Kleiman and Anders Thorup, Editors,** Enumerative algebraic geometry, 1991

122 **D. H. Sattinger, C. A. Tracy, and S. Venakides, Editors,** Inverse scattering and applications, 1991

121 **Alex J. Feingold, Igor B. Frenkel, and John F. X. Ries,** Spinor construction of vertex operator algebras, triality, and $E_8^{(1)}$, 1991

120 **Robert S. Doran, Editor,** Selfadjoint and nonselfadjoint operator algebras and operator theory, 1991

119 **Robert A. Melter, Azriel Rosenfeld, and Prabir Bhattacharya, Editors,** Vision geometry, 1991

118 **Yan Shi-Jian, Wang Jiagang, and Yang Chung-chun, Editors,** Probability theory and its applications in China, 1991

117 **Morton Brown, Editor,** Continuum theory and dynamical systems, 1991

116 **Brian Harbourne and Robert Speiser, Editors,** Algebraic geometry: Sundance 1988, 1991

115 **Nancy Flournoy and Robert K. Tsutakawa, Editors,** Statistical multiple integration, 1991

(*Continued in the back of this publication*)

CONTEMPORARY MATHEMATICS

144

Banach Spaces

Proceedings of an International Workshop
on Banach Space Theory, held January 6–17, 1992
with financial support from the National Science Foundation,
Consejo Nacional de Investigaciones Cientificas
y Tecnologicas of Venezuela, International Mathematical Union,
Institute of Mathematics and its Applications, University of Iowa,
Universidad de Los Andes, Fundacion Polar and Fundacite-Merida

Bor-Luh Lin
William B. Johnson
Editors

American Mathematical Society
Providence, Rhode Island

The International Workshop on Banach Space Theory was held at the Universidad de Los Andes in Merida, Venezuela, on January 6–17, 1992 with financial support from the National Science Foundation, Grant INT-9100331, Consejo Nacional de Investigaciones Cientificas y Tecnologicas of Venezuela, the International Mathematical Union, the Institute of Mathematics and its Applications, the University of Iowa, the Universidad de Los Andes, Fundacion Polar and Fundicate-Merida.

1991 *Mathematics Subject Classification.* Primary 46-06, 46B03, 46B04, 46B07, 46B09, 46B10, 46B15, 46B20, 46B22, 46B26, 46E15, 46E25, 46E30, 46E50, 46G15, 46G20.

Library of Congress Cataloging-in-Publication Data

International Workshop on Banach Space Theory (1992: Universidad de Los Andes)
Banach spaces: proceedings of an International Workshop on Banach Space Theory, January 6–17, 1992/Bor-Luh Lin, William B. Johnson, editors.
p. cm.—(Contemporary mathematics; v. 144)
Held at the Universidad de Los Andes, Merida, Venezuela.
ISBN 0-8218-5157-8
1. Banach spaces—Congresses. I. Lin, Bor-Luh. II. Johnson, W. B. (William B.), 1944– . III. Title. IV. Series: Contemporary mathematics (American Mathematical Society); v. 144.
QA322.2.I58 1992 92-46467
515′.732—dc20 CIP

This volume was printed directly from copy typeset by the authors using AMS-TeX, the American Mathematical Society's TeX macro system.

10 9 8 7 6 5 4 3 2 1 98 97 96 95 94 93

Contents

Preface

This volume contains the proceedings of an International Workshop on Banach Space Theory held at the Universidad de Los Andes in Merida, Venezuela, January 6–17, 1992. The articles contain original research and high quality survey papers, some of which were presented at the workshop. All papers in this volume are in final form and no version of them will be submitted for publication elsewhere.

The first international research conference at Merida workshop received financial support from the National Science Foundation, Consejo Nacional de Investigaciones Cientificas y Tecnologicas of Venezuela, the International Mathematical Union, the Institute of Mathematics and its Applications, the University of Iowa, the Universidad de Los Andes, Fundacion Polar and Fundacite–Merida. The members of the organizing committee were Bor-Luh Lin, Chair; W. B. Johnson, J. Lindenstrauss, J. Palis Jr., A. Pełczyński, H. P. Rosenthal and J. Vielma. Lin did essentially all of the hard preparatory work (after appropriate consultation with the rest of the organizing committee) as well as most of the editorial work associated with this volume, while Vielma was responsible for the local arrangements.

Mornings were reserved for the informal discussions among participants. A number of participants (P. Aiena, M. Aquirre, R. Aron, G. Bennett, O. Blasco, G. Bombal, P. Enflo, J. Jarchow, J. Lindenstrauss, P. Papini, J. Sanchez and A. Tonge) were kind enough to give introductory expositions on Banach space theory for the many non-specialists and graduate students in attendance. In the afternoons, there were lectures for experts on recent progress in Banach space theory and related fields.

I was struck by the breadth of mathematics presented and how little of it I had seen before. The highlight of the workshop was the announcement by T. Schlumprecht of the solution by E. Odell and himself of the distortion problem for Hilbert space. In addition to his formal address, Schlumprecht gave several hours of informal talks on this research.

W. B. Johnson

Taller Internacional en Espacios de Banach
Workshop on Banach Space Theory

Monday, Jan. 6

6:00 p.m.	Welcome to the Workshop

Tuesday, Jan. 7

10:00 a.m.	Registration, Science Center

Tuesday, Jan. 7
Session 1: Chair (moderador): J. Lindenstrauss

2:00 – 3:00 p.m.	G. Bennett Factorizing the Classical Inequalities
3:15 – 4:15 p.m.	O. Blasco Multipliers on Spaces of Analytic Functions
4:30 – 5:00 p.m.	M. Smith Sufficient Geometrical Conditions for Normal Structure in Banach Spaces
5:15 – 5:45 p.m.	K. Jarosz Isometries of Bloch Spaces

Wednesday Jan. 8
Session 2: Chair (moderador): W.B. Johnson

2:00 – 3:00 p.m.	P. Enflo A Nonlinear Problem of Operators

3:15 – 4:15 p.m.	P. Dowling	Complex Geometry of Vector–Valued H_ϕ Spaces
4:30 – 5:00 p.m.	H. Jarchow	Remarks on Self–Induced Compactness in Banach Spaces
5:15 – 5:45 p.m.	Z. Hu	When is RNP and CPCP Equivalent in Lebesgue–Bochner Function Spaces?

Thursday, Jan. 9
Session 3: Chair (moderador): P. Enflo

2:00 – 3:00 p.m.	B. Kashin	Polynomials with Integer Coefficients Least Deviating From Zero, A Survey
3:15 – 4:15 p.m.	Y. Gordon	Local Theory of Quasi–Normed Spaces
4:30 – 5:00 p.m.	D. Alspach	ℓ_1 Preduals
5:15 – 5:45 p.m.	W. Zhang	Banach Space–Valued Lorentz Sequence Spaces and Strong WCG Spaces

Friday, Jan. 10
Session 4: Chair (moderador): H. Jarchow

2:00 – 3:00 p.m.	N. Tomczak–Jaegermann	Defining Weak Hilbert Spaces by Basis Constants of Large Subspaces

3:15 – 4:15 p.m. R. Komorowski
Weak Hilbert Spaces with No Unconditional Basis

4:30 – 5:00 p.m. G. Schlutermann
Parameterized Operators and Semigroups

5:15 – 5:45 p.m. F.L. Hernández
Representation of Uncountable Symmetric Basic Sets

Monday, Jan. 13
Session 5: Chair (moderador): A. Pelczynski

2:00 – 3:00 p.m. R. Aron
Boundaries for Analytic Functions in Infinite Dimensions

3:15 – 4:15 p.m. T. Schlumprecht
A Solution of the Distortion Problem in Hilbert Spaces

4:30 – 5:00 p.m. W.B. Johnson
Analytic Disc in Fibers Over the Unit Ball of a Banach Space

5:15 – 5:45 p.m. R. Alencar
An Application of Singer's Theorem to Homogeneous Polynomials

Tuesday, Jan. 14
Session 6: Chair (moderador): R. Aron

2:00 – 3:00 p.m. V. Milman
Remarks on Geometry of Infinite Dimensional Convex Bodies

3:15 – 4:15 p.m.	P. Terenzi On Basis Problem
4:30 – 5:00 p.m.	A. Tonge Generalized Roundness
5:15 – 5:45 p.m.	J. Farmer On the Primarity of Tensor Products of ℓ_p Spaces

Wednesday, Jan. 15
Session 7: Chair (moderador): N. Tomczak–Jaegermann

2:00 – 3:00 p.m.	K. Wozniakowski Polynomial Bases in Function Spaces
3:15 – 4:00 p.m.	A. Pełczyński* and M. Mojciechowski Stratification Decomposition of the Sobolev Space $L^1_{(1)}(R^d)$
4:15 – 5:00 p.m.	M. Wojciechowski Ornstein Type "Non–Inequalities" Between L^1–Norms of Partial Derivatives

*Banquet time to be announced.

Thursday, Jan. 16
Session 8: Chair (moderador): G. Bennett

2:00 – 3:00 p.m.	K. Grobel Retraction Problems on Banach Spaces
3:15 – 4:15 p.m.	P.G. Casazza Approximation Properties for Non–Separable Banach Spaces

4:30 – 5:00 p.m. W.A. Kirk
Property (β) and Edelstein's Algorithms for Constructing Nearest and Farthest Points

5:15 – 5:45 p.m. R. Asthagiri
Weak Continuity of Non–Linear Maps Between Banach Spaces

Friday, Jan. 17
Session 9: Chair (moderador): K. Gobel

9:30 – 10:00 a.m. L. Rodriguez–Piazza
Vector Measures Determined by their Ranges

10:15 – 10:45 a.m. Z. Hu
Strongly Exterme Points and RNP

11:00 – 12:00 p.m. S. Szarek
Norms of Functions of Random Matrices

Participants

<u>U.S. Participants</u>

Dale E. Alspach, Oklahoma State University, Stillwater, Oklahoma

Richard M. Aron, Kent State University, Kent, Ohio

Rajappa K. Asthagiri, Miami University, Middletown, Ohio

Grahame Bennett, Indiana University, Bloomington, Indiana

Peter G. Casazza, University of Missouri, Columbia, Missouri

Patrick N. Dowling, Miami University, Oxford, Ohio

Per Enflo, Kent State University, Kent, Ohio

Jefferey D. Farmer, Texas A&M University, College Station, Texas

Maria K. Girardi, University of South Carolina, Columbia, South Carolina

Zhibao Hu, Miami University, Oxford, Ohio

Krzysztof Jarosz, University of Southern Illinois, Edwardsville, Illinois

William B. Johnson, Texas A&M University, College Station, Texas

William A. Kirk, University of Iowa, Iowa City, Iowa

Helmut Knaust, University of Texas, El Paso, Texas

Bor–Luh Lin, University of Iowa, Iowa City, Iowa

Vitali D. Milman, The Ohio State University, Columbus, Ohio/Tel Aviv University, Tel Aviv, Israel

Thomas Schlumprecht, Louisiana State University, Baton Rouge, Louisiana

Mark A. Smith, Miami University, Oxford, Ohio

Stanislaw J. Szarek, Case Western Reserve University, Cleveland, Ohio

Andrew M. Tonge, Kent State University, Kent, Ohio

Wenyao Zhang, Northern Illinois University, DeKalb, Illinois

Participants from Europe, Canada, and Israel

Pietro Aiena, University of Palermo, Palermo, Italy

Yoav Benyamini, Technion–Israel Institute of Technology, Haifa, Israel

Oscar Blasco, University of Zaragoza, Zaragoza, Spain

Gordon F. Bombal, University of Complutense, Madrid, Spain

Benedetto Bongiorno, University of Palermo, Palermo, Italy

Catherine Finet, University of Mons, Mons, Belgium

Kazimierz Goebel, Marie–Curie Sklodowska University, Lubin, Poland

Yehoram Gorden, Technion–Israel Institute of Technology, Haifa, Israel

Francisco L. Hernández, University of Complutense, Madrid, Spain

Hans Jarchow, University of Zurich, Zurich, Switzerland

Boris Kashin, Moscow State University, Moscow, Russia

Ryszard Komorowski, University of Alberta, Edmonton, Canada

Joram Lindenstrauss, Hebrew University at Jerusalem, Jerusalem, Israel

Pier L. Papini, University of Bologna, Bologna, Italy

Aleksander Pełczyński, Polish Academy of Sciences, Warsaw, Poland

Luis Rodriguez–Piazza, University of Sevilla, Sevilla, Spain

Gideon Schechtman, The Weizmann Institute of Science, Rehovot, Israel

Georg Schluchtermann, University of Munich, Munich, Germany

Paolo Terenzi, Polytech Institute of Milan, Milan, Italy

Nicole Tomczak–Jaegermann, University of Alberta, Edmonton, Canada

Michal Wojciechowski, Polish Academy of Sciences, Warsaw, Poland

Krzysztof Wozniakowski, Polish Academy of Sciences, Warsaw, Poland

Clemente Zanco, University of Milan, Milan, Italy

Participants from Latin America

Manuel Aguirre, University of Tandil, Tandil, Argentina

Raymundo Alencar, University of Campinas, Campinas, Brazil

Jose Saldana, National University of San Marcos, Lima, Peru

Jorge Samur, National University of LaPlata, LaPlata, Argentina

Jose A. Sanchez, University of Concepcion, Concepcion, Chile

Participants from Venezuela other than Universidad de Los Andes

Alexis S. Araujo, Universidad del Zulia, Maracaibo

Maria M. Bolivar, Universidad Nacional Alierta, Maracay

Ramon J. Bruzual, Universidad Central de Venezuela, Caracas

Mario Calvani, Universidad de Oriente, Cumanó

Marisela Dominguez, Universidad Central de Venezuela, Caracas

Ventura Echandia, Universidad Central de Venezuela, Caracas

Carlos Finol, Universidad Central de Venezuela, Caracas

Jose J. Finol, Universidad del Zulia, Maracaibo

Eliecer Guevara, Universidad del Tachira, San Cristobal

Jose Z. Lopez, Universsidad Central de Venezuela, Caracas

Justo R. Marquez, Universidad del Zulia, Maracaibo

Ramon Mirabal, Universidad del Tachira, San Cristobal

Maria D. Moran, Universidad Central de Venezuela, Caracas

Wilson R. Pacheco, Universidad del Zulia, Maracaibo

Alexi Quevedo, Universidad Central de Venezuela, Caracas

Sael N. Romero, Universidad de Oriente, Cumaná

Oscar G. Rondon, Universidad de Guayana, Ciudad Bolivar

Ennis R. Rosas, Universidad de Oriente, Cumaná

Wilfredo Urbina, Universidad Central de Venezuela, Caracas

Participants from Universidad de Los Andes

Carlos Alvarez

Bartolo Barreto

Wilman Brito

Jose L. Chacon

Jesus J. Cols

Carlos E. Cova

Hernando Gaitan

Jose P. Gimenez

Luis Gonzalez

Gilberto Gonzalez

Maria T. Gonzalez

Edgar Iturriaga

Luis R. Leon

Glauco Lopez

Elio Marquez

Osmin A. Monsalve

Luis L. Montilla

Jose Morales

Victor Padron

Thiruvaiyaru Panchapagesan

Jesus A. Perez

Jesus Rivero

Luisa S. de Rosales

Cesar A. Sanchez

Mauro J. Silva

Gaetano Tepedino

Antonio Tineo

Carlos Uzcategui

Contemporary Mathematics
Volume **144**, 1993

AN APPLICATION OF SINGER'S THEOREM TO HOMOGENEOUS POLYNOMIALS

RAYMUNDO ALENCAR

ABSTRACT. With the help of Singer's theorem we characterize the dual of $\mathcal{P}_c({}^m E; F)$, the compact type m–homogeneous polynomials. Using known results on nuclear and integral polynomials we establish that the reflexivity of $\mathcal{P}({}^m E; F)$ is equivalent to the coincidence of $\mathcal{P}({}^m E; F)$ with its subspace $\mathcal{P}_c({}^m E; F)$, when E and F are reflexive Banach spaces with the approximation property.

0. INTRODUCTION

In [1] we established a similar condition for the reflexivity of $\mathcal{P}({}^m E)$, the space of m–homogeneous polynomials scalar valued, using a characterization of the dual $\mathcal{P}_c^*({}^m E)$ due to S. Dineen [7].

For vector valued polynomials, we can follow the model of the scalar case by introducing the Bartle integral, where for the same vector measure we consider two kind of integrands, scalar valued functions and vector valued functions. This allows us to apply the Riesz representation theorem and the Singer theorem to characterize the dual of $\mathcal{P}_c({}^m E, F)$, thereby establishing a preliminary result. The main result follows from the fact that the spaces of nuclear and integral polynomials coincide when the dual space E^* has the Radon Nikodym property.

In the last paragraph we give some examples illustrating our results.

1. NOTATION AND DEFINITIONS

E and F will represent Banach spaces, E^* and F^* their duals, and B_E the closed unit ball of E.

$L({}^m E, F)$ the space of continuous m–linear mappings from $E \times \cdots \times E$ (m-fold) to F. $\mathcal{P}({}^m E, F)$ the space of continuous m-homogeneous polynomials from E to F with the sup norm on the unit ball B_E. $\mathcal{P}_f({}^m E, F)$ is the subspace of $\mathcal{P}({}^m E, F)$ of finite type polynomials, i.e., if $P \in \mathcal{P}_f({}^m E, F)$ there are finite sequences $\varphi_j \in E^*$, $y_j \in F$, $j = 1, \dots, r$ such that

$$P(x) = \sum_{j=1}^{r} \varphi_j^m(x) y_j \quad \text{for all} \quad x \in E .$$

$\mathcal{P}_c({}^m E, F)$ denotes the closure of $\mathcal{P}_f({}^m E, F)$ in $\mathcal{P}({}^m E, F)$, called the compact type polynomials.

1991 *Mathematics Subject Classification.* 46G20 (46B10).

Partially supported by CNPq - 301122/83-0 and FAPESP - 91/4486-4

$\mathcal{P}_N({}^mE,F)$ is the space of all $P \in \mathcal{P}({}^mE,F)$, that can be represented by

$$P(x) = \sum_{j=1}^{\infty} \varphi_j^m(x) y_j \quad \text{where} \quad \varphi_j \in E^* \ , \ y_j \in F$$

with $\sum_{j=1}^{\infty} ||\varphi_j||^m ||y_j|| < \infty$. $\mathcal{P}_N({}^mE,F)$ endowed with the

norm $||P||_N = \inf\left\{\sum_{j=1}^{\infty} ||\varphi_j||^m ||y_j|| : \text{over all representations}\right\}$,

is called the space of nuclear polynomials.

$\mathcal{P}_I({}^mE,F)$ is the space of all $P \in \mathcal{P}({}^mE,F)$ that can be represented by

$$P(x) = \int_{B_{E^*}} \varphi^m(x) d\mu(\varphi)$$

where μ is an F–valued regular σ–additive Borel measure of bounded variation, on the Borel sets of $(B_{E^*}, \sigma(E^*,E))$. $\mathcal{P}_I({}^mE,F)$ endowed with the norm,

$$||P||_I = \inf\{|\mu|(B_{E^*}) : \ \text{over all representations } \mu \text{ of } P\} \ ,$$

is called the space of integral polynomials.

As usual, if K is a compact Hausdorff space, $C(K)$ and $C(K,F)$ denote the Banach space of continuous scalar–valued and F–valued functions respectively, endowed with the sup norm.

2. Integration with respect to a vector measure

We shall limit ourselves to some basic facts on the Bartle integral in two special situations that fulfill our needs.

Let Ω be a set and $\sum$ a σ–algebra of subsets of Ω and F a Banach space.

In the following we give the notion of Bartle integral $\int f \, d\mu$ in two special cases (see for example [5]):

- f is a scalar function and μ is an F–valued measure.
- f is a F–valued function and μ is an F^*–valued measure.

Let μ be a vector measure from $\sum$ into F (or F^*) of bounded semivariation. Let f be a scalar (or F–valued) simple function defined on Ω represented by

$$f = \sum_{i=1}^{n} \alpha_i \chi_{A_i}$$

where α_i are nonzero scalars (or F–elements) and $A_1, \dots, A_n$ are pairwise disjoint members of $\sum$, then we define:

$$\int_\Omega f \, d\mu = \sum_{i=1}^{n} \alpha_i \mu(A_i) \quad \left(\text{or} \quad \int_\Omega \langle f, d\mu\rangle = \sum_{i=1}^{n} \langle \alpha_i, \mu(A_i)\rangle\right)$$

The above integral induces a continuous linear operator T_μ on the space of simple functions normed by the sup norm, given by

$$T_\mu(f) = \int_\Omega f\, d\mu \in F \quad \left(\text{or} \quad \int_\Omega \langle f, d\mu\rangle \in I\!K\right)$$

If $\mathcal{B}(\Omega)$ is the Borel σ–algebra, of subsets of a compact Hausdorff space Ω, then every continuous scalar– valued (or F–valued) function is an uniform limit of a sequence of simple functions. Therefore we can extend the above definitions for f a continuous, scalar valued (or F–valued) function, and we have also a continuous operator

$$T_\mu(f) = \int_\Omega f\, d\mu, \quad f \in C(\Omega) \quad \left(\text{or} \quad \int_\Omega \langle f, d\mu\rangle, \quad f \in C(\Omega, F)\right)$$

Some basic properties are the following.

1 – $||\int_\Omega f\, d\mu|| \leq ||f||\,||\mu||(\Omega)$
2 – μ, F^*–valued, $f \in C(\Omega)$

$$\text{for every} \quad x^{**} \in F^{**}, \quad x^{**}\left(\int_\Omega f\, d\mu\right) = \int_\Omega f d(x^{**}\mu)$$

3 – μ, F^*–valued, $f(x) = \sum_{i=1}^n g_i(x) y_i$, $g_i \in C(\Omega)$, $y_i \in F$

$$\int_\Omega \langle f, d\mu\rangle = \sum_{i=1}^n \int_\Omega \langle g_i(x) y_i, d\mu\rangle$$

Now we recall two representation theorems.

The Singer Theorem. *"The dual $C^*(K, F)$ is isomorphic and isometric to the space of all F^*–valued regular, σ–additive Borel measures μ of bounded variation defined on the Borel sets of K, endowed with the variation norm."*

The isomorphism is given by $T \in C^*(K, F) \longleftrightarrow \mu$, where $T(f) = \int_K \langle f(x), d\mu\rangle$.

The Riesz Representation Theorem. *"Let T be a bounded linear operator from $C(K)$ into F. Then there is a weak* σ–additive measure μ defined on the Borel sets of K with values in F^{**} such that:*

(i) *$\mu(\cdot)x^*$ is a regular σ–additive Borel measure for each $x^* \in F^*$.*
(ii) *the mapping $x^* \in F^* \longrightarrow \mu(\cdot)x^*$, from F^* into $C^*(K)$ is weak*– to–weak* continuous.*
(iii) *$x^*T(f) = \int_K f d(x^*\mu)$, for each $f \in C(K)$ and each $x^* \in F^*$.*
(iv) *$||T|| = ||\mu||(K)$*

Conversely, if μ is an F^{**}–valued measure defined on the Borel sets of K for which (i) and (ii) hold, then (iii) defines a bounded linear operator from $C(K)$ to F which satisfies (iv)."

Definition. Let T be a bounded linear operator from $C(K)$ into F, then the measure μ satisfying (i) to (iv) will be called the representing measure of T.

The next theorem strengthens the utility of the Riesz Representation Theorem if we know more about T or its representing measure.

The Bartle–Dunford–Schwartz Theorem. *"Let $T : C(K) \longrightarrow F$ be a bounded linear operator with representing measure μ. Any one of the following statements implies all the others.*

(a) *the operator T is weakly compact.*
(b) *the measure μ takes all its values in F.*
(c) *the measure μ is σ–additive."*

For the above results and definitions on vector measures, the Riesz Representation theorem and the Bartle–Dunford– Schwartz theorem we refer to the book of Diestel and Uhl ([6]). For the Singer Theorem we refer to the Singer's paper ([10]) and also to the book of Jean Schmets ([9]).

3. Preliminary Result (Application of Singer's Theorem)

We start with the following remark: if μ is an F–valued regular σ–additive Borel measure of bounded variation, defined on the Borel sets of a compact Hausdorff space K, then it defines a bounded linear operator T from $C(K)$ into F given by

$$T(f) = \int_K f d\mu$$

and μ is a representing measure of T. This is straightforward if we look at the Riesz Representation theorem and the Bartle–Dunford–Schwartz theorem.

Lemma. *Let E and F be reflexive Banach spaces. Then $\mathcal{P}_I({}^mE^*, F^*)$ and $\mathcal{P}_c^*({}^mE, F)$ are isometrically isomorphic.*

Proof. First of all we notice that if $P \in \mathcal{P}_f({}^mE, F)$, then P is $\sigma(E, E^*)$–continuous. As a consequence we have that the inclusion $\mathcal{P}_c({}^mE, F) \hookrightarrow C(B_E, F)$ is continuous, where B_E is endowed with the weak–topology. Since the norm of $\mathcal{P}_f({}^mE, F)$ coincides with the norm induced by $C(B_E, F)$, it follows that the closure of $\mathcal{P}_f({}^mE, F)$, the compact type polynomials $\mathcal{P}_c({}^mE, F)$ is a closed subspace of $C(B_E, F)$. Therefore, if $T \in \mathcal{P}_c^*({}^mE, F)$, T can be extended via Hahn–Banach to some $\widetilde{T}$ on $C(B_E, F)$, preserving the norm.

Now, by Singer's theorem we can find an F^*–valued measure μ defined on the Borel sets of B_E such that

$$\widetilde{T}(f) = \int_{B_E} \langle f, d\mu \rangle \quad \text{for all} \quad f \in C(B_E, F) \quad \text{and} \quad \|\widetilde{T}\| = |\mu|(B_E).$$

Now, by the previous remark, μ is the representing measure of the operator $\tau : C(B_E) \longrightarrow F^*$, given by

$$\tau(g) = \int_{B_E} g(x) d\mu(x), \quad \text{with} \quad \|\tau\| = \|\mu\|(B_E).$$

If $\varphi \in E^*$, then $\varphi^m(x) = (\varphi(x))^m$ defines a continuous function on B_E. Therefore

$$Q_T(\varphi) = \int_{B_E} \varphi^m(x) d\mu(x)$$

defines a m–homogeneous polynomial and it is clear that $Q_T \in \mathcal{P}_I({}^mE^*, F^*)$. Furthermore, we have,

$$||Q_T||_I \leq |\mu|(B_E) = ||\widetilde{T}|| = ||T||.$$

Conversely, let Q be in $\mathcal{P}_I({}^mE^*, F^*)$, so that Q can be represented by

$$Q(\varphi) = \int_{B_E} \varphi^m(x) d\mu(x).$$

Since $Q \in \mathcal{P}({}^mE^*, F^*)$ there is an unique symmetric $B \in L({}^mE^*, F^*)$ such that $\widehat{B} = Q$.

Now, using the isometric isomorphism between $L({}^mE^*, F^*)$ and $L({}^mE^* \times F)$ (the $m+1$–linear scalar valued mappings), given $B \in L({}^mE^*, F^*)$ we can find $\widetilde{B} \in L({}^mE^* \times F)$ such that

$$\widetilde{B}(\varphi_1, \dots, \varphi_m, y) = \langle y,\ B(\varphi_1, \dots, \varphi_m)\rangle.$$

By the universal property of the tensor product there is an unique linear form S on $E^* \otimes \cdots E^* \otimes F$ such that

$$\widetilde{B}(\varphi_1, \dots, \varphi_n, y) = S(\varphi_1 \otimes \cdots \otimes \varphi_n \otimes y).$$

Now, let $P \in \mathcal{P}_f({}^mE, F)$ have the representation

$$P(x) = \sum_{j=1}^{r} \varphi_j^m(x) y_j, \quad \varphi_j \in E^* \quad \text{and} \quad y_j \in F, \quad j = 1, \dots, r, \ .$$

There is an unique symmetric $A \in L_f({}^mE, F)$ such that $\widehat{A} = P$. It is clear that A can be represented by

$$A(x_1, \dots, x_m) = \sum_{j=1}^{r} \varphi_j(x_1) \cdots \varphi_j(x_m) y_j.$$

On the other hand A can be viewed as an element of $E^* \otimes \cdots \otimes E^* \otimes F$, represented by

$$A = \sum_{j=1}^{r} \varphi_j \otimes \cdots \otimes \varphi_j \otimes y_j.$$

We define a linear functional T on $\mathcal{P}_f({}^mE,F)$ by

$$\begin{aligned}T(P) = S(A) &= \sum_{j=1}^{r} S(\varphi_j \otimes \cdots \otimes \varphi_j \otimes y_j) = \sum_{j=1}^{r} \widetilde{B}(\varphi_j,\dots,\varphi_j,y_j)\\ &= \sum_{j=1}^{r} \langle y_j, B(\varphi_j,\dots,\varphi_j)\rangle = \sum_{j=1}^{r} \langle y_j, Q(\varphi_j)\rangle\\ &= \sum_{j=1}^{r} \langle y_j, \int_{B_E} \varphi_j^m(x)d\mu(x)\rangle = \int_{B_E} \langle \sum_{j=1}^{r} \varphi_j^m(x)y_j, d\mu(x)\rangle\\ &= \int_{B_E} \langle P(x), d\mu(x)\rangle.\end{aligned}$$

This shows that T is well defined and we have

$$||T(P)|| \leq ||P|| |\mu|(B_E),$$

for every μ that represents Q. Therefore $T \in \mathcal{P}_f^*({}^mE,F)$ and $||T|| \leq ||Q||_I$.

Now, T can be extended in an unique way to $\mathcal{P}_c({}^mE,F)$ and the lemma follows.

4. The Reflexivity of $\mathcal{P}({}^mE,F)$

We recall some specific results on nuclear and integral polynomials.

Theorem 1. *[1]. If E^* has the Radon Nikodym property then $\mathcal{P}_N({}^mE,F)$ and $\mathcal{P}_I({}^mE,F)$ are isomorphic for $m = 1,2,\cdots$ and all F.*

Theorem 2. *[8]. If E and F have the approximation property, then the spaces $\mathcal{P}_N^*({}^mE^*,F)$ and $\mathcal{P}({}^mE,F^*)$ are isometrically isomorphic.*

Now we prove the result:

Theorem 3. *Let E and F be reflexive Banach spaces, both with the approximation property. Then $\mathcal{P}({}^mE,F)$ is reflexive if and only if $\mathcal{P}({}^mE,F)$ coincides with its subspace $\mathcal{P}_c({}^mE,F)$.*

Proof. First of all we notice that E has the Radon Nikodym property since E is reflexive. Therefore by Theorem 1 we have that $\mathcal{P}_N({}^mE^*,F^*) = \mathcal{P}_I({}^mE^*,F^*)$ isomorphically. Now, combining this with Theorem 2 and the lemma we get the identities:

$$\mathcal{P}({}^mE,F) = \mathcal{P}_N^*({}^mE^*,F^*) = \mathcal{P}_I^*({}^mE^*,F^*) = \mathcal{P}_c^{**}({}^mE,F).$$

To conclude it is enough to observe that $\mathcal{P}_c({}^mE,F)$ is a closed subspace of $\mathcal{P}({}^mE,F)$.

Remark. If $I : \mathcal{P}_c^{**}({}^mE,F) \to \mathcal{P}({}^mE,F)$ is this isometry, it is a simple matter to see that $I_0\chi_{\mathcal{P}_c}$ ($\chi_G : G \hookrightarrow G^{**}$ the canocial injection) is the inclusion mapping $\mathcal{P}_c \hookrightarrow \mathcal{P}$.

5. Examples

1 – In [2] and [3] we showed that if T^* is the space discovered by B. S. Tsirelson then the spaces $\mathcal{P}(^nT^*)$ and $\mathcal{P}(^nT^*, l_p)$ are reflexive for all $n = 1, 2, \ldots$ and all p, $1 < p < \infty$. Therefore by Theorem 3 we have that the elements of those spaces are in fact polynomials of compact type.

2 – Now we apply theorem 3, in negative form, to some concrete space of polynomials. We consider only real l_p spaces.

We recall the following result, whose proof is a simple adaptation of the proof of prop. 5.3 in [4].

Proposition. *Let $m, n \in \mathbb{N}, m \leq n$. Then $\mathcal{P}(^mE, F)$ is isomorphic to a complemented subspace of $\mathcal{P}(^nE, F)$.*

For real numbers $p, q > 1$ and m a positive integer such that $mq = p$ we have that $\mathcal{P}(^mlp, lq)$ is not a reflexive space. In fact, let $x = (x_i) \in l_p$ and define a m-homogeneous polynomial P by

$$P(x) = (x_i^m)$$

It is clear that $P(x) \in l_q$ and $P \in \mathcal{P}(^ml_p, l_q)$. Now if $Q \in \mathcal{P}_f(^ml_p, l_q)$, Q has the form $Q(x) = \sum_{j=1}^{k} a_j \gamma_j^m(x)$ where $\gamma_j \in l_p^*$ and $a_j \in l_q, j = 1 \ldots k$. We can choose $y = (y_i) \in \cap_{j=1}^k \gamma_j^{-1}(0) \in l_p$ such that $y \neq 0$, and therefore $Q(y) = 0$. On the other hand, if $z = \dfrac{y}{||y||_p}$ from the definition of P we have that

$$\begin{aligned}(||P(z)||_q)^q = \sum_{i=1}^{\infty} \left| \frac{y_i}{||y||_p} \right|^{mq} &= \frac{1}{(||y||_p)^{mq}} \sum_{i=1}^{\infty} |y_i|^{mq} \\ = \left(\frac{||y||_{mq}}{||y||_p} \right)^{mq} &= 1\end{aligned}$$

Since $||P - Q|| \geq ||P(z) - Q(z)||_q = ||P(z)||_q = 1$ and Q is an arbitrary finite type polynomial it follows that $\mathcal{P}_c(^ml_p, l_q) \neq \mathcal{P}(^ml_p, l_q)$. By theorem 3 $\mathcal{P}(^ml_p, l_q)$ is not a reflexive space. Now, if $n \in \mathbb{N}$ $n \geq m$, by the proposition we have in particular that $\mathcal{P}(^ml_p, l_q)$ is a closed subspace of $\mathcal{P}(^nl_p, l_q)$. Therefore $\mathcal{P}(^nl_p, l_q)$ cannot be a reflexive space.

Remark. The same argument can be applied to $\mathcal{P}(^nl_p)$, to obtain that $\mathcal{P}(^nl_p)$ is not reflexive for $n \in \mathbb{N}, n \geq p$.

The author acknowledges useful conversations with professors K. Kloret, M. Matos and J. Mujica.

References

1. Alencar, R., *On reflexivity and basis for $\mathcal{P}(^mE)$*, Proc. R. Ir. Acad. **85A** (1985), no. 2, 131–138.
2. Alencar, R., Aron, R. and Fricke, G., *Tensor Products of Tsirelson's space*, Illinois J. of Math. **31** (1987), no. 1.

3. Alencar, R., Aron, R. and Dineen, S., *A reflexive space of holomorphic functions in infinitely many variables*, Proc. Am. Math. Soc. **90** (1984), no. 3.
4. Aron, R. and Schottenloher, M., *Compact Holomorphic Mappings on Banach Spaces and the Approximation Property*, Journal of Functional Analysis **21** (1976), no. 1.
5. Bartle, R. G., *A general bilinear vector integral*, Studia Math. **15** (1956), 377–352.
6. Diestel, J. and Uhl, J. J., *Vector Measures*, Math. Surveys No. 15, Am. Math. Soc. (1977).
7. Dineen, S., *Holomorphy types on a Banach space*, Studia Math. **39** (1971), 241–288.
8. Dwyer, T., *Convolutions equations for vector–valued entire functions of bounded nuclear type*, Trans. Am. Math. Soc. **217** (1976), 105–119.
9. Schmets, J., *Spaces of vector–valued continuous functions*, Lecture Notes in Mathematics, Vol. 1003, 1983.
10. Singer, I., *Sur les applications lineàires intègrales des espaces de fonctions continues*, I, Rev. Roumaine Math. Pures et Appl. **4** (1949), 391–401.

INSTITUTO DE MATEMÁTICA, UNIVERSIDADE ESTADUAL DE CAMPINAS, CAIXA POSTAL 6065, 13.081 - CAMPINAS - S.P., BRASIL

Contemporary Mathematics
Volume **144**, 1993

A ℓ_1-Predual which is not Isometric to a Quotient of $C(\alpha)$

DALE E. ALSPACH

ABSTRACT. About twenty years ago Johnson and Zippin showed that every separable $L_1(\mu)$-predual was isometric to a quotient of $C(\Delta)$, where Δ is the Cantor set. In this note we will show that the natural analogue of the theorem for ℓ_1-preduals does not hold. We will show that there are many ℓ_1–preduals which are not isometric to a quotient of any $C(K)$-space with K a countable compact metric space. We also prove some general results about the relationship between ℓ_1-preduals and quotients of $C(K)$-spaces with K a countable compact metric space.

About 20 years ago Johnson and Zippin [**J-Z**] proved the following theorem.

THEOREM. *Suppose that X is a separable $L_1(\mu)$-predual, then there is subspace Y of $C(\Delta)$, where Δ is the Cantor set, such that X is isometric to $C(\Delta)/Y$.*

Because the space X might be $C(\Delta)$, $C(\Delta)$ is the smallest $L_1(\mu)$-predual that one could use for such a result. If we consider the class of ℓ_1-preduals then it is conceivable that a smaller space might be sufficient although the space would need to depend on some measurement of the size of the ℓ_1-predual. A natural class of spaces to consider is the spaces $C(\alpha)$ where α is a countable ordinal. (This is the same as the class of $C(K)$-spaces with K a countable compact Hausdorff space by a classical result of Mazurkiewicz and Sierpiński, [**M-S**].) Thus one can consider the following question.

This paper is in final form and no version of it will be submitted for publication elsewhere.

1991 *Mathematics Subject Classification.* Primary 46E10, Secondary 46B04..

Research supported in part by National Science Foundation grant DMS-8902327

QUESTION. *If X is a ℓ_1–predual is there a countable ordinal α such that X is isometric (isomorphic) to a quotient of $C(\alpha)$?*

We will show that the isometric question has a negative answer and prove some technical results which are useful for deciding whether an ℓ_1–predual is isomorphic to a quotient of $C(\alpha)$. The isomorphic question remains open and at the end of the paper we discuss some variants of the isomorphic problem. Note also that we consider only X for which X^* is isometric to ℓ_1 because even the Johnson–Zippin result is false for isomorphic ℓ_1-preduals, [**B-D**].

Throughout this paper ℓ_1-predual will mean a Banach space with dual isometric to ℓ_1. If α is an ordinal $C(\alpha)$ will denote the space of continuous functions on the ordinals less than or equal to α with the order topology. If A is a subset of a Banach space, $[A]$ is the norm closed linear span of A. Notation and standard results from Banach space theory may be found in the books of Lindenstrauss and Tzafriri, [**L-T**], Dunford and Schwartz, [**D-S**], and Diestel, [**D**].

We begin with some technical results about basic sequences equivalent to the usual ℓ_1 basis which are contained in dual spaces.

LEMMA 1. *Let Z be a Banach space with separable dual and let Y be a subspace of Z^* which is isomorphic to ℓ_1 with normalized ℓ_1-basis (y_n). If $\overline{\{y_n\}}^{w^*} \subset Y$ then Y is w^*-closed in Z^*.*

PROOF. By the Krein-Smulian Theorem, [**D-S, Theorem V.5.7**], it is sufficient to show that $B_Y = \{y \in Y : \|y\| \leq 1\}$ is w*-closed. Let (z_n^*) be a sequence in B_Y with w*-limit z^*. We will show that $z \in Y$. Let $K=\overline{\{y_n\}}^{w^*}$ and let $\lambda \geq 1$ satisfy $\lambda\|\sum a_n y_n\| \geq \sum |a_n|$ for all finite sequences (a_n). We know that $\overline{\text{co}(\lambda K \cup (-\lambda)K)}^{w^*} \supset \overline{\text{co}\{\pm\lambda y_n\}}^{w^*} \supset B_Y$.

Because $z^* \in \overline{B_Y}^{w^*} \subset \overline{\text{co}(\lambda K \cup (-\lambda)K)}^{w^*}$, by the Choquet, [**D, p 154**], and Milman Theorems, [**D-S, Lemma V.8.5**], z^* is represented by a probability measure μ supported on

$$\text{Ext}\ \overline{\text{co}(\lambda K \cup (-\lambda)K)}^{w^*} \subset \lambda K \cup (-\lambda)K.$$

If $z^* \notin Y$ then by the Hahn-Banach Theorem there exists $z^{**} \in Z^{**}$ such that $z^{**}_{|Y} = 0$, $\|z^{**}\| = 1$, and $z^{**}(z^*) > 0$. z^{**} is the w*-limit of a sequence in Z, (z_j), with $\|z_j\| \leq 1$ for all j. Thus

$$z^{**}(z^*) = \lim_j z^*(z_j) = \lim_j \int_{\lambda K \cup (-\lambda)K} z_j d\mu.$$

Since $K \subset Y$, $z^{**}(k) = 0$ for all $k \in K$. Hence $z_j(\lambda k) \to 0$ for all $k \in K$. Therefore (z_j) is a uniformly bounded sequence converging pointwise to 0 on $\lambda K \cup (-\lambda)K$ and it follows from the Bounded Convergence Theorem that $z^{**}(z^*) = 0$. □

The next lemma tells us that the w*-closure of the ℓ_1-basis is the only thing that is important.

LEMMA 2. *Suppose that X and Y are separable Banach spaces and that (x_n^*) and (y_n^*) are normalized sequences in X^* and Y^*, respectively, which are equivalent to the usual unit vector basis of ℓ_1 and for which $\overline{[x_n^*]}^{w^*} = [x_n^*]$ and $\overline{[y_n^*]}^{w^*} = [y_n^*]$. Suppose that the basis to basis map ϕ of $[x_n^*]$ onto $[y_n^*]$, i.e., $\phi(\sum a_n x_n^*) = \sum a_n y_n^*$, is a w^*-homeomorphism of the w^*-closure of $\{x_n^*\}$ onto the w^*-closure of $\{y_n^*\}$. Then ϕ is a w^*-continuous isomorphism of $[x_n^*]$ onto $[y_n^*]$.*

PROOF. Only the w*-continuity needs to be proved. By passing to quotients we may assume that (x_n^*) and (y_n^*) are bases for the duals of X and Y, respectively. If we identify X and Y with the w*-continuous affine symmetric functions on B_{X^*} and B_{Y^*}, respectively, we need only show that the linear extension of a w*-continuous linear function on $K(X) = \lambda^{-1}\overline{\{x_n^*\}}^{w^*} \cup \{0\}$, where (x_n^*) is λ equivalent to the usual basis of ℓ_1, is w*-continuous on B_{X^*}. Let Φ be the map defined by $(\Phi f)(x^*) = f(\phi(x^*))$ for $x^* \in K(X)$ and let J_X and J_Y be the evaluation maps from X into $C(K(X)$ and from Y into $C(K(Y)$, respectively. ($K(Y)$ is defined analogously to $K(X)$.) We have the following diagram.

$$\begin{array}{ccc} Y & & X \\ {\scriptstyle J_Y}\big\downarrow & & \big\downarrow{\scriptstyle J_X} \\ C(K(Y)) & \xrightarrow{\ \Phi\ } & C(K(X)) \end{array}$$

In order to invert J_X we need to show that any $x^{**} \in C(K(X)) \cap X^{**}$ is actually in X. Any such x^{**} is the w*–limit of a sequence $(x_n) \in \|x^{**}\| B_X$. Because x^{**} is continuous on $K(X)$, (x_n) converges to x^{**} weakly in $C(K(X))$. Hence $x^{**}_{|K(X)} \in \overline{\text{co}\ \{x_{n|K(X)}\}}^{\|\cdot\|}$ and thus $x^{**} \in X$ because $K(X)$ is norming. □

If we reformulate these results for the special case where X is a ℓ_1-predual and Y is some $C(\alpha)$, we get

PROPOSITION 3. *Suppose that X is an ℓ_1-predual and let (x_n^*) be the ℓ_1-basis of the dual. Then X is isometric to a quotient of $C(\alpha)$, for some $\alpha < \omega_1$, if and only if there is a sequence (μ_n) of norm 1, mutually singular measures on α such that $K = \overline{\{\mu_n^*\}}^{w^*} \subset [\mu_n]$ and the map ϕ defined by $\phi(x_n^*) = \mu_n$ extends to a w^*-continuous map from $\overline{\{x_n^*\}}^{w^*}$ onto K.*

PROOF. It is well known that a sequence of measures in the dual of a $C(K)$-space is isometric to a ℓ_1-basis if and only if the measures are norm one and mutually singular. The two lemmas complete the proof. □

To our knowledge this proposition has not appeared previously but for a long time it was used on an intuitive level by the author. This can be seen in the construction of the example in [**A**] and in the treatment of the Benyamini-Lindenstrauss example [**B-L**] given in [**A-B**]. To see the utility of the proposition consider the following standard example of an ℓ_1-predual which is not isometric to a $C(K)$-space.

EXAMPLE 1. Let (t_n) be a sequence of points in the open unit square in $\mathbb{R}^2$ such that $K = \overline{\{t_n\}} = \{t_n\} \cup \{(t,0) : 0 \le t \le 1\}$. Let $X = \{f \in C(K) : f(t,0) = tf((0,0)) + (1-t)f((1,0))\}$. Clearly X is isomorphic to c_0 and the ℓ_1-basis for the dual is the evaluations at the t_n's and at (0,0) and (1,0). We claim that X is isometric to a quotient of $C(\omega \cdot 2)$. Indeed, let $t_{n,1}$ denote the first coordinate of t_n for all n. Define $\phi(\delta_{t_n}) = t_{n,1}\delta_n + (1 - t_{n,1})\delta_{\omega+n}$, $\phi(\delta_{(0,0)}) = \delta_\omega$, and $\phi(\delta_{(1,0)}) = \delta_{\omega \cdot 2}$. It is easy to verify that ϕ extends to be a w*-continuous map from the closure of the basis of X^* onto the closure of the image. Clearly the span of the range of ϕ contains the w*-closure of the range of ϕ. Thus X is a quotient of $C(\omega \cdot 2)$ by the proposition.

Next we will note some simple facts about basic sequences in the dual of a $C(K)$-space which are 1-equivalent to the usual basis of ℓ_1.

LEMMA 4. *If (μ_n) is a sequence of mutually singular (non-zero) measures in $C(K)^*$, K compact metric, which converges w^* to μ then*

$$\textit{supp } \mu \subset \{k \in K : \textit{there exist } k_n \in \textit{supp } \mu_n \textit{with } \lim k_n = k\}.$$

If the measures in Lemma 4 are all atomic then we may replace the support of the measures by the set of points where the measure is non-zero and the result remains true. The proof of the lemma is straight-forward so we leave it to the reader.

COROLLARY 5. *If K is a countable compact metric space and (μ_n) is a sequence of mutually singular measures in $C(K)^*$ with w^*-limit μ, then supp $\mu \subset K^{(1)}$, the first derived set of K.*

PROOF. This follows immediately from Lemma 4. □

Now let X be a ℓ_1-predual and let (e_n) be the ℓ_1-basis for the dual. We want to define a system of derived sets $N^{(\alpha)} = N^{(\alpha)}((e_n)_{n\in\mathbb{N}})$ of $\mathbb{N}$ that has properties similar to those of the Szlenk sets for the w*-closure K of the sequence (e_n). (See [S] for the definition and properties of the Szlenk sets.)

DEFINITION. Let $N^{(0)} = \mathbb{N}$ and if $N^{(\alpha)}$ has been defined, let

$$N^{(\alpha+1)} = \{n \in N^{(\alpha)} : \text{there exists an infinite } M \subset N^{(\alpha)} \text{ such that } (\text{w}^* \lim_{j\in M} e_j)(n) \neq 0\}.$$

If α is a limit ordinal define $N^{(\alpha)} = \cap_{\beta<\alpha} N^{(\beta)}$.

PROPOSITION 6. *Let K be a countable compact metric space. If (μ_n) is a sequence of norm one measures in $C(K)^*$, K a countable compact metric space, which are 1-equivalent to the unit vector basis of ℓ_1 and $[\mu_n]$ is w^*-closed, then $\{N^{(\alpha)}((\mu_n)_{n\in\mathbb{N}}) : \alpha < \alpha_0\}$ is a strictly decreasing family of subsets of $\mathbb{N}$, where α_0 is the smallest ordinal α such that $N^{(\alpha)} = \emptyset$.*

PROOF. By Corollary 5 we have that if $(\mu_j)_{j\in M}$ converges to a non-zero measure $\mu = \sum a_k\mu_k$, then supp $\mu \subset K^{(1)}$. Therefore if $a_n \neq 0$ then $n \in N^{(1)}$, and supp $\mu \subset K^{(1)}$ implies that supp $\mu_n \subset K^{(1)}$ because the measures μ_k are mutually singular. Hence for all $n \in N^{(1)}$, supp $\mu_n \subset K^{(1)}$. A simple induction argument completes the proof. □

Using this proposition it is very easy to find ℓ_1-preduals which are not isometric to a quotient of any $C(\alpha)$.

EXAMPLE 2. Let $X = \{(a_n) \subset \mathbb{R} : \lim_{j\to\infty} a_j = \sum_{n\in\mathbb{N}} \frac{a_n}{2^n}\}$. Note that X is a codimension 1 subspace of $c = C(\omega)$. Thus X is isomorphic to c_0. We claim that the evaluation functionals defined by $e_n((a_j)) = a_n$ are a basis for the dual 1-equivalent to the usual ℓ_1-basis. To do this for each $k \in \mathbb{N}$ we will exhibit a copy of ℓ_∞^k which norms the span of the first k of these functionals. Fix $K > k$ and let

$$f_j(n) = \begin{cases} 1, & \text{if } n = j \\ 0, & \text{if } n < K \text{ and } n \neq j \\ -2^{-j}, & \text{if } n = K \\ 2^{-j}, & \text{if } n > K. \end{cases}$$

for $j = 1, 2, \ldots, k$. Obviously $\lim_{n\to\infty} f_j(n) = 2^{-j} = \sum_{n\in\mathbb{N}} 2^{-n} f_j(n)$. Thus $(f_j)_{j\leq k}$ is a sequence in X 1-equivalent to the unit vectors in ℓ_∞^k and biorthogonal to the first k evaluation functionals.

We claim that X is not isometric to a quotient of $C(\alpha)$. Indeed, the w*-limit in X^* of (e_j) is $\sum_{n\in\mathbb{N}} \frac{e_n}{2^n}$. Hence $N^{(\alpha)}((e_n)_{n\in\mathbb{N}}) = \mathbb{N}$ for all α. By Propositions 3 and 6, X^* is not isometric to a w*-closed subspace of $C(K)^*$ for any countable compact metric space K, consequently, X is not isometric to a quotient of any $C(\alpha)$.

The point here is that in this case one really needs the Cantor set or some other uncountable compact metric space to support the measures. If we relax the requirement from isometric to a quotient to $(1+\epsilon)$-isomorphic to a quotient of $C(\alpha)$, then we can embed the basis of X^* in $C(\omega^2)^*$.

Fix N and define $\phi(e_n) = \delta_{\omega\cdot n}$ for $n \leq N$ and

$$\begin{aligned}\phi(e_n) &= (1 - \frac{1}{2^{n-1}})^{-1}[\sum_{j\leq N} \frac{\delta_{\omega\cdot(j-1)+n}}{2^j} + \sum_{N+1\leq j<n} \frac{\phi(e_j)}{2^j}] \\ &= (1 - \frac{1}{2^{n-1}})^{-1} \sum_{1\leq j<n} \frac{\phi(e_j)}{2^j}\end{aligned}$$

for $n > N$. It follows from Lemma 2 that ϕ can be extended to a w*-isomorphism of X^* into $C(\omega^2)$.

The example provides a trivial way of producing more such examples. Indeed, take any ℓ_1-predual and construct an isomorph by adding the example as a ℓ_∞ direct summand to get another ℓ_1-predual which is not isometric to a quotient of any $C(\alpha)$. There are probably much more interesting and subtle examples that can be constructed by using the behavior of the w*-topology in the example.

As we noted earlier it is unknown whether every ℓ_1-predual is isomorphic to a quotient of $C(\alpha)$. In addition there are several variants of this question which are open. Below X is a ℓ_1-predual.

QUESTION 1. Given $\epsilon > 0$ does there exist $\alpha < \omega_1$ and a w*-continuous into isomorphism S of X^* into $C(\alpha)^*$ with $\|S\|\|S^{-1}_{|S(X^*)}\| < 1+\epsilon$? In other words is X $(1+\epsilon)$-isomorphic to a quotient of $C(\alpha)$?

QUESTION 2. Is there an $\alpha < \omega_1$ and a quotient Y of $C(\alpha)$ such that Y^* is isometric to ℓ_1 and X is isomorphic to Y?

A positive answer to Question 1 or 2 would of course answer the isomorphic question. A positive answer to Question 1 would imply that there is an α that works for all ϵ simultaneously, that is, there is an α so that X is almost isometric to a quotient of $C(\alpha)$. Such a result should provide very useful information of the structure of ℓ_1-preduals. Let us also note that very little is known about quotients of $C(\alpha)$. In particular it is unknown whether they must be c_0-saturated. (Y is c_0-saturated if every subspace contains c_0.)

REFERENCES

[A] D. E. Alspach, *A quotient of $C(\omega^\omega)$ which is not isomorphic to a subspace of $C(\alpha), \alpha < \omega_1$*, Israel J. Math **35** (1980), 49–60.

[A-B] D. E. Alspach and Y. Benyamini, *A geometrical property of $C(K)$ spaces*, Israel J. Math. **64** (1988), 179–194.

[B-L] Y. Benyamini and J. Lindenstrauss, *A predual of ℓ_1 which is not isomorphic to a $C(K)$ space*, Israel J. Math **13** (1972), 246–259.

[B-D] J. Bourgain and F. Delbaen, *A class of special $\mathcal{L}^\infty$-spaces*, Acta Math. **145** (1980), 155–176.

[D] J. Diestel, *Sequences and Series in Banach Spaces*, Graduate Texts in Mathematics 92, Springer-Verlag, New York, 1984.

[D-S] N. Dunford and J.T. Schwartz, *Linear Operators: General Theory*, Pure and Applied Mathematics, vol.7, Interscience, New York, 1958.

[J-Z] W.B. Johnson and M.Zippin, *Every separable predual of an L_1-space is a quotient of $C(\Delta)$*, Israel J. Math **16** (1973), 198–202.

[L-T] J. Lindenstrauss and L. Tzafriri, *Classical Banach spaces I, Sequence spaces*, Springer-Verlag, Berlin, 1977.

[M] S. Mazur, *Über konvexe Mengen in linearen normierte Räumen*, Studia Math. **4** (1933), 70–84.

[M-S] S. Mazurkiewicz and W. Sierpinski., *Contribution à la topologie des ensembles dénombrables*, Fund. Math. **1** (1920), 17–27.

[S] W. Szlenk, *The non-existence of a separable reflexive Banach space universal for all separable reflexive Banach spaces*, Studia Math. **30** (1968), 53–61.

DEPARTMENT OF MATHEMATICS, OKLAHOMA STATE UNIVERSITY, STILLWATER, OK 74078-0613

E-mail address: alspach@hilbert.math.okstate.edu

Contemporary Mathematics
Volume **144**, 1993

BOUNDARIES FOR ALGEBRAS OF ANALYTIC FUNCTIONS ON INFINITE DIMENSIONAL BANACH SPACES

R. M. ARON,[1] Y. S. CHOI,[2] M. L. LOUREN,[3] O. W. PAQUES,[4]

ABSTRACT. Following Globevnik [7], we study boundaries for infinite dimensional analogues of the classical disc algebra $A(D)$ where D is replaced by the open unit ball of a Banach space E. We consider only the situation of $E = \ell_p$, concentrating on what seems to be the most difficult case of $p = \infty$.

1. INTRODUCTION

Let $E = \ell_p$ where $1 \leq p \leq \infty$ and let B be the open unit ball of E. Globevnik [7] (see also [3]) defined and studied the following analogues of the classical disc algebra:

$$A_U(B_{\ell_p}) = \{f : B_{\ell_p} \to \mathbb{C} : f \quad \text{is holomorphic on} \quad B_{\ell_p}$$
$$\text{and uniformly continuous on} \quad \overline{B}_{\ell_p}\},$$
$$A^\infty(B_{\ell_p}) = \{f : B_{\ell_p} \to \mathbb{C} : f \quad \text{is holomorphic on} \quad B_{\ell_p}$$
$$\text{and continuous and bounded on} \quad \overline{B}_{\ell_p}\}.$$

Both $A_U(B)$ and $A^\infty(B)$ are Banach algebras when given the natural norm, $\|f\| = \sup_{\|x\|\leq 1}|f(x)|$. In [3], it is shown that $A_U(B)$ is a proper subset of $A^\infty(B)$ as long as B is the open unit ball of any infinite dimensional Banach space.

Our interest here will be on what Globevnik in [7] called "boundaries" for such algebras A. Specifically, a subset F of $\overline{B}$ is called a *boundary* for A if for all $f \in A$, $\|f\| = \|f\|_F$ where $\|f\|_F$ denotes $\sup_{x\in F}|f(x)|$. We will discuss three separate situations. Our main interest will be the most difficult case of $E = \ell_\infty$, where we are only able to obtain incomplete information about boundaries which is similar to work done by Globevnik [7] on $E = c_o$. We will also treat the substantially simpler situation when E is uniformly convex, obtaining the expected result that the unit sphere is a minimal closed boundary. In addition, we will examine some algebras of special interest in the case $E = \ell_1$and $E = \ell_\infty$. More particularly, we will study the boundaries of the following type of algebra: Let X be any fixed Banach space and let $A_{w^*}(B_{X^*}) = \{f : B_{X^*} \to \mathbb{C} : f$ is holomorphic on B_{X^*} and weak*- continuous on $\overline{B}_{X^*}\}$. As we will see, the condition of being a boundary for $A_{w^*}(B_{\ell_1})$ does not depend on the choice of predual X of ℓ_1.

For general background on holomorphic functions, we refer to [4] and [10].

1991 *Mathematics Subject Classification.* 46B20, 46E25.

The authors gratefully acknowledge many helpful suggestions from the referee.

[1]The research of this author was done while he was a visitor at the Univ. de Campinas (Brazil) and the Univ. Complutense de Madrid (Spain)

[2] Supported by KOSEF and POSTECH Research Grant.

[3]Partially supported by CNPq, FINEP, and USP (Brazil).

[4]Partially supported by CNPq, FINEP and UNICAMP (Brazil)

2. Boundaries

Let K be a Hausdorff topological space and A a function algebra on K. If the intersection of all closed boundaries for A is again a boundary for A, then it is called the *Shilov boundary* for A, denoted by ∂A. Since A is not a uniform algebra in general, the existence of the Shilov boundary for A is not guaranteed. In [7], Globevnik obtained various characterizations of boundaries F for $A_U(B_{c_o})$, and showed that the Shilov boundary for $A_U(B_{c_o})$ does not exist. Using what is known as the *Littlewood-Bogdanowicz-Pelczynski* property of c_0, namely that every scalar-valued polynomial on c_o is uniformly continuous on B_{c_o} for the weak topology, boundaries for $A_U(B_{c_o})$ can be described as follows [3]: A holomorphic function $f : B_{c_o} \to \mathbb{C}$ is in $A_U(B_{c_o})$ if and only if f is uniformly continuous on B_{c_o}, when this set is endowed with the weak topology. As a result, each $f \in A_U(B_{c_o})$ can be extended as a weak* continuous function to $\overline{B}_{\ell_\infty}$. Consequently, $F \subset B_{c_o}$ is a boundary for $A_U(B_{c_o})$ if and only if F is weak*-dense in the set T of extreme points of B_{ℓ_∞}. See [3] for a more complete discussion. On the other hand, there is no complete description of boundaries of $A^\infty(B_{c_o})$.

We recall some definitions and results about function algebras on a set K (see e.g. [9]). A point $x \in K$ is called a *peak point* for A if there is some $f \in A$ such that $f(x) = 1$, and $|f(y)| < 1, y \neq x$ in K. A point $x \in K$ is called a *strong peak point* for A if there is some $f \in A$ such that $f(x) = 1$ and such that given any open neighborhood V of x there is some $\delta > 0$ satisfying $|f(y)| < 1 - \delta$ for all $y \in K \backslash V$. If A is a uniform algebra on a compact Hausdorff space K, it is well known that the Shilov boundary for A exists. The set of all peak points for A is called the *Bishop boundary* for A and denoted by ρA. If M is a convex subset of a Banach space E, then the set of all real extreme points of M is denoted by $\mathrm{Ext}_{\mathbb{R}}(M)$. A point e of M is called a *complex extreme point* of M if there is no nonzero x in E with $\{e + \lambda x : |\lambda| \leq 1, \lambda \in \mathbb{C}\} \subset M$. The set of all complex extreme points of M is denoted by $\mathrm{Ext}_{\mathbb{C}}(M)$. Let A^* be the dual Banach space of A and let S_1^* be the intersection of the unit sphere S_{A^*} of A^* with the hyperplane $\{x^* \in A^* : x^*(1) = 1\}$. To each $x \in K$ corresponds the element $\delta_x \in S_1^*$, where $\delta_x(f) = f(x)$ for all $f \in A$. The set $\chi A = \{x \in K : \delta_x \in \mathrm{Ext}_{\mathbb{R}}(S_1^*)\}$ is called the *Choquet boundary* for A. Since for $1 \leq p \leq \infty$ each predual of ℓ_p is separable, $\overline{B}_{\ell_p}$ is metrizable and compact with respect to the weak*-topology, and hence we have the well known results (see [9]) that $\rho A_{w^*}(B_{\ell_p}) = \chi A_{w^*}(B_{\ell_p})$ and the weak*-closure of $\rho A_{w^*}(B_{\ell_p})$ in $\overline{B}_{\ell_p}$ is $\partial A_{w^*}(B_{\ell_p})$.

The connection between these various notions of boundary is described in our first result, which holds for the dual of a separable Banach space.

Theorem 1. *Let E be the dual of a separable Banach space. Then $\rho A_{w^*}(B) = \chi A_{w^*}(B) = S$, the unit sphere of E, and $\partial A_{w^*}(B) = \overline{B}$, independent of the predual of E.*

Proof. Let $P(B_E)$ be the closed subalgebra of $A_{w^*}(B_E)$ generated by the constants and all weak*-continuous linear functionals of E. By Arenson [1] $\chi P(B_E) = \mathrm{Ext}_{\mathbb{C}}(\overline{B}_E) = S_E$. Since $\overline{B}_E$ is metrizable and compact with respect to the weak*-topology, we have $\rho P(B_E) = \chi P(B_E)$. Since $\rho P(B_E) \subset \rho A_{w^*}(B_E)$, it is clear that $\mathrm{Ext}_{\mathbb{C}}(\overline{B}_E) = S_E$ is contained in $\rho A_{w^*}(B_E)$. Following Globevnik [6] it is easy to see that $\rho A_{w^*}(B_E) \subset \mathrm{Ext}_{\mathbb{C}}(\overline{B}_E)$. Hence $\rho A_{w^*}(B_E) = S_E = \chi A_{w^*}(B_E)$ and $\partial A_{w^*}(B_E) = \overline{B}_E$. Q.E.D.

Theorem 2. *The set $T = Ext_{\mathbb{C}}(\overline{B}_{\ell_\infty}) = \{x = (x_i) : |x_i| = 1 \text{ for all } i\}$ is a closed boundary for $A^\infty(B_{\ell_\infty})$, and hence for $A_U(B_{\ell_\infty})$.*

Proof. The proof uses an argument modeled on ideas of Harris [8]. Let $f \in A^\infty(B_{\ell_\infty})$ and $\|f\| = 1$. Given $\epsilon > 0$, there is $y = (y_i) \in \overline{B}_{\ell_\infty}$ with $|f(y)| > 1 - \epsilon/2$. Since f is continuous at y, there exists $r < 1$ and a point $x \in B_{\ell_\infty}$ such that $||x|| \leq r$ and $|f(x)| > 1 - \epsilon$.

Let $\Delta_{1/r} = \{\lambda \in \mathbb{C} : |\lambda| < 1/r\}$, and define $g : \Delta_{1/r} \to \ell_\infty$ by $g(\lambda) = (g_i(\lambda))$, where $g_i(\lambda) = \frac{\lambda + x_i}{1 + \overline{x_i}\lambda}$. By [2], g is holomorphic on Δ. Define $F : \overline{\Delta} \to \mathbb{C}$ by $F(\lambda) = (f \circ g)(\lambda), (\lambda \in \overline{\Delta})$. Then F is holomorphic on Δ and continuous on $\overline{\Delta}$. By the maximum modulus theorem there exists $\lambda_\circ, |\lambda_\circ| = 1$, such that

$$1 - \epsilon < |f(x)| = |F(0)| \leq |F(\lambda_\circ)| = |f(g(\lambda_\circ))|.$$

Since $g(\lambda_\circ) \in T$ and ϵ is arbitrary, we conclude that $\sup_{x \in T}|f(x)| = 1$. Hence T is a boundary for $A^\infty(B_{\ell_\infty})$. Q.E.D.

The proof of the following result, which is needed in Theorem 4, is included for the sake of completeness.

Lemma 3. *Let (x_j) be a bounded and pairwise disjointly supported sequence in ℓ_∞. Then (x_j) converges weakly to 0 in ℓ_∞.*

Proof. Since (x_j) is a bounded sequence in ℓ_∞, there exists $M > 0$ such that $\|x_j\| \leq M$ for all j. If (x_j) does not converge weakly to 0 in ℓ_∞, then there is a continuous linear functional ϕ on ℓ_∞ such that $(\phi(x_j))$ does not converge to 0. There is no loss of generality if we assume that $\phi(x_j) > 0$ for all j, and so by passing to a subsequence of (x_j) we may assume that there exists $\delta > 0$ such that $\phi(x_j) \geq \delta$ for all j. Then $\phi(\sum_{j=1}^n x_j) \geq n\delta$ for all n and so $\lim_{n\to\infty}\phi(\sum_{j=1}^n x_j) = \infty$. Since the sequence (x_j) has pairwise disjoint support, it follows that $\| \sum_{j=1}^n x_j\| \leq M$ for all n. This contradicts the continuity of ϕ. Q.E.D.

Theorem 4. *Let (ξ_j) be a sequence of points in T. The set $F \equiv T \setminus \cup_{j=1}^\infty B_2(\xi_j)$ is a boundary for $A_U(B_{\ell_\infty})$.*

Proof. Since T is a boundary, it suffices to show that for every $f \in A_U(B_{\ell_\infty}), y = (y_j) \in T \cap (\cup_{j=1}^\infty B_2(\xi_j))$, and $\epsilon > 0$, there is a point $w \in F$ such that $|f(y) - f(w)| < 5\epsilon$. Since f is uniformly continuous, there is $\delta > 0$ such that if $s, t \in \overline{B}_{\ell_\infty}, \|s - t\| < 3\delta$, then $|f(s) - f(t)| < \epsilon$. Since the Taylor series of f converges uniformly to f on any smaller ball, there is a polynomial P such that $|f(x) - P(x)| < \epsilon$ whenever $\|x\| \leq 1 - \delta$. Let $z = (1 - 2\delta)y$ and let $\mathbb{N} = \cup_{j=1}^\infty \mathbb{A}_j$ be a decomposition of the natural numbers $\mathbb{N}$ into a sequence of infinite, pairwise disjoint sets. Write each $\mathbb{A}_j$ as $\{n_{j,1}, n_{j,2}, \dots\}$. Let $z_j \equiv \sum_{k=1}^\infty (2\delta - 1)\xi_{k,n_{j,k}} e_{n_{j,k}} + \sum_{k \in \mathbb{N}\setminus A_j}(1 - 2\delta)y_k e_k$, where $\xi_j = (\xi_{j,1}, \xi_{j,2}, \dots)$. Here, of course, the summation notation is taken formally. Then $z - z_j = \sum_{k=1}^\infty (1 - 2\delta)(y_{n_{j,k}} + \xi_{k,n_{j,k}})e_{n_{j,k}} \to 0$ weakly by Lemma 3.

Since $\|y - z\| = 2\delta, |f(y) - f(z)| < \epsilon$. Since $\|z\| = 1 - 2\delta < 1 - \delta, |f(z) - P(z)| < \epsilon$. Since ℓ_∞ has the Dunford-Pettis property, it also has the polynomial

Dunford-Pettis property [11]. Thus, we can conclude that $|P(z)-P(z_m)| < \epsilon$ for m sufficiently large, and $\|z_m\| = 1-2\delta$. Finally, if we set $w = \sum_{k=1}^{\infty}(-\xi_{k,n_{m,k}})e_{n_{m,k}} + \sum_{k\in\mathbb{N}\setminus A_n} y_k e_k$, we see that $\|z_n - w\| = 2\delta$ and $\|w-\xi_j\| = 2$ for all j. Thus, $w \in F$ and $|f(y) - f(w)| < 5\epsilon$. Q.E.D.

We do not know whether the set F in Theorem 4 is also a boundary for $A^{\infty}(B_{\ell_\infty})$. On the other hand, Theorem 4 shows in particular that the intersection of all boundaries for $A_U(B_{\ell_\infty})$ is empty.

The following shows that being a boundary for algebras of analytic functions on B_{ℓ_∞} is closely connected with the weak topology of ℓ_∞.

Proposition 5. *If F is a boundary for $A_U(B_{\ell_\infty})$, then F is weak* dense in the "distinguished boundary" T of $\overline{B}_{\ell_\infty}$. If F is a weakly sequentially dense in T, then F is a boundary for $A_U(B_{\ell_\infty})$.*

Proof. If F is a boundary for $A_U(B_{\ell_\infty})$, then it is also a boundary for $A_{w^*}(B_{\ell_\infty})$. By our introductory remarks, the latter space is a uniform algebra with Shilov boundary T, and so it follows that $T \subset \overline{F}^{\text{weak}^*}$.

Next, suppose that F is weakly sequentially dense in T. If $y \in T, f \in A_U(B_{\ell_\infty})$, and $\epsilon > 0$, we argue as in the preceding theorem to find first $\delta > 0$ such that $|f(s) - f(t)| < \epsilon/2$ whenever s and t are in $\overline{B}_{\ell_\infty}, \|s-t\| < \delta$. Since the Taylor series of f converges uniformly to f on any ball centered at 0 and with radius less than 1, we can find a polynomial P such that $|f(z) - P([1-\delta]z)| < \epsilon$, for all $z \in \overline{B}_{\ell_\infty}$. By assumption, y is the weak limit of a sequence (x_j) of points of F, so that $|P([1-\delta]y) - P([1-\delta]x_j)| < \epsilon$ for large j, by the (polynomial) Dunford-Pettis property [11]. It follows that $|f(y) - f(x_j)| < 3\epsilon$, which implies that F is a boundary. Q.E.D.

Unlike the case of functions in $A_U(B_{c_0})$ which was described earlier, not every $f \in A_U(B_{\ell_\infty})$ is weakly uniformly continuous on B_{ℓ_∞}. Indeed, there are 3-homogeneous polynomials on ℓ_∞ which are not weakly uniformly continuous. (However, every 2- homogeneous polynomial is in fact weakly uniformly continuous, since it corresponds to a continuous linear operator from ℓ_∞ to ℓ_∞^*. Also, by the Dunford-Pettis property, every scalar-valued polynomial on ℓ_∞ is weakly sequentially continuous ([11]).) We remark that there exist subsets F of $\overline{B}_{\ell_\infty}$ which satisfy the following properties: F is a boundary for $A_{w^*}(B_{\ell_\infty})$, the weak closure of F contains T, and F is not weakly sequentially dense in T. Although the convex hull of such a set F is necessarily a boundary for $A^{\infty}(B_{\ell_\infty})$, we do not know if F itself is a boundary for either $A^{\infty}(B_{\ell_\infty})$ or $A_U(B_{\ell_\infty})$. See also Remark 8 below.

Our next two results give us more information about boundaries for $A^{\infty}(B_{\ell_\infty})$. For $\gamma > 0$ and $\xi \in E$ let $B_\gamma(\xi)$ denote the open ball in E with center ξ and radius γ.

Theorem 6. *Fix $\gamma, 0 < \gamma < 2$, and let $(\xi_i) \subset T$ be an arbitrary sequence. Then the set $S = \overline{B}_{\ell_\infty} \setminus \cup_{i=1}^{\infty} B_\gamma(\xi_i)$ is a closed boundary for $A^{\infty}(B_{\ell_\infty})$, and hence for $A_U(B_{\ell_\infty})$.*

Proof. Let $\epsilon > 0$ and $y \in T \bigcap B_\gamma(\xi_i)$, for some i, be given. Given $f \in A^{\infty}(B_{\ell_\infty})$, there exists $\delta, 0 < \delta < 1 - \frac{\gamma}{2}$, such that $|f(x) - f(y)| < \epsilon/4$ if $\|x-y\| < 3\delta, x \in \overline{B}_{\ell_\infty}$. As in the proof of Theorem 4, let $z = (1-2\delta)y$ and write the natural numbers $\mathbb{N}$ as a disjoint union of infinite subsets, $\mathbb{N} = \cup_{j=1}^{\infty} \mathbb{A}_j$, where each $\mathbb{A}_j = \{n_{j,1}, n_{j,2}, \cdots\}$. For each positive integer j, let $z_j \equiv \sum_{k=1}^{\infty}(2\delta - 1)\xi_{k,n_{j,k}} e_{n_{j,k}} + \sum_{k\in\mathbb{N}\setminus\mathbb{A}_j}(1-2\delta)y_k e_k$. Then $\|\xi_k - z_j\| = 2 - 2\delta, \|z\| = 1 - 2\delta = \|z_j\|$ for each j and k, and (z_j) converges weakly to z in ℓ_∞.

Since the Taylor series of f at 0 converges uniformly on $\|x\| \leq 1-\delta$, there exists a continuous polynomial P on ℓ_∞ such that $|f(x) - P(x)| < \epsilon/4, \|x\| < 1 - \delta$. Also, $(P(z_n))$ converges to $P(z)$ and so there exists a positive integer N such that $|P(z_n) - P(z)| < \epsilon/4$ for all $n \geq N$. Hence we have $|f(y) - f(z_N)| \leq |f(y) - f(z)| + |f(z) - P(z)| + |P(z) - P(z_N)| + |P(z_N) - f(z_N)| \leq \epsilon/4 + \epsilon/4 + \epsilon/4 + \epsilon/4 = \epsilon$. Since $\|\xi_k - z_j\| = (2 - 2\delta) > \gamma$ for all j and $k, z_N \in S$. Since ϵ and $y \in T \bigcap B_\gamma(\xi)$ are arbitrary, we have that $\sup_{x\in S}|f(x)| = \sup_{x\in T}|f(x)|$ for all $f \in A^\infty(B_{\ell_\infty})$. Hence S is a closed boundary for $A^\infty(B_{\ell_\infty})$. Q.E.D.

Of course, Theorem 6 is of interest only in the A^∞-case, in light of Theorem 4. It also shows that the intersection of all boundaries for $A^\infty(B_{\ell_\infty})$ is empty, although the following result shows that much more is true.

Theorem 7. *There exists a closed boundary for $A^\infty(B_{\ell_\infty})$ (hence for $A_U(B_{\ell_\infty})$), disjoint from T.*

Proof. Let F be the set of all points $x = (x_j) \in \overline{B}_{\ell_\infty}$ such that there exists only one positive integer i satisfying $x_i = 0$ and $|x_j| = 1$ for all $j \neq i$. Then F is closed and disjoint from T. We show that F is a boundary for $A^\infty(B_{\ell_\infty})$.

Let $\xi = (\xi_j) \in T$ and $\epsilon > 0$ be given. Given $f \in A^\infty(B_{\ell_\infty})$ there exists $\delta, 0 < \delta < 1$, such that $|f(x) - f(\xi)| < \epsilon/4$ if $\|x - \xi\| < 3\delta, x \in \overline{B}_{\ell_\infty}$. By the same argument as in Theorem 6, there is some point $z_N = \sum_{j\neq N}(1 - 2\delta)\xi_j e_j$ such that $|f(z_N) - f(\xi)| \leq \epsilon$. Since f is continuous at z_N, there exist $\tau, 0 < \tau < 1$, such that if $\|x - z_N\| \leq \tau, x \in \overline{B}_{\ell_\infty}$, then $|f(x) - f(z_N)| \leq \epsilon$. Let $y = (y_j) \in \overline{B}_{\ell_\infty}$ be defined as follows: If $\delta > \tau/2$, then let $y = z_N$. If $\delta \leq \tau/2$, then let $y_N = 0$ and $y_j = \xi_j$ if $j \neq N$.

If $\delta \leq \tau/2$, then $y \in F, \|y - z_N\| \leq \tau$ and so $|f(\xi)| - 2\epsilon \leq |f(y)|$. If $\delta > \tau/2$, then choose $\tau', 1 < \tau' < \frac{1}{1-\tau}$. Define $g : \Delta_{\tau'} \to \ell_\infty$ by $g(\lambda) = (g_j(\lambda))$, where

$$g_j(\lambda) = \begin{cases} 0 & \text{if } j = N \\ \frac{\lambda + y_j}{1 + \overline{y_j}\lambda} & \text{if } j \neq N\ . \end{cases}$$

As before, g is analytic on $\Delta_{\tau'}$, and the maximum modulus theorem implies that for some $\lambda_0 \in C, |\lambda_0| = 1, \|g(0)\| \leq \|g(\lambda_0)\|$. Hence, we can find a point $\zeta \in F$ such that $|f(\xi)| - \epsilon \leq |f(z_N)| = |f(y)| \leq |f(\zeta)|$.

Since $\xi \in T$ and ϵ are arbitrary, the above arguments show that in either case, F is a boundary for $A^\infty(B_{\ell_\infty})$. Q.E.D.

In fact, combining the argument given above with that of the preceding results, one can show the following. Let $\mathbb{N} = \cup_{i=1}^{\infty} \mathbb{A}_i$ be a decomposition of the natural numbers into an infinite number of infinite sets. Let F be the set of all points $x \in \overline{B}_{\ell_\infty}$ such that for some positive integer i, $x_m = 0$ if $i \in A_m$ and $|x_m| = 1$ if $i \notin A_m$. Then F is a closed boundary for $A^\infty(B_{\ell_\infty})$.

Remark 8. Since $A_{w^*}(B_{\ell_\infty})$ is contained in $A_U(B_{\ell_\infty})$, if a set $F \subset \overline{B}_{\ell_\infty}$ is a boundary for $A_U(B_{\ell_\infty})$, then the weak*-closure of F must contain the set T. However this is not a sufficient condition for a set to be a boundary for $A_U(B_{\ell_\infty})$. In fact the set F of all finite vectors in S_{ℓ_∞} is not a boundary for $A_U(B_{\ell_\infty})$, although it is a boundary for $A_U(B_{c_o})$ by [7], and its weak*-closure contains the set T. Indeed, let c be the subspace of ℓ_∞ consisting of all convergent sequences and let Φ be a norm-preserving extension to ℓ_∞ of the continuous linear functional ϕ on c defined by $\phi(x) = \lim_{i\to\infty} x_i$. It is easily seen that $\|\Phi\|_F = 0$.

We turn now to boundaries for $A_U(B)$ and $A^\infty(B)$ where B is the open unit ball of a Banach space E which is either uniformly convex or the dual of a separable Banach space. Here the situation is much clearer. The proof for the uniformly convex case is easy: Let $\xi \in E$ be a fixed unit vector and let $\phi \in E^*$ be such that $\|\phi\| = \phi(\xi) = 1$. The uniform convexity of E implies that the function $f \in A_U(B_{\ell_p})$ defined by $f(x) = \frac{1}{2}(1+\phi(x)), (x \in B)$ attains its maximum at only the point ξ and that ξ is a strong peak point. Theorem 10 below shows that the Shilov boundary for both $A_U(B)$ and $A^\infty(B)$ exists for $E = \ell_1$. The proof will make use of the following simple result.

Lemma 9. *Let $0 < a < 1$. Then $f(x) = (1+\frac{x}{1-a})(1+\frac{1-x}{a})$, $(x \in \mathbb{R})$ has maximum at $x = a$.*

Theorem 10. *The set of all strong peak points for $A_U(B_{\ell_1})$ (hence for $A^\infty(B_{\ell_1})$) is a dense subset of the unit sphere S_{ℓ_1}.*

Proof. It suffices to show that every finite vector in the unit sphere S_{ℓ_1} is a strong peak point for $A_U(B_{\ell_1})$. Let $\xi = (\xi_i)$ be a finite vector in S_{ℓ_1}. For simplicity we may assume that $\xi = (\xi_1, \xi_2, ..., \xi_n, 0, ...)$ for some $n \geq 1$, where $\xi_i \neq 0, i = 1, ..., n$. When $n = 1$, define $f(x) = \dfrac{1}{2}(1 + \bar{\xi}_1 x_1), (x = (x_i) \in \overline{B}_{\ell_1})$. Then $f \in A_U(B_{\ell_1})$, $f(\xi) = 1$ and $|f(x)| < 1, x \neq \xi$ in $\overline{B}_{\ell_1}$. Hence we have that ξ is a strong peak point for $A_U(B_{\ell_1})$ by the same argument as the case above.

When $n > 1$, define

$$f(x) = \sum_{k=1}^{n} \left(1 + \frac{1}{1-|\xi_k|} e^{i\theta_k} x_k\right) \left(1 + \frac{1}{|\xi_k|} \sum_{\substack{j=1 \\ j\neq k}}^{n} e^{i\theta_j} x_j\right)$$

where $e^{i\theta_k}\xi_k = |\xi_k|, \theta_k \in \mathbb{R}$, $k = 1, \ldots, n$. Then $f \in A_U(B_{\ell_1})$. Let $x = (x_i) \in \overline{B}_{\ell_1}$ and $x \neq \xi$. Suppose that $\sum_{k=1}^{n} |x_k| = 1$. Since $x \neq \xi$, x satisfies either $|x_i| = |\xi_i| (i = 1, \ldots, n)$ but $x_{i_0} \neq \xi_{i_0}$ for some $i_0 (1 \leq i_0 \leq n)$ or $|x_{i_0}| \neq |\xi_{i_0}|$ for some $i_0 (1 \leq i_0 \leq n)$.

If $|x_i| = |\xi_i| (i = 1, \ldots, n)$ but $x_{i_0} \neq \xi_{i_0}$ for some $i_0 (1 \leq i_0 \leq n)$, then

$$|f(x)| \leq \sum_{\substack{k=1 \\ k\neq i_0}}^{n} \left(1 + \frac{1}{1-|\xi_k|}|x_k|\right) \left(1 + \frac{1}{|\xi_k|} \sum_{\substack{j=1 \\ j\neq k}}^{n} |x_j|\right)$$
$$+ \left|\left(1 + \frac{1}{1-|\xi_{i_0}|} e^{i\theta_{i_0}} x_{i_0}\right)\right| \left(1 + \frac{1}{|\xi_{i_0}|} \sum_{\substack{j=1 \\ j\neq i_0}}^{n} |x_j|\right).$$

Since $\left|1 + \frac{1}{1-|\xi_{i_0}|} e^{i\theta_{i_0}} x_{i_0}\right| < 1 + \frac{1}{1-|\xi_{i_0}|}|\xi_{i_0}|$, we have that

$$|f(x)| < \sum_{k=1}^{n} \left(1 + \frac{1}{1-|\xi_k|}|\xi_k|\right) \left(1 + \frac{1}{|\xi_k|} \sum_{\substack{j=1 \\ j\neq k}}^{n} |\xi_j|\right) = f(\xi).$$

If $|x_{i_0}| \neq |\xi_{i_0}|$ for some $i_0 (1 \leq i_0 \leq n)$, then

$$\begin{aligned} |f(x)| &\leq \sum_{k=1}^{n} \left(1 + \frac{1}{1-|\xi_k|}|x_k|\right) \left(1 + \frac{1}{|\xi_k|} \sum_{\substack{j=1 \\ j\neq k}}^{n} |x_j|\right) \\ &= \sum_{k=1}^{n} \left(1 + \frac{1}{1-|\xi_k|}|x_k|\right) \left(1 + \frac{1}{|\xi_k|}(1 - |x_k|)\right) . \end{aligned}$$

By Lemma 9

$$\begin{aligned} &\left(1 + \frac{1}{1-|\xi_{i_0}|}|x_{i_0}|\right) \left(1 + \frac{1}{|\xi_{i_0}|}(1 - |x_{i_0}|)\right) \\ &< \left(1 + \frac{1}{1-|\xi_{i_0}|}|\xi_{i_0}|\right) \left(1 + \frac{1}{|\xi_{i_0}|}(1 - |\xi_{i_0}|)\right) . \end{aligned}$$

Hence $|f(x)| < f(\xi)$.

In the case that $\sum_{k=1}^{n} |x_k| < 1$, let

$$x' = \left(e^{-i\theta_1}|x_1|, \dots , e^{-i\theta_{n-1}}|x_{n-1}|, e^{-i\theta_n}(1 - \sum_{k=1}^{n-1} |x_k|), 0, \dots \right) .$$

Then $\sum_{k=1}^{n} |x'_k| = 1$. Since $|x_n| < 1 - \sum_{k=1}^{n-1} |x_k| = |x'_n|$, we obtain that $|f(x)| < f(x') \leq f(\xi)$.

Hence $|f(x)| < f(\xi)$ for all $x \neq \xi$ in $\overline{B}_{\ell_1}$ and we have that ξ is a strong peak point for $A_U(B_{\ell_1})$ by the same argument as in the case $1 < p < \infty$. Q.E.D.

As the above proof shows, it is easy to construct peaking functions in the case $1 < p < \infty$. Let $\xi = (\xi_i)$ be a complex extreme point of $\overline{B}_{\ell_\infty}$. Then the function $f(x) = \sum_i 2^{-i-1}(1 + \overline{\xi}_i x_i)$, $(x = (x_i) \in \overline{B}_{\ell_\infty})$ is a peaking function for ξ in $A_{w^*}(B_{\ell_\infty})$. However, we do not have an explicit peaking function in $A_{w^*}(B_{\ell_1})$ for an infinite vector in S_{ℓ_1}.

Corollary 11. *$\partial A_U(B_{\ell_p}) = \partial A^\infty(B_{\ell_p}) = S_{\ell_p}$ where $1 \leq p < \infty$. In particular, a set $F \subset \overline{B}_{\ell_p}$ is a boundary for $A_U(B_{\ell_p})$ if and only if it is a boundary for $A^\infty(B_{\ell_p})$, where $1 \leq p < \infty$.*

Proof. Let F be a closed boundary for $A_U(B_{\ell_p})$. If F does not contain some strong peak point ξ for $A_U(B_{\ell_p})$, then there exists an $\epsilon > 0$ such that $F \bigcap B_\epsilon(\xi) = \emptyset$. Since ξ is a strong peak point for $A_U(B_{\ell_p})$, there exist some $f \in A_U(B_{\ell_p})$ and $\delta > 0$ such that $f(\xi) = 1$ and $|f(x)| < 1 - \delta$ for all $x \in F$, which is a contradiction. Hence F contains all strong peak points for $A_U(B_{\ell_p})$. By Theorem 10, F must contain S_{ℓ_p}, which is a closed boundary for both $A_U(B_{\ell_p})$ and $A^\infty(B_{\ell_p})$ by the maximum modulus theorem. Hence $\partial A_U(B_{\ell_p}) = \partial A^\infty(B_{\ell_p}) = S_{\ell_p}$. Q.E.D.

References

1. E.L. Arenson, *Gleason parts and the Choquet boundary of a function algebra on a convex compactum*, Otdel. Mat. Inst. Steklov. (LOMI), Zap. Nauchn. Sem. Leningrad. **113** (1981), 204-207.
2. R. Aron and J.A. Cima, *A theorem on holomorphic mappings into Banach spaces with basis*, Proc. A.M.S., vol. 36, 1972, pp. 289-292.
3. R. Aron, B. Cole and T.W. Gamelin, *The spectra of algebras of analytic functions associated with a Banach space*, (Crelle) (to appearin J.Reine Angew.Math.).
4. S. Dineen, *Complex Analysis in Locally Convex Spaces*, Math. Studies(57), North-Holland, Amsterdam, 1981.
5. N. Dunford and J. Schwartz, *Linear operators*, part I, Interscience Publishers, 1966.
6. J. Globevnik, *On interpolation by analytic maps in infinite dimensions*, Math. Proc. Cambridge Philos. Soc. **83** (1978), 243-262.
7. J. Globevnik, *Boundaries for polydisc algebras in infinite dimensions*, Math. Proc. Cambridge Philos. Soc. **85** (1979), 291-303.
8. L. Harris, *Schwarz's lemma in normed linear spaces*, Proc. Nat. Acad. Sc. **62** (1969), 1014-1017.
9. R. Larsen, *Banach Algebras*, Marcel-Dekker, 1973.
10. J. Mujica, *Complex Analysis in Banach Spaces*, Math. Studies, vol. 120, North-Holland, Amsterdam, 1986.
11. R. Ryan, *Dunford-Pettis properties*, Bull. Acad. Polon. Sci. Math. **27** (1979), 373-379.

R. M. Aron, Department of Mathematics, Kent State University, Kent, Ohio 44242, USA

Y. S. Choi, Department of Mathematics, Pohang Institute of Science and Technology, P.O. Box 125, Pohang, Korea 790

M. L. Louren, Departamento de Matemática, Universidade de São Paulo, 01000 São Paulo, Brazil

O. W. Paques, Instituto de Matemática, Universidade Estadual de Campinas, 13081 Campinas, S.P., Brazil

Contemporary Mathematics
Volume **144**, 1993

MULTIPLIERS ON WEIGHTED BESOV SPACES OF ANALYTIC FUNCTIONS

OSCAR BLASCO

Departamento de Matemáticas
Universidad de Zaragoza

ABSTRACT. We characterize the space of multipliers between certain weighted Besov spaces of analytic functions. This extend and give a new proof of a result of Wojtaszczyk about multipliers between Bergman spaces.

INTRODUCTION.

P. Wojtaszczyk [W], using certain factorization theorems due to Maurey and Grothendieck, proved the following results:

Let $\alpha > 0$, $0 < p \leq 2 \leq q < \infty$ and $\frac{1}{r} = \frac{1}{p} - \frac{1}{q}$.

(0.1) $$(B_q, B_p) = \{\lambda_n : \sup_{2^n \leq k < 2^{n+1}} (k^{-1/r}|\lambda_k|) \in l^r\}$$

(0.2) $$(X_\alpha, B_p) = \{\lambda_n : \sup_{2^n \leq k < 2^{n+1}} (k^{\alpha-1/p}|\lambda_k|) \in l^p\}$$

where B_p and X_α stand for the spaces

$$B_p = \{f : D \to \mathbb{C} \text{ analytic } : \Big(\int_D |f(z)|^p d\sigma(z)\Big)^{1/p} < \infty\} \text{ and}$$

$$X_\alpha = \{f : D \to \mathbb{C} \text{ analytic } : |f(z)| = O(\frac{1}{(1-|z|)^\alpha})\}.$$

The main objective of this paper is to extend such results to a much more general situation of general weighted Bergman and Besov spaces. We shall present a proof based simply on Kintchine's inequality for the analogue to (0.1) and then we shall use the previous case combined with some duality arguments to get the analogue to (0.2).

1991 *Mathematics Subject Classification.* 42A45.
Key words and phrases. Besov spaces, multipliers.
This research has been partially supported by the Spanish DGICYT, Proyecto PS89-0106.

Given be a nondecreasing function on $(0,1)$, say ρ, such that $\frac{\rho(t)}{t} \in L^1((0,1))$, we denote by $B_p(\rho)$, $0 < p < \infty$, the space of analytic function f on the unit disc such that

$$\|f\|_{p,\rho} = \Big(\int_D \frac{\rho(1-|z|)}{(1-|z|)}|f(z)|^p d\sigma(z)\Big)^{1/p} < \infty.$$

For certain weights the spaces $B_p(\rho)$ have been extensively studied in the literature. They can be regarded as extensions of the classical Bergman spaces ($\rho(t) = t$). Although the condition appearing in the case $p = 1$ and $\rho(t) = t^{1/q-1}$ for $q < 1$ goes back to the work of Hardy and Littlewood (see [HL1, Theorem 3.1], the corresponding spaces were first studied as Banach spaces by P. Duren, B.W. Romberg and A.L. Shield in [DRS] and by A.L. Shields and D.L. Williams [SW] for certain weights. Later J. Shapiro [Sh] considered the spaces, denoted by A^p_α and called weighted Bergman spaces, for $0 < p \leq \infty$ and $\rho(t) = t^{\alpha+1}$.

The extension of (0.1) and (0.2) for weighted Bergman spaces can be achieved from similar arguments as those used by Wojtaszczyk, as it was pointed out in [W], when we deal with weights of the type considered in [SW]. In fact they can be obtained, as we shall prove, not only for these weights but for bit more general weights, defined by Janson [J], under Dini and b_α condition (see definitions below).

In [B1] the author gave a procedure to characterize multipliers acting on $B_p(\rho)$ when $0 < p \leq 1$ and ρ verifies the previous conditions (see [B1, section 3]). Here we shall complete the cases when $2 \leq p < \infty$, and not only for weighted Bergman spaces but for weighted Besov classes of analytic functions.

For $0 < p \leq \infty$, $0 < q < \infty$, we denote by $H(p,q,\rho)$ the spaces formed by analytic functions on the unit disc D satisfying

$$\|f\|_{p,q,\rho} = \Big(\int_0^1 \frac{\rho(1-r)}{1-r} M_p^q(f,r)dr\Big)^{1/q} < \infty,$$

and by H^p_ρ the spaces of analytic functions such that

$$M_p(f,r) = O\Big(\frac{\rho(1-r)}{1-r}\Big) \quad (r \to 1).$$

The definition of these classes for the particular case of $\rho(t) = t^\alpha$, goes back to Hardy and Littlewood (see [HL1] [HL2]) and they were extensively studied for different reasons and by different authors (see [F1, F2, M, MP1, MP2, S]).

Given two sequence spaces X, Y we denote by (X,Y) the space of multipliers from X to Y, that is $(X,Y) = \{(\lambda_n) : (\lambda_n a_n) \in Y \text{ for every } (a_n) \in X\}$.

We identify $H(p,q,\rho)$ and H^p_ρ with sequence spaces by associating with each analytic function the sequence of its Taylor coefficients. Hence λ will stand for either the sequence (λ_n) or the function $\lambda(z) = \sum_{n=0}^\infty \lambda_n z^n$.

Theorem 1. *Let $0 < p_2 \leq 2 \leq p_1 < \infty$, $0 < q_1, q_2 < \infty$, and ρ_1, ρ_2 weight functions verifying Dini and b_{α_1} and b_{α_2} respectively for some $\alpha_1, \alpha_2 > 0$. Let $\frac{1}{r} = \frac{1}{min(q_1,q_2)} - \frac{1}{q_1}$. Then*

$$\Big(H(p_1,q_1,\rho_1), H(p_2,q_2,\rho_2)\Big) = \{\lambda_n : \sup_{2^n \leq k < 2^{n+1}} (\rho_2(k^{-1})^{1/q_2}\rho_1(k^{-1})^{-1/q_1}|\lambda_k|) \in l^r\}.$$

Theorem 2. *Let $2 \leq p_1 \leq \infty$, $1 < p_2 \leq 2$, $1 < q_2 < \infty$ and ρ_1, ρ_2 weight functions verifying Dini and b_1 and b_{α_2} respectively for some $\alpha_2 > 0$. Then*

$$\left(H^{p_1}_{\rho_1}, H(p_2, q_2, \rho_2)\right) = \{\lambda_n : \sup_{2^n \leq k < 2^{n+1}} \left(k\rho_2(k^{-1})^{1/q_2}\rho_1(k^{-1})|\lambda_k|\right) \in l^{q_2}\}.$$

The reader is referred to [B2] for these and more results about multipliers when $\rho(t) = t^\alpha$.

The key point for these results is that for $p \geq 2$ multipliers on $H(p, q, \rho)$ depend on those on $H(2, q, \rho)$ and this space is isomorphic to $l(2, q)$ which makes them very easy to deal with.Then some duality arguments allow us to get Theorem 2 from Theorem 1.

Throughout the paper all functions f will be analytic on the unit disc and $M_p(f, r)$ stands for $\left(\int_0^{2\pi} |f(re^{i\theta})|^p \frac{d\theta}{2\pi}\right)^{1/p}$. A weight function ρ will be a nondecreasing function on $(0, 1)$ with $\rho(0^+) = 0$ such that $\frac{\rho(t)}{t} \in L^1\left((0, 1)\right)$ and C will be a numerical constant not necessarily the same in each instance.

§1. Definitions and Lemmas.

There are several types of weight functions that have been considered to extend those results valid for $\rho(t) = t^\alpha$ to more general situations (see [SW, BS, BTS, MP2, J]). We shall be dealing with conditions considered in [J], which cover the other cases and have been found adequated to many other situations.

Definition 1. *A weight function will be a non negative non-decreasing function on $(0, 1)$ such that $\rho(t)/t \in L^1\left((0, 1)\right)$. ρ is said to be a Dini-weight if*

$$\int_0^s \frac{\rho(t)}{t} dt \leq C\rho(s) \tag{1.1}$$

Given $0 < \alpha < \infty$, ρ is said to be a b_α-weight, $\rho \in b_\alpha$, if

$$\int_s^1 \frac{\rho(t)}{t^{\alpha+1}} dt \leq C\frac{\rho(s)}{s^\alpha} \tag{1.2}$$

Remark 1. The main example of a Dini-weight ρ such that $\rho \in b_{\alpha+\varepsilon}$ for any $\varepsilon > 0$ is

$$\rho(t) = C_{\alpha,\beta}t^\alpha(log\frac{e}{t})^\beta.$$

Remark 2. It is not difficult to see that if $\rho \in b_\alpha$ for some $\alpha > 0$, then the Dini condition is equivalent to

$$\int_0^s \frac{\rho(t)}{t} dt = O\left(\rho(s)\right) \qquad (s \to 0) \tag{1.1$'$}$$

It is clear that b_α condition can be written

$$\int_s^1 \frac{\rho(t)}{t^{\alpha+1}} dt \approx \frac{\rho(s)}{s^\alpha}, \tag{1.2$'$}$$

which implies that there is a constant $C < 1$ such that

$$C\rho(2s) \leq \rho(s) \leq \rho(2s) \qquad (0 < s < \frac{1}{2}). \tag{1.3}$$

Let us include now some results on weights to be used later on.

Lemma 1. *Let ρ be a Dini weight such that $\rho \in b_\alpha$. Then*

$$(1.4) \qquad \rho(2^{-n}) \in l^1 \text{ and } \quad \sum_{m=n}^{\infty} \rho(2^{-m}) = O\big(\rho(2^{-n})\big) \quad (n \to \infty)$$

$$(1.5) \quad \int_0^1 \frac{\rho(1-r)}{(1-r)} r^n dr = O(\rho(\frac{1}{n})) \qquad (n \to \infty) \qquad (\text{see [B1, Lemma 1.1]})$$

Proof. Note that from (1.3)

$$\rho(2^{-n}) \approx \int_{1-2^{-n}}^{1-2^{-(n+1)}} \frac{\rho(1-r)}{1-r} dr.$$

Therefore $\rho(2^{-n}) \in l^1$ follows from $\frac{\rho(t)}{t} \in L^1\big((0,1)\big)$ and

$$\sum_{m=n}^{\infty} \rho(2^{-m}) \approx \int_{1-2^{-n}}^{1} \frac{\rho(1-r)}{1-r} dr \approx \int_0^{2^{-n}} \frac{\rho(t)}{t} dt = O\big(\rho(2^{-n})\big). \quad \square$$

Definition 2. *Let $0 < p \le \infty$, $0 < q < \infty$ and ρ a weight function. $H(p,q,\rho)$ will denote the space of analytic functions on the unit disc D satisfying*

$$\|f\|_{p,q,\rho} = \Big(\int_0^1 \frac{\rho(1-r)}{1-r} M_p^q(f,r) dr\Big)^{1/q} < \infty.$$

Let ρ be a weight function with $\inf \frac{\rho(1-r)}{1-r} > 0$. H_ρ^p will denote the space of analytic functions on the unit disc D satisfying

$$M_p(f,r) = O(\frac{\rho(1-r)}{1-r}).$$

We define the norm by

$$\|f\|_{p,\rho} = \inf\{C : M_p(f,r) \le C\frac{\rho(1-r)}{1-r}, 0 < r < 1\}.$$

Definition 3. *Let $0 < p \le \infty, 0 < q < \infty$ and $\bar{\gamma} = (\gamma_n)$ a sequence of positive real numbers. Denote by $I_n = \{k \in \mathbb{N} : 2^{n-1} \le k < 2^n\}$ and $I_0 = \{0\}$.*

$$l(p,q,\bar{\gamma}) = \{(a_n) \in \mathbb{C} : \|(a_n)\|_{p,q,\bar{\gamma}} = \Big(\sum_n \big(\sum_{k \in I_n} |a_n|^p\big)^{q/p} \gamma_n\Big)^{1/q} < \infty\}$$

We denote by $l(p,q)$, with the obvious modification for $q = \infty$, the case where $\gamma_n = 1$ for all $n \in \mathbb{N}$.

Let us recall a very useful example of multipliers to be used later on.

Lemma 2. *([K]) Let* $0 < p_1, q_1, p_2, q_2 \leq \infty$ *Then*

$$\Big(l(p_1, q_1), l(p_2, q_2)\Big) = l(p, q)$$

where $\frac{1}{p} = \frac{1}{min(p_1,p_2)} - \frac{1}{p_1}$ *and* $\frac{1}{q} = \frac{1}{min(q_1,q_2)} - \frac{1}{q_1}$.

Note that Plancherel's theorem implies

$$\|f\|_{2,2,\rho} = \Big(\int_0^1 \frac{\rho(1-r)}{1-r} \sum_{n=0}^{\infty} |a_n|^2 r^{2n} dr\Big)^{1/2}$$
$$= \Big(\sum_{n=0}^{\infty} |a_n|^2 \int_0^1 \frac{\rho(1-r)}{1-r} r^{2n} dr\Big)^{1/2} \approx \Big(\sum_{n=0}^{\infty} |a_n|^2 \rho(n^{-1})\Big)^{1/2}.$$

Hence, for ρ verifying (1.3), we have

$$H(2,2,\rho) = \{\lambda_n : \lambda_n \in l\big(2, 2, \rho(2^{-n})\big)\} \tag{1.6}$$

To extend this to other values of $0 < q < \infty$ (see [MP1, S] for the case $\rho(t) = t^\alpha$) we shall use the following lemma.

Lemma 3. *Let* $0 < q < \infty$, ρ *a Dini weight with* $\rho \in b_\alpha$ *for some* $\alpha > 0$ *and* $\alpha_n \geq 0$. *Then*

$$\int_0^1 \frac{\rho(1-r)}{1-r} (\sum_{n=0}^{\infty} \alpha_n r^n)^q dr \approx \sum_{n=0}^{\infty} (\sum_{k \in I_n} \alpha_k)^q \rho(2^{-n})$$

Proof.

$$\int_0^1 \frac{\rho(1-r)}{1-r} (\sum_{n=0}^{\infty} \alpha_n r^n)^q dr = \sum_{n=0}^{\infty} \int_{1-2^{-n}}^{1-2^{-(n+1)}} \frac{\rho(1-r)}{1-r} (\sum_{n=0}^{\infty} \alpha_n r^n)^q dr$$
$$\geq C \sum_{n=0}^{\infty} \int_{1-2^{-n}}^{1-2^{-(n+1)}} \frac{\rho(1-r)}{1-r} (\sum_{k \in I_n} \alpha_k)^q r^{2^n q} dr$$
$$\geq C \sum_{n=0}^{\infty} (\sum_{k \in I_n} \alpha_k)^q \rho(2^{-n}).$$

Let us now show the converse inequality. Assume first $0 < q \leq 1$.

$$\int_0^1 \frac{\rho(1-r)}{1-r} (\sum_{n=0}^{\infty} \alpha_n r^n)^q dr \leq \int_0^1 \frac{\rho(1-r)}{1-r} \Big(\sum_{n=0}^{\infty} (\sum_{k \in I_n} \alpha_k) r^{2^n - 1}\Big)^q dr$$
$$\leq \int_0^1 \frac{\rho(1-r)}{1-r} \Big(\sum_{n=0}^{\infty} (\sum_{k \in I_n} \alpha_k)^q r^{(2^n - 1)q}\Big) dr$$
$$\leq \sum_{n=0}^{\infty} (\sum_{k \in I_n} \alpha_k)^q \int_0^1 \frac{\rho(1-r)}{1-r} r^{(2^n-1)q} dr$$
$$\leq C \sum_{n=0}^{\infty} (\sum_{k \in I_n} \alpha_k)^q \rho(2^{-n})$$

where the last inequality follows from (1.5).

Let us then asumme $q > 1$.

$$\int_0^1 \frac{\rho(1-r)}{1-r}(\sum_{n=0}^{\infty}\alpha_n r^n)^q dr = \sum_{n=0}^{\infty}\int_{1-2^{-n}}^{1-2^{-(n+1)}} \frac{\rho(1-r)}{1-r}(\sum_{n=0}^{\infty}\alpha_n r^n)^q dr$$
$$\leq C\sum_{n=0}^{\infty}\int_{1-2^{-n}}^{1-2^{-(n+1)}} \frac{\rho(1-r)}{1-r}(\sum_{m=0}^{n}\sum_{k\in I_m}\alpha_k)^q dr$$
$$+ C\sum_{n=0}^{\infty}\int_{1-2^{-n}}^{1-2^{-(n+1)}} \frac{\rho(1-r)}{1-r}(\sum_{m=n+1}^{\infty}(\sum_{k\in I_m}\alpha_k)r^{2^{m-1}})^q dr$$
$$= (I) + (II).$$

In order to estimate (I) note that for any sequence $\beta_n \geq 0$ we have

$$\sum_{n=0}^{\infty}(\sum_{m=0}^{n}\beta_m)^q\rho(2^{-n}) \leq C\sum_{n=0}^{\infty}\beta_n^q\rho(2^{-n}) \tag{1.7}$$

Indeed, put $\gamma_{n+1} = \sum_{m=0}^{n}\beta_m$, $\gamma_0 = 0$ and $\mu(A) = \sum_{n\in A}\rho(2^{-n})$, then

$$\sum_{n=0}^{\infty}(\sum_{m=0}^{n}\beta_m)^q\rho(2^{-n}) = q\int_0^{\infty}\lambda^{q-1}\mu(\{n : \sum_{m=0}^{n}\beta_m > \lambda\})d\lambda$$
$$= q\sum_{n=0}^{\infty}\int_{\gamma_n}^{\gamma_{n+1}}\lambda^{q-1}\mu(\{k : \sum_{m=0}^{k}\beta_m > \lambda\})d\lambda$$
$$\leq C\sum_{n=0}^{\infty}(\sum_{m=0}^{n}\beta_m)^{q-1}(\sum_{m=n}^{\infty}\rho(2^{-m}))(\gamma_{n+1}-\gamma_n)$$
$$\leq C\sum_{n=0}^{\infty}(\sum_{m=0}^{n}\beta_m)^{q-1}\beta_n\rho(2^{-n})$$
$$\leq C\Big(\sum_{n=0}^{\infty}(\sum_{m=0}^{n}\beta_m)^q\rho(2^{-n})\Big)^{1/q'}\Big(\sum_{n=0}^{\infty}\beta_m^q\rho(2^{-n})\Big)^{1/q}$$

This gives (1.7), what allows us to get by writing $\beta_n = \sum_{k\in I_n}\alpha_k$

$$(I) \leq C\sum_{n=0}^{\infty}\sum_{m=0}^{n}(\sum_{k\in I_m}\alpha_k)^q\rho(2^{-n}) \leq C\sum_{n=0}^{\infty}(\sum_{k\in I_n}\alpha_k)^q\rho(2^{-n}).$$

In order to estimate (II), first observe that for any $\beta_n \geq 0$

$$(\sum_{m=n+1}^{\infty}\beta_m r^{2^{m-1}})^q \leq (\sum_{m=n+1}^{\infty}2^{-m\alpha}\beta_m^q)(\sum_{m=n+1}^{\infty}2^{m\alpha(q'-1)}r^{(2^{m-1})q'})^{q-1}$$
$$\leq C(\sum_{m=n+1}^{\infty}2^{-m\alpha}\beta_m^q)(\sum_{m=2^n}^{\infty}m^{\alpha(q'-1)-1}r^{mq'})^{q-1}$$
$$\leq C(\sum_{m=n+1}^{\infty}2^{-m\alpha}\beta_m^q)\frac{1}{(1-r)^{\alpha}}.$$

Hence

$$\begin{aligned}
(II) &\leq C\sum_{n=0}^{\infty}\int_{1-2^{-n}}^{1-2^{-(n+1)}}\frac{\rho(1-r)}{1-r}\Big(\sum_{m=n+1}^{\infty}\big(\sum_{k\in I_m}\alpha_k\big)r^{2^{m-1}}\Big)^q dr\\
&\leq C\sum_{n=0}^{\infty}\Big(\sum_{m=n+1}^{\infty}2^{-m\alpha}\big(\sum_{k\in I_m}\alpha_k\big)^q\Big)\int_{1-2^{-n}}^{1-2^{-(n+1)}}\frac{\rho(1-r)}{(1-r)^{\alpha+1}}dr\\
&\leq C\sum_{m=0}^{\infty}\big(\sum_{k\in I_m}\alpha_k\big)^q 2^{-m\alpha}\sum_{n=0}^{m}\int_{1-2^{-n}}^{1-2^{-(n+1)}}\frac{\rho(1-r)}{(1-r)^{\alpha+1}}dr\\
&= C\sum_{m=0}^{\infty}\big(\sum_{k\in I_m}\alpha_k\big)^q 2^{-m\alpha}\int_{0}^{1-2^{-(m+1)}}\frac{\rho(1-r)}{(1-r)^{\alpha+1}}dr\\
&\leq C\sum_{m=0}^{\infty}\big(\sum_{k\in I_m}\alpha_k\big)^q\rho(2^{-m}). \quad \square
\end{aligned}$$

Corollary 1. *Let $0 < q < \infty$, ρ a Dini weight, $\rho \in b_\alpha$ and $f(z) = \sum_{n=1}^{\infty} a_n z^n$. Then*

$$f \in H(2,q,\rho) \text{ if and only if } a_n \in l\big(2,q,\rho(2^{-n})\big).$$

Corollary 2. *Let $1 < q < \infty$, $\frac{1}{q}+\frac{1}{q'}=1$, ρ a Dini weight and $\rho \in b_\alpha$. Then*

$$\Big(H(2,q,\rho)\Big)^* = H(2,q',\rho).$$

under the duality pair $< f,g >= \sum_{n=0}^{\infty} a_n\bar{b}_n\rho(\frac{1}{n})$.

Remark 3. For other values of $1 \leq p \leq \infty$ using (1.5)

$$\begin{aligned}
\sum_{n=0}^{\infty} a_n\bar{b}_n\rho(n^{-1}) &\approx \int_0^1\frac{\rho(1-r)}{1-r}\sum_{n=0}^{\infty}a_n\bar{b}_n r^n dr\\
&= \int_0^1\frac{\rho(1-r)}{1-r}\int_0^{2\pi}f(re^{it})\bar{g}(re^{it})\frac{dt}{2\pi}\\
&\leq C\int_0^1\frac{\rho(1-r)}{1-r}M_p(f,r)M_{p'}(g,r)dr\\
&\leq C\|f\|_{p,q,\rho}\|g\|_{p',q',\rho}.
\end{aligned}$$

This shows that for $1 \leq p \leq \infty$ and $1 < q < \infty$

$$H(p',q',\rho) \subset \Big(H(p,q,\rho)\Big)^*. \tag{1.8}$$

Actually for the case $1 < p < \infty$ it can be shown that $H(p',q',\rho)$ is the dual of $H(p,q,\rho)$. Altough we shall not need the duality result the reader is referred to [S] for a proof that can be easily extended to the weighted case.

Let us finally mention the duality for the case $q = 1$.

Theorem A. *(See* [BS, Theorem 3.2]. *Let $1 \le p < \infty$ and ρ a Dini weight such that $\rho \in b_1$. Then*

$$\Big(H(p,1,\rho)\Big)^* = H^{p'}_{\rho}$$

under the pairing $< f,g >_\alpha = lim_{r\to 1}\sum_{n=0}^{\infty}(n+1)^{-1}a_n b_n r^n$.

Remark 4. This can be extended to weights having b_α condition by means of fractional derivatives (see [B2]).

§2 The Theorems and their proofs.

Next result follows easily from Kintchine's inequality.

Lemma 4. *Let $0 < p,q < \infty$. Let $f(z) = \sum_{n=0}^{\infty} a_n z^n$. Let us denote by $f_t(z) = \sum_{n=0}^{\infty} r_n(t) a_n z^n$ where r_n are the Rademacher functions.*

$$M_2^q(f,r) \approx \int_0^1 M_p^q(f_t,r)dt \tag{2.1}$$

Proposition 1. *Let $0 < p_2 \le 2 \le p_1 < \infty$, $0 < q_1, q_2 < \infty$ and ρ_1, ρ_2 weight functions. Then*

$$\Big((H(p_1,q_1,\rho_1), H(p_2,q_2,\rho_2)\Big) = \Big((H(2,q_1,\rho_1), H(2,q_2,\rho_2)\Big).$$

Proof. It is immediate that

$$\Big((H(2,q_1,\rho_1), H(2,q_2,\rho_2)\Big) \subset \Big((H(p_1,q_1,\rho_1), H(p_2,q_2,\rho_2)\Big).$$

Take now $\lambda_n \in \Big((H(p_1,q_1,\rho_1), H(p_2,q_2,\rho_2)\Big)$ and $f \in H(p_1,q_1,\rho_1)$. Denote by $(f*\lambda)(z) = \sum_{n=0}^{\infty} a_n\lambda_n z^n$ and $(f*\lambda)_t(z) = \sum_{n=0}^{\infty} r_n(t) a_n \lambda_n z^n$ where $r_n(t)$ stand for the Rademacher functions.

An application of (2.1) and Fubini's Theorem give

$$\begin{aligned}
\int_0^1 \frac{\rho_2(1-r)}{1-r} M_2^{q_2}(f*\lambda,r)dr &\le C\int_0^1 \frac{\rho_2(1-r)}{1-r}\int_0^1 M_{p_2}^{q_2}\big((f*\lambda)_t,r\big)dt\,dr \\
&= C\int_0^1\int_0^1 \frac{\rho_2(1-r)}{1-r} M_{p_2}^{q_2}\big((f*\lambda)_t,r\big)dr\,dt \\
&\le C\int_0^1 \Big(\int_0^1 \frac{\rho_1(1-r)}{1-r} M_{p_1}^{q_1}(f_t,r)dr\Big)^{q_2/q_1} dt = A
\end{aligned}$$

Now if $q_2 \le q_1$ then from Fubini and (2.1) again

$$\begin{aligned}
A &\le \Big(\int_0^1\int_0^1 \frac{\rho_1(1-r)}{1-r} M_{p_1}^{q_1}(f_t,r)dr\,dt\Big)^{q_2/q_1} \\
&\le \Big(\int_0^1 \frac{\rho_1(1-r)}{1-r}\int_0^1 M_{p_1}^{q_1}(f_t,r)dt\,dr\Big)^{q_2/q_1} \\
&\le C\Big(\int_0^1 \frac{\rho_1(1-r)}{1-r} M_2^{q_1}(f,r)dr\Big)^{q_2/q_1}
\end{aligned}$$

If $q_2 > q_1$ put $s = \frac{q_2}{q_1}$ and apply duality

$$\begin{aligned} A &= \sup_{\|h\|_{s'}=1} \Big(\int_0^1 \big(\int_0^1 \frac{\rho_1(1-r)}{1-r} M_{p_1}^{q_1}(f_t, r) dr \big) h(t) dt \Big)^{q_2/q_1} \\ &= \sup_{\|h\|_{s'}=1} \Big(\int_0^1 \Big(\int_0^1 M_{p_1}^{q_1}(f_t, r) h(t) dt \Big) \frac{\rho_1(1-r)}{1-r} dr \Big)^{q_2/q_1} \\ &\le \Big(\int_0^1 \Big(\int_0^1 M_{p_1}^{q_2}(f_t, r) dt \Big)^{q_1/q_2} \frac{\rho(1-r)}{1-r} dr \Big)^{q_2/q_1} \\ &\le C \Big(\int_0^1 \frac{\rho(1-r)}{1-r} M_2^{q_1}(f, r) dr \Big)^{q_2/q_1}. \quad \square \end{aligned}$$

Theorem 1. *Let $0 < p_2 \le 2 \le p_1 < \infty$, $0 < q_1, q_2 < \infty$, and ρ_1, ρ_2 weight functions verifying Dini and b_{α_1} and b_{α_2} respectively for some $\alpha_1, \alpha_2 > 0$. Let $\frac{1}{r} = \frac{1}{min(q_1, q_2)} - \frac{1}{q_1}$. Then*

$$\Big(H(p_1, q_1, \rho_1), H(p_2, q_2, \rho_2)\Big) = \{\lambda_n : \sup_{2^n \le k < 2^{n+1}} \big(\rho_2(k^{-1})^{1/q_2} \rho_1(k^{-1})^{-1/q_1} |\lambda_k|\big) \in l^r\}.$$

Proof. Using Proposition 1 and Corollary 1 we simply have to find

$$\Big(l\big(2, q_1, \rho_1(2^{-n})\big), l\big(2, q_2, \rho_2(2^{-n})\big)\Big).$$

Note that $(\lambda_n) \in \Big(l\big(2, q_1, \rho_1(2^{-n})\big), l\big(2, q_2, \rho_2(2^{-n})\big)\Big)$ is equivalent to

$$\rho_2(n^{-1})^{1/q_2} \lambda_n \rho_1(n^{-1})^{-1/q_1} \in \Big(l(2, q_1), l(2, q_2)\Big).$$

Hence the proof is completed by invoking Lemma 2. $\square$

Lemma 5. *(see* [B1, Lemma 5.1 and 5.2]*) Let ρ be a Dini weight and $\rho(t) \in b_\alpha$ for some $\alpha > 0$. Let $(\alpha_n) \ge 0$. Then*

$$\sum_{n=0}^{\infty} \alpha_n r^n = O\big(\frac{\rho(1-r)}{(1-r)^\alpha}\big) \qquad \text{if and only if} \qquad \sum_{m \in I_n} n^{-\alpha} \alpha_n = O\big(\rho(2^{-n})\big).$$

Theorem 2. *Let $2 \le p_1 \le \infty$, $1 < p_2 \le 2$, $1 < q_2 < \infty$ and ρ_1, ρ_2 weight functions verifying Dini and b_1 and b_{α_2} respectively for some $\alpha_2 > 0$. Then*

$$\Big(H_{\rho_1}^{p_1}, H(p_2, q_2, \rho_2)\Big) = \{\lambda_n : \sup_{2^n \le k < 2^{n+1}} \big(k \rho_2(k^{-1})^{1/q_2} \rho_1(k^{-1}) |\lambda_k|\big) \in l^{q_2}\}.$$

Proof. It is clear that $\Big(H_{\rho_1}^2, H(2, q_2, \rho_2)\Big) \subset \Big(H_{\rho_1}^{p_1}, H(p_2, q_2, \rho_2)\Big)$.

Let $(\lambda_n) \in \Big(H^{p_1}_{\rho_1}, H(p_2, q_2, \rho_2)\Big)$. It is not hard to see from duality (see (1.8) and Theorem A) that then

$$(n+1)\lambda_n \rho_2(n^{-1}) \in \Big(H(p_2', q_2', \rho_2), H(p_1', 1, \rho_1)\Big).$$

Using Proposition 1 and duality again this implies $(\lambda_n) \in \Big(H^2_{\rho_1}, H(2, q_2, \rho_2)\Big)$, which, from Corollary 1 and Lemma 5, is equivalent to

$$\rho_2(n^{-1})^{1/q_2} \lambda_n \rho_1(n^{-1})(n+1) \in \Big(l(2,\infty), l(2, q_1)\Big).$$

The proof is again finished from invoking Lemma 2. $\square$

References

[AS] J.M. Anderson, A.L. Shields, *Coefficient multipliers on Bloch functions*, Trans. Amer. Math. Soc. **224** (1976), 256-265.

[BST] G. Bennett, D.A. Stegenga, R.M. Timoney, *Coefficients of Bloch and Lipschitz functions*, Illinois J. of Math. **25** (1981), 520-531.

[B1] O. Blasco, *Operators on weighted Bergman spaces and applications*, Duke Math. J. (to appear).

[B2] O. Blasco, *Multipliers on spaces of analytic functions* (to appear).

[BS] O. Blasco, G.S. de Souza, *Spaces of analytic functions on the disc where the growth of $M_p(F,r)$ depends on a weight*, J. Math. Anal. and Appl. **147** (1990), 580-598.

[BlS] S. Bloom, G.S. de Souza, *Atomic decomposition of generalized Lipschitz spaces*, Illinois J. Math. **33** (1989), 181-189.

[D] P. Duren, *Theory of H_p-spaces* (1970), Academic Press, New York.

[DS1] P. Duren, A.L. Shields, *Coefficient multipliers of H^p and B^p spaces*, Pacific J. Math. **32** (1970), 69-78.

[DS2] P. Duren, A.L. Shields, *Properties of H^p $(0 < p < 1)$ and its containing Banach space*, Trans. Amer. Math. Soc. **141** (1969), 255-262.

[DRS] P.L. Duren, B.W. Romberg, A.L. Shields, *Linear functionals on H_p-spaces with $0 < p < 1$*, J. Reigne Angew. Math. **238** (1969), 32-60.

[F1] T.M. Flett, *The dual of an inequality of Hardy and Littlewood and some related inequalities*, J. Math. Anal. and Appl. **38** (1972), 746-765.

[F2] T.M. Flett, *Lipschitz spaces of functions on the circle and the disc*, J. Math. Anal. and Appl. **39** (1972), 125-158.

[F3] T.M. Flett, *On the rate of growth of mean valeus of holomorphic and harmonic functions*, Proc. London Math. Soc. **20** (1970), 749-768.

[HL1] G.H. Hardy, J.E. Littlewood, *Some properties of fractional integrals II*, Math. Z. **34** (1932), 403-439.

[HL2] G.H. Hardy, J.E. Littlewood, *Theorems concerning mean values of analytic or harmonic functions*, Quart. J. Math. **12** (1941), 221-256.

[K] C. N. Kellogg, *An extension of the Hausdorff-Young Theorem*, Michigan. Math. J. **18** (1971), 121-127.

[J1] S. Janson, *Generalization on Lipschitz spaces and applications to Hardy spaces and bounded mean oscillation*, Duke Math. J. **47** (1980), 959-982.

[M] M. Marzuq, *Linear functionals on some weighted Bergman spaces*, Bull. Austral. Math. Soc. **42** (1990), 417-425.

[MP1] M. Mateljevic, M. Pavlovic, *L^p behaviour of power series with positive coefficients and Hardy spaces*, Proc. Amer. Math. Soc. **87** (1983), 309-316.

[MP2] M. Mateljevic, M. Pavlovic, *L^p behaviour of the integral means of analytic functions*, Studia Math. **77** (1984), 219-237.

[M] A. Matheson, *A Multipliers theorem for analytic functions of slow mean growth*, Proc. Amer. Math. Soc. **77** (1979), 53-57.

[S] W. Sledd, *Some results about spaces of analytic functions introduced by Hardy and Littlewood*, J. London Math. Soc. **2** (1974), 328-336.

[Sh] J.H. Shapiro, *Mackey topologies, reproducing kernels and diagonal maps on the Hardy and Bergman spaces*, Duke Math. J. **43** (1976), 187-202.

[SW] A.L.Shields, D.L. Williams, *Bounded projections, duality and multipliers in spaces of analytic functions*, Trans. Amer. Math. Soc. **162,** (1971), 287-302.

[W] P. Wojtaszczyk, *Multipliers into Bergman spaces and Nevalinna class*, Canad. Math. Bull. **33** (1990), 151-161.

[Z] K. Zhu, *Operator theory in function spaces*, Marcel Dekker, Inc., New York, 1990.

Departamento de Matemáticas, Universidad de Zaragoza, 50009 Zaragoza, Spain.
Current address:
E-mail address: Blasco@cc.unizar.es

Contemporary Mathematics
Volume **144**, 1993

SOME QUESTIONS ARISING FROM THE HOMOGENEOUS BANACH SPACE PROBLEM

PETER G. CASAZZA

ABSTRACT. We review the current state of the homogeneous Banach space problem. We then formulate several questions which arise naturally from this problem, some of which seem to be fundamental but new. We give many examples defining the bounds on the problem. We end with a simple construction showing that every infinite dimensional Banach space contains a subspace on which weak properties have become stable (under passing to further subspaces). Implications of this construction are considered.

1. INTRODUCTION

A Banach space is said to be homogeneous if it is isomorphic to all of its infinite dimensional subspaces.

The Homogeneous Banach Space (HBS) Problem

Is every homogeneous Banach space isomorphic to a Hilbert space?

This problem has frequently been referred to as "Banach's problem" since it is stated in Banach's book [B]. However, it was pointed out in [MTJ2] that in the original polish version of his book, Banach attributes this problem to Mazur. The objective of this paper is to generate renewed interest in the HBS problem and a host of other interesting problems surrounding it.

Essentially no progress was made on this problem of mazur until recently when J. Bourgain [B*] solved the finite dimensional version and W. B. Johnson [J1] solved a special case of the general problem. Both papers rely on advances made in the local theory of Banach spaces in the 1980's. These results are discussed in section 3.

In section2, we consider the long list of immediate questions we can also ask about a homogeneous Banach space, most of which are still unanswered. In section 4, we will look at a series of problems which are stronger than the homogeneous Banach space problem, in the sense that a positive answer to one of these would yield a positive answer to the HBS problem. In section 5 we look at some fundamental questions which arise from the HBS problem, some of which do not seem to have

---

1991 *Mathematics Subject Classification.* 46B07, 46B20, 46B25.

Supported by NSF DMS 9001344 and a grant from the University of Missouri Research Council

Part of this research was conducted while the author was a visiting scholar at the University of Cambridge, ENGLAND been asked before. Finally, in section 6, we give a simple construction that can be carried out in any Banach space to yield a subspace which is stable for weak properties, for passing to further subspaces. We will also show that this is the strongest subspace that we can expect to find inside of every Banach space.

The author expresses his deepest gratitude to W. B. Johnson for many interesting discussions related to the material in this paper.

2. Some Immediate Questions

There are several immediate questions which we can ask about a homogeneous Banach space X: (1) Is X reflexive?; (2) Is X suprreflexive?; (3) Does X have a separable dual space? None of these questions has been answered yet. However, we do not believe that any of these questions will be important for the solution of the problem. But these questions do give rise to some interesting problems in the general theory of Banach spaces which are discussed in section 5. Even the isometric version of the HBS problem is still open.

The Homogeneous Banach Space (HBS) Problem (Isometric Version)

If X is isometric to all of its infinite dimensional subspaces, is X isometric (or even isomorphic) to a Hilbert space?

With so little progress having been made on this problem, we might hope to get some movement on it by strengthening our hypotheses. Unfortunately, assuming that X is homogeneous and has a unconditional basis, or even a symmetric basis, does not seem to help. These hypotheses do yield that X is superreflexive and that X is weak cotype 2 [J1] but do not seem to lead to any serious breakthrough on the problem.

Another obvious question is: Is every homogeneous Banach space X isomorphic to a square? (i.e. Is $X \approx X \oplus X$?) Even such an elementary question was unresolved until recently and requires some heavy machinery of W. T. Gowers and B. Maurey [GoM]. Recall that a projection P on a Banach space is said to be nontrivial if $\text{rnk}P = \text{rnk}(I - P) = \infty$.

Theorem 2.1 ($\mathbf{G_0M}$). *If an infinite dimensional Banach space X has no nontrivial projections on any infinite dimensional subspace, then every bounded operator on X is a strictly singular pertubation of the identity.*

This means that every bounded operator T on such a space can be written as: $T = aI + S$, where S is a strictly singular operator (i.e. S is not an isomorphism when restricted to any infinite dimensional subspace of X.) We also need the notion of Fredholm index. If $T : X \to Y$ is a bounded operator with closed range, put $\alpha(T) = \dim \ker T$, $\beta(T) = \dim Y/TX$. If either $\alpha(T) < \infty$ or $\beta(T) < \infty$, we define the Fredholm index $i(T)$ by: $i(T) = \alpha(T) - \beta(T)$. If $i(T)$ is defined and is finite, then T is called a Fredholm operator. (See chapter 2.c of [LT1], for basic information on $i(T)$). We are now ready for:

Result 2.2. *If X is homogeneous then $X \approx X \oplus X$.*

Proof. If not, then X has no non–trivial projection on any infinite dimensional subspace. Let T be an isomorphism of X onto a hyperplane in X. By Theorem 2.1, $T = aI + S$. But, T has Fredholm index -1 while $aI + S$ has Fredholm index 0 by Proposition 2.c.10 of [LT1]. This contradiction completes the proof. □

There are some immediate questions which would be useful for the solution to the HBS problem. One such question is,

Question 2.3. *If X is homogeneous, is X^* homogeneous?*

In section 3, we will see a case where W. B. Johnson [J1] made use of such a hypothesis. If X is homogeneous then, since X has a subspace with a basis, every subspace of X has a basis. In particular, X is separable and every subspace of X has the approximation property. It follows ([LT2], Theorem 1.g.6) that;

$$\sup\{p | X \text{ is type } p\} = 2 = \inf\{q | X \text{ is cotype } q\}. \tag{2.1}$$

This fact gives rise to some interesting open questions in section 4.

We also feel that the following open question will be important to the solution of the HBS problem.

Question 2.4. *If X is homogeneous, is X uniformly isomorphic to all of its infinite dimensional subspaces?*

That is, does there exist a $K \geq 1$ so that X is K–isomorphic to all of its infinite dimensional subspaces? One reason for the importance of question 2.4 is that, perhaps the HBS problem has a positive answer in the uniform case but a negative answer in the general case. We can prove, with some relatively soft infinite dimensional theory, the following result:

Proposition 2.5. *If X is a homogeneous Banach space, then there is a constant $K \geq 1$ so that X K–embeds into every infinite dimensional subspace of X.*

Proposition 2.5 will follow from a general result on minimal Banach spaces. A Banach space is minimal if it embeds into every one of its infinite dimensional subspaces. We discuss such spaces in more detail in section 6.

Proposition 2.6. *If X is a minimal Banach space then there is a $K \geq 1$ so that X K–embeds into every infinite dimensional subspace of X.*

Proof. Assume, for the sake of contradiction, that X is not uniformly embeddable into all of its infinite dimensional subspaces. Then no subspace of X has this property either. So there are infinite dimensional subspaces $Y_1 \supset Y_2 \supset \cdots$ so that X does not $2^{f(n)}$–embed into Y_n, where $f(n) = n(1+\sqrt{n})^2$. Choose $y_n \in Y_n$ so that (y_n) is a basic sequence in X and let $E_n = \text{span}_{n \leq i}\, y_i$. By assumption, there is a subspace $Z \subset E_1$ and a $K \geq 1$ so that Z is K–isomorphic to X. Without loss of generality, we may assume that $\dim E_1/Z = \infty$. For each n let $H_n = Z \cap E_n$ and choose $F_n \subset E_n$ with $H_n \subset F_n$ and $k = \dim E_n/F_n = \dim Z/H_n \leq n$. There are projections $P : E_n \to F_n$ and $Q : Z \to H_n$ with $\|p\| \leq 1+\sqrt{n}$ and $\|Q\| \leq 1+\sqrt{n}$. Hence, there are k–dimensional spaces $W \subset E_n$ and $W' \subset Z$ so that $W' \oplus_1 H_n$ is $(1+\sqrt{n})$–isomorphic to Z while *span*(W, H_n) is $(1+\sqrt{n})$–isomorphic to $W \oplus_1 H_n$.

Since W and W' are n–isomorphic, it follows that $V = \text{span}(W, H_n)$ is $n(1+\sqrt{n})^2$–isomorphic to Z and hence V is $Kn(1+\sqrt{n})^2$–isomorphic to X. But $V \subset E_n$ implies X is $Kn(1+\sqrt{n})^2$–embeddable into E_n. For large n, this contradicts our assumption that X is not $2^{f(n)}$–embeddable into E_n, for $f(n) = n(1+\sqrt{n})^2$. □

Proposition 2.6 easily gives a subspace $Y \subset X$ with a finite dimensional decomposition $\Sigma \oplus E_n$ so that for every k, $(E_n)_{n=k}^{\infty}$ is dense in the family of all finite dimensional subspaces of X. Hence, there is a constant $K \geq 1$ so that every finite dimensional subspace of X is K–isomorphic to a K–complemented subspace of every finite codimensional subspace of X.

Another consequence of a positive answer to 2.4 comes from Krivine's Theorem [Kr], [MS]. If X is homogeneous, a positive answer to 2.4, combined with 2.1 and Krivine's Theorem implies:

Theorem 2.7. *If X is K–isomorphic to all of its infinite dimensional subspaces, then for every n, there is a basis (x_i) of X with basis constant $\leq K$ and for every finite set of natural numbers I with $|I| = n$, $(x_i)_{i\in I} \approx_K (e_i^2)_{i\in I}$, where (e_i^2) is the unit vector basis of l_2.*

This says that X has a sequence of bases $(x_n^k)_{n=1}^{\infty}$ with uniformly bounded basis constants so that every k elements of $(x_n^k)_{n=1}^{\infty}$ are uniformly the l_2^k unit vector basis. This property itself is very strong and we might hope that it already characterizes a Hilbert space. Although, as we will see, it does not characterize Hilbert space, it does give rise to the notion of sequences of successively better bases for a Banach space.

Definition 2.8. We say that a Banach space X has basis property p $(1 \leq p \leq \infty)$ if there is a $K \geq 1$ so that for every n, there is a basis (x_i) of X with basis constant $\leq K$ and for every n–element subset I of the natural numbers, $(x_i)_{i\in I} \approx_K (e_i^p)_{i\in I}$, where (e_i^p) is the unit vector basis of l_p (or c_0 if $p = \infty$). To simplify notation, we write $(x_n) \leq (y_n)$, for two sequences in Banach spaces X, Y respectively, if there is a constant $K \geq 1$ so that for every sequence of scalars (a_n),

$$\left\| \sum_{n=1}^{\infty} a_n x_n \right\| \leq K \left\| \sum_{n=1}^{\infty} a_n y_n \right\| .$$

We now have,

Proposition 2.9. *Let X be a Banach space with a normalized basis (x_n) satisfying $(x_n) < (e_n^p)$. Then $X \oplus l_p$ has basis property p.*

Proof. Let (x_n) be the normalized basis of X satisfying $(x_n) < (e_n^p)$. Fix a natural number m. Define $y_n \in X \oplus l_p$ by:

$$y_{k2m+(k+1)} = \left(x_{k+1}, \frac{1}{(2m)^{1/p}} \sum_{i=k2m+1}^{(k+1)2m} e_i^p \right)$$

$$y_{k2m+(k+1)} + j = \left(0, e_{k2m+j}^p\right), \qquad \text{for } 1 \leq j \leq 2m,$$

and $k = 0, 1, 2, \ldots$. For any sequence of scalars (a_i) and any $t < s$, choose k_1, k_2, j_1, j_2 so that $t = k_1 2m + (k_1+1) + j_1$ and $s = k_2 2m + (k_2+1) + j_2$. Then, treating $X \oplus l_p$ as an l_p–sum and letting b be the basis constant of (x_n) we have,

$$
\begin{aligned}
\left\| \sum_{i=1}^{t} a_i\, y_i \right\| = \\
& \left[\left\| \sum_{i=0}^{k_1} a_{i2m+(i+1)}\, x_{i+1} \right\|^p + \right. \\
& \sum_{i=0}^{k_1} \sum_{j=1}^{2m} \left| a_{i2m+(i+1)+j} - \frac{1}{(2m)^{1/p}}\, a_{i2m+(i+1)} \right|^p \\
& + \sum_{j=1}^{j_1} \left| a_{i2m+(i+1)+j} - \frac{1}{(2m)^{1/p}}\, a_{i2m+(i+1)} \right|^p + \\
& \left. \frac{2m - j_1}{2m} \left| a_{k_1 2m+(k_1+1)} \right|^p \right]^{1/p} \\
& \le \left[b \left\| \sum_{i=0}^{k_1} a_{i2m+(i+1)}\, x_{i+1} \right\|^p + \right. \\
& \sum_{i=0}^{k_1} \sum_{j=1}^{2m} \left| a_{i2m+(i+1)+j} - \frac{1}{(2m)^{1/p}}\, a_{i2m+(i+1)} \right|^p \\
& \left. + \sum_{j=1}^{j_1} \left| a_{i2m+(i+1)+j} - \frac{1}{(2m)^{1/p}}\, a_{i2m+(i+1)} \right| \right]^{1/p} \\
& \le \left[b^2 \left\| \sum_{i=0}^{k_1} a_{i2m+(i+1)}\, x_{i+1} \right\|^p + \right. \\
& \sum_{i=0}^{k_2} \sum_{j=1}^{2m} \left| a_{i2m+(i+1)+j} - \frac{1}{(2m)^{1/p}}\, a_{i2m+(i+1)} \right|^p \\
& + \sum_{j=1}^{j_2} \left| a_{i2m+(i+1)+j} - \frac{1}{(2m)^{1/p}}\, a_{i2m+(i+1)} \right|^p + \\
& \left. \frac{2m - j_2}{2m} \left| a_{k_2\, 2m+(k_2+1)} \right|^p \right]^{1/p} \\
& \le b^2 \left\| \sum_{i=1}^{s} a_i\, y_i \right\| .
\end{aligned}
$$

So the basis constant of (y_n) is b^2 where b is the basis constant of (x_n).

Since $(x_n) < (e_n^p)$, there is a constant $K \ge 1$ so that for every sequence of scalars

(a_n) we have,

$$\left\|\sum_{n=1}^{\infty} a_n x_n\right\| \le K\left(\sum_{n=1}^{\infty} |a_n|^p\right)^{1/p}. \tag{2.2}$$

On the other hand, if $I \subset \{k2m+(k+1)+1, k2m+(k+1)+2, \dots, (k+1)2m+(k+1)\}$ with $|I| \le m$,

$$\left[\sum_{j\in I}\left|a_{k2m+(k+1)+j} - \frac{1}{(2m)^{1/p}} a_{k2m+(k+1)}\right|^p + \frac{2m-|I|}{2m}\left|a_{k2m+(k+1)}\right|^p\right]^{1/p} \tag{2.3}$$

$$\ge \frac{1}{4}\left[\sum_{j\in I}\left|a_{k2m+(k+1)+j}\right|^p + \left|a_{k2m+(k+1)}\right|^p\right]^{1/p}. \tag{2.4}$$

Now, if $I \subset \mathbb{N}$, $|I| = m$, for each $k = 0, 1, \dots$ let $I_k = I \cap \{k2m+(k+1), k2m+(k+1)+1, \dots, (k+1)2m+(k+1)\}$ and let P be the natural projection of $X \oplus l_p$ onto l_p. From 2.4 we have,

$$\left\|\sum_{i\in I} a_i y_i\right\| \ \ge \ \left(\sum_{k=0}^{\infty}\left\|\sum_{i\in I_k} a_i P y_i\right\|^p\right)^{1/p} \tag{2.5}$$

$$\ge \ \frac{1}{4}\left(\sum_{k=0}^{\infty}\sum_{i\in I_k} |a_i|^p\right)^{1/p}. \tag{2.6}$$

By 2.2 and 2.6, we have that $(y_i)_{i\in I} \approx_{8K} (e_i^p)_{i\in I}$.
This completes the proof of the proposition. □

For $1 < p \le 2$, $L_p[0,1] \approx L_p[0,1] \oplus l_2$, and the Walsh system (W_n) is a basis of $L_p[0,1]$ satisfying $(W_n) < (e_n^2)$. So we have,

Corollary 2.10. *For $1 < p \le 2$, $L_p[0,1]$ has basis property 2.*

Since any space with basis property p contains uniformly complemented l_p^n's, it follows that $L_1[0,1]$ fails basis property 2. It is immediate that every basis (x_n) for a Banach space X satisfies $(e_n^0) < (x_n) < (e_n^1)$. So for example, $L_1[0,1]$ has basis property 1. That is, $L_1[0,1]$ has bases with uniformly bounded basis constants every n–elements of which are the unit vector basis of l_1^n (and hence are well unconditional) despite the fact that $L_1[0,1]$ has no unconditional basis. It can be shown that $T^{(2)}$–Tsirelson's space [CS] has basis property 2 despite its not containing a subspace isomorphic to c_0 or l_p, for any p. Also, $(\sum \oplus T^{(2)})_{l_2}$ has basis property 2.

We could strengthen this notion to "unconditional" basis property p by requiring the unconditional basis constant to be $\le K$ in definition 2.9. Although $l_p \oplus l_2$ has basis property p, for $2 < p < \infty$ (by proposition 2.10) it fails to have unconditional basis property p by the quantitative version of a result of Edelstein and Wojtaszczyk [EW] (see also [W], [CKT]). Also, $T^{(2)}$ fails to have unconditional basis property 2 by the uniqueness, up to a permutation, of the unconditional basis for $T^{(2)}$ [BCLT]. We do not know of a non–Hilbert space with unconditional basis property 2. Probably, such examples exist in the class of Orlicz spaces (Probably even spaces with "symmetric" basis property 2). This whole idea could warrant further study if we could first find a good use for this concept.

Recently, V. Mascioni [Ma] has introduced an interesting variant of the HBS problem.

Problem 2.11. [Ma] *If X is an infinite dimensional Banach space, and every infinite dimensional subspace of X is isomorphic to its dual space, is X isomorphic to a Hilbert space?*

We could also ask the isometric version of problem 2.11. Mascioni [Ma] then finite dimensionalizes the problem.

Definition 2.12. [Ma] A Banach space is locally self dual (LSD) if there is a constant c such that every finite dimensional subspace of X is c–isomorphic to its dual space.

Problem 2.13. [Ma] *Are LSD spaces isomorphic to Hilbert spaces?*

We will discuss these results further in section 4.

3. Some Positive Results

The first major advance in this area was due to J. Bourgain [B^*], who solved the finite dimensional version of the HBS problem. Later, N. Tomczak–Jaegermann and P. Mankiewicz [MTJ1] gave the best constants for the finite dimensional homogeneous Banach space problem. V. D. Milman first posed this problem in its finite dimensional form. In the finite dimensional setting, we cannot ask for X to be isomorphic to its subspaces. Instead, we assume that $\dim X = n$ and for some $1 \le k \le n$, we assume that all k–dimensional subspaces of X are K–isomorphic. Now we must ask for a quantitative answer: what is the smallest constant $f(K)$ so that X is $f(K)$–isomorphic to a Hilbert space? Clearly, $n = 1$ will give no information about the Banach space X. Yet, we have a quite exact answer to this problem from [B^*], and [MJT1].

Theorem 3.1. [MTJ1] *If an n–dimensional Banach space X has the property that all its $[\alpha n]$–dimensional subspaces are K–isomorphic, for some $0 < \alpha < 1$, then X is $f(\alpha, K)$–isomorphic to a Hilbert space, where $f(\alpha, K) = CK^{3/2}$, if $0 < \alpha < 2/3$ and $f(\alpha, K) = CK^2$, if $2/3 < \alpha < 1$, and $C = C(\alpha)$ depends on α only.*

We will come back to this theorem in a moment. In the meantime, we consider the only solution of a special case of the infinite dimensional problem due to W. B. Johnson [J1]. Recall that a Banach space X is said to have the GL–property (Gordon–Lewis property) if every absolutely summing operator from X to L_2 factors through L_1. Y. Gordon and D. Lewis [GL] in a landmark paper showed that every Banach space with an unconditional basis (or even LUST) has the GL–property. W. B. Johnson [J1] conjectured that every Banach space has a subspace with the GL–property. This conjecture does not seem to have been tested yet on the new examples of Banach spaces without unconditional bases due to W. T. Gowers and B. Maurey [GoM]. For a full understanding of Johnson's result, we need to recall some definitions.

Definition 3.2. (1) A Banach space X is a weak cotype q space if there is a constant $wc_q(X)$ so that for all n and all operators $u : l_2^n \longrightarrow X$, we have

$$\sup_{k\geq 1} k^{1/q}\, a_k\,(u) \leq we_q(X)\, l(u),$$

where

$$l(u) = \left(\int_{\mathbb{R}} \|u(x)\|^2\, \gamma_n(dx) \right)^{1/2}$$

and γ_n is the canonical Gaussian probability measure on $\mathbb{R}^n$, and

$$a_k(u) = \inf\{\, \|u - s\| \,|s : l_2^n \longrightarrow X,\ \text{rank}\ s \leq k\,\}.$$

(2) A Banach space X is a weak type p space if there is a constant $wt_p(X)$ so that for all n and all operators $V : X \longrightarrow l_2^n$ we have

$$\sup_{k\geq 1} k^{1/p}\, a_k(V) \leq wt_p(X)\, l^*(V)$$

where

$$l^*(V) = \sup\{tr(uv)\,|u : l_2^n \longrightarrow X, l(u) \leq 1\,\}.$$

(3) A Banach space X is a weak Hilbert space if it is weak type 2 and weak cotype 2 space.

(4) A Banach space is as–Hilbertian (asymptotically Hilbertian) if there is a constant $K \geq 1$ so that for every n, there is a finite codimensional subspace H of X with the property that every n–dimensional subspace of H is K–isomorphic to l_2^n.

See G. Pisier [P] for a complete treatment of these notions. Johnson observed;

Theorem 3.3. [J1] *A homogeneous weak Hilbert space is isomorphic to a Hilbert space.*

The main result from [J1] is:

Theorem 3.4. [J1] *If X and X^* are homogeneous and if X has the GL–property, then X is isomorphic to a Hilbert space.*

Returning to Theorem 3.1, we might naturally formulate an alternative infinite dimensional HBS problem as:

If X is K–isomorphic to all of its finite codimensional subspaces, is X $f(K)$–isomorphic to a Hilbert space?

It is easily seen that this property is so strong that even c_0, l_p, $1 \leq p < \infty$, fail it. In particular, a space X with this property also satisfies the property that all finite dimensional subspaces of X are K–isomorphic to K–complemented subspaces of X. Also, the proof of Theorem 3.4 shows that whenever X and X^* have this property and X has the GL–property, then X is isomorphic to a Hilbert space. However, this property does not characterize a Hilbert space as our next proposition shows.

Proposition 3.5. (1) *If X is a Banach space with a separable dual X^*, and $\varepsilon > 0$, then there is a Banach space Y with a separable dual containing X and Y is $1 + \varepsilon$–isomorphic to every subspace of Y of finite codimension.*

(2) *If in (1), X^{**} is separable, we may construct our Y satisfying (1) and so that Y^* is also $1+\varepsilon$–isomorphic to every subspace of Y^* of finite codimension.*

Proof. Let $X_1 = (\sum_{k=1}^{\infty} \oplus X)_{l_2}$ and choose $f_n^1 \in X_1^*$ so that $\{f_n^1\}$ is dense in X_1^* and each f_n^1 appears infinitely many times in the sequence. Let J be the family of all finite subsets of $\mathbb{N}$. For each $A \in N$ let $Z(A) = \operatorname{span}_{n \in A} f_n^1$ and let

(1) $X_2 = X_1 \oplus_2 \left(\sum_{A \in J} \oplus Z(A)_\perp\right)_{l_2}$
It follows that

(2) $X_2 \approx_{(1)} X_1 \oplus X_2$,
and if $A \in J$, then

(3) $X_2 \approx_{(1)} X_2 \oplus_2 Z(A)_\perp$.
Now choose (f_n^2) dense in X_2^* with each f_n^2 appearing infinitely many times in the sequence and repeat the construction to get X_3. By induction, we construct $X_1, X_2, \ldots$. Now let

(4) $Y = (\sum_{n=1}^{\infty} \oplus X_n)_{l_2}$.
If follows that

(5) $Y \approx_{(1)} (\sum_{k=n}^{\infty} \oplus X_k)_{l_2}$, for every n, and

(6) if H is a subspace of $(\sum_{i=1}^{n} \oplus X_i)_{l_2}$ of finite codimension, and $H = (\operatorname{span}_{k \in A} f_k^n)_\perp$, for some $A \in J$, then

$$\left(\sum_{k=n+1}^{\infty} \oplus X_k\right)_{l_2} \approx_{(1)} \left(\sum_{k=n+1}^{\infty} \oplus X_k\right)_{l_2} \oplus_2 H.$$

Now let $H \subset Y$ be a subspace of finite codimension and $H^\perp = \operatorname{span}_{1 \le i \le n} f_i$, where $f_i = \sum_{j=1}^{\infty} f_{ij}$, $f_{ij} \in X_j^*$ and the basis constant of $\{f_i\}_{i=1}^n$ is $\le n^2$. Choose $\varepsilon_0 < 1 + \varepsilon / 2\sqrt{n} n^2$ and an m so that

(7) $\|f_i - \sum_{j=1}^{m} f_{ij}\| < \varepsilon_0$, for all $i = 1, 2, \ldots, n$.
Let $g_i = \sum_{j=1}^{m} f_{ij}$, $1 \le i \le n$ and choose $f_{k_i}^{\ m} \in \left(\sum_{j=1}^{m} \oplus X_j^*\right)_{l_2}$ so that

(8) $\|f_{k_i}^{\ m} - g_i\| < \varepsilon_0$.
It follows that

(9) $\|f_{k_i}^{\ m} - f_i\| < 2\varepsilon_0$.
Choose a w^*–continuous projection P^* of Y^* onto $\operatorname{span}_{1 \le i \le n} f_{k_i}^{\ m}$ with $\|P^*\| < 2\sqrt{n}$. Finally, define $T : Y^* \longrightarrow Y^*$ by: Given $f \in Y^*$, let $P^*(f) = \sum_{i=1}^{n} a_i f_{k_i}^{\ m}$ and define

(10) $T(f) = \sum_{i=1}^{n} a_i f_i + (I - P^*)(f)$.

We now have, for all $f \in Y^*$,

(11) $\|(I-T)(f)\| \le \sum_{i=1}^{n} |a_i| \|f_i - f_{K_i^m}\| < 2\sqrt{n} n^2 \varepsilon_0 < 1 + \varepsilon$.
So T is a w^* continuous isomorphism on Y^* with $\|T\| \, \|T^{-1}\| \le (1+\varepsilon)^2$. Let $S : T \longrightarrow Y$ be the isomorphism satisfying $T = S^*$. Finally, $T\left(\text{span}_{i \le i \le n} f_{k_i^n}\right) = H^\perp$ implies $S(H) \subset \left(\text{span}_{1 \le i \le n} f_{k_i^m}\right)_\perp$, and dimension considerations yields that S is a $(1+\varepsilon)^2$ isomorphism of H onto $\left(\text{span}_{1 \le i \le n} f_{k_i^n}\right)_\perp$. Now, $H \approx_{(1+\varepsilon)^2} \left(\text{span}_{1 \le i \le n} f_{k_i^n}\right)_\perp$ and so $H \approx_{(1+\varepsilon)^2} H_1 \oplus \left(\sum_{j=m+1}^{\infty} \oplus X_j\right)_{l_2}$ where H_1 has codimension n in $\left(\sum_{j=1}^{n} \oplus X_j\right)_{l_2}$.
So by (5) and (6),

$$H \approx_{(1+\varepsilon)^2} \left(\sum_{j=m+1}^{\infty} \oplus X_j\right)_{l_2} \approx_{(1)} Y.$$

For part (2) of the proposition, we alternate the above construction between X_n and X_n^*. □

A variation of this construction carried out transfinitely would yield an alternate proof of a result in the literature. (This result definitely exists in the literature despite our inability to locate it at this time). One can view this as a counterexample to the "non–separable" HBS problem.

Theorem 3.6. *There is a non–separable Banach space X which is not isomorphic to a Hilbert space, but such that X is isometric to every subspace of the same density as X.*

4. Some Stronger Problems

At this time, weak Hilbert space theory seems to be closing in on the HBS problem. Several natural questions from that area would yield a positive solution to the HBS problem. Since every subspace of a homogeneous Banach space has a basis, a positive answer to the following question, combined with Theorem 3.3, would yield a positive answer to the HBS problem:

Question 4.1. *If every subspace of X has a basis, is X a weak Hilbert space?*

The converse of 4.1 is also an open problem. That is; Does every separable weak Hilbert space have a basis? A result of B. Maurey and G. Pisier (which appears for the first time in [Ma]) states that a separable weak Hilbert space has a finite dimensional decomposition. R. Komorowski [K] has constructed the first weak Hilbert spaces which fail to have unconditional bases (and they are even unconditional sums of two dimensional subspaces).

Johnson's result 3.3 actually asserts that a homogeneous Banach space which is as–Hilbertian is isomorphic to a Hilbert space. And an earlier result of Johnson (see [P]) states that every weak Hilbert space is as–Hilbertian. So a weaker formulation of 4.1 would be:

Question 4.2. *If every subspace of X has the approximation property, is X as–Hilbertian?*

It is easily seen that there are as–Hilbertian spaces which fail the approximation property. Also, Johnson [J2] has exhibited a class of Banach spaces which are as–Hilbertian, every subspace has the approximation property (even the finite dimensional decomposition property) but they have subspaces without bases. There is a possible counterexample to question 4.2. That is, the symmetric convexified Tsirelson space [CS]. This is a Banach space with a symmetric basis for which all n–dimensional subspaces are within a fixed iterated logarithm of l_2^n. It is possible, however, that every non Hilbert space with a symmetric basis has a subspace which fails the approximation property.

N. Tomczak–Jaegermann and P. Mankiewicz [MTJ1] (see [MTJ3]) obtained the following corollary while proving some deep results in the local theory of Banach spaces:

Theorem 4.3. [MTJ1] *Let X be a Banach space for which there exists a constant K such that every finite dimensional subspace F of X satisfies $bc(F) \leq K$. Then X is of weak cotype 2.*

This Theorem gives added importance to the following question which has been around for quite some time.

Question 4.4. *Are the following equivalent for a Banach space X?*

(1) *Every subspace of X has a basis.*
(2) *There is a constant $K \geq 1$ so that every finite dimensional subspace of X has a basis with basis constant $\leq K$.*

The importance of question 4.4 is that a positive solution to (1) $\Longrightarrow$ (2) would yield that every homogeneous Banach space is a weak cotype 2 space. This might then be combined with a positive solution to question 2.3 to solve the whole problem.

The argument of Johnson [J2] is "local" and shows that convexified Tsirelson's space has both properties (1) and (2) of question 4.4. N. Nielsen and N. Tomczak–Jaegermann [NTJ] have shown that very weak Hilbert space with LUST also has these properties.

Recently, P. Mankiewicz and N. Tomczak–Jaegermann [MTJ3] made an important step in resolving the above questions:

Theorem 4.5. [MTJ3] *If a Banach space X has the property that every subspace of every quotient space of $l_2(X)$ has a Schauder basis, then X is isomorphic to a Hilbert space.*

Neither implication in question 4.4 is known, which points out a serious gap in the available techniques in Banach space theory. Namely, we have no reasonable way of passing results back and forth between local theory and infinite dimensional theory. The paper [MTJ3] should be read not only for the main result, but also because it is the first serious integration of local theory and infinite dimensional theory.

Mascioni [Ma] has proved the corresponding result for LSD–spaces.

Theorem 4.6. [Ma] *If $l_2(X)$ is LSD, then X is isomorphic to a Hilbert space.*

Another possible counterexample to question 4.2 is $X = \left(\sum \oplus T^{(2)}\right)_{l_2}$, where $T^{(2)}$ is convexified Tsirelson space (see [CS]). This space is of type 2 but fails to be weak cotype 2, while still satisfying:

$$\inf\{q|X \text{ is cotype } q\} = 2.$$

Results of N. Tomczak–Jaegermann and P. Mankiewicz [MTJ1] show that X fails the finite dimensional basis property (i.e. (2) of question 4.4). Also, results from [MTJ3] show that this space has a subspace of a quotient space which fails to have a basis. It is possible, however, that every subspace of X has the approximation property.

Finally, a positive answer to the following question (plus its dual formulation for weak cotype) would yield a positive solution to the HBS problem:

Question 4.7. *If* $\sup\{p|X \text{ is Type } p\} = p_0$, *does* X *contain a subspace of weak type* p_0*?*

5. Some Basic Questions

Let us return to some of the open questions of section 2. It seems strange that we do not know if a homogeneous Banach space is reflexive, especially in light of formula (1) of section 2. R. C. James [Ja] has given us a class of Banach spaces with type which fail to be reflexive. But a positive answer to the following question would show that homogeneous Banach spaces are reflexive.

Question 5.1. *If a Banach space* X *is of type* p, *for some* $1 < p \leq 2$, *must* X *contain a reflexive subspace?*

If a Banach space has a subspace with LUST, the answer is "yes" (see [LT2]). Actually, a stronger conclusion could hold:

Question 5.2. *Does every Banach space of type* p *for some* $1 < p \leq 2$, *contain a superreflexive subspace?*

The cotype version of question 5.1 also seems to be unasked and unanswered:

Question 5.3. *If* X *is of cotype* q, *for some* $2 \leq q \leq \infty$, *must* X *contain a separable dual space?*

An equivalent formulation of question 5.3 would be to ask if X must contain a boundedly complete basic sequence. These questions are special cases of the old problem: Does every Banach space contain either a reflexive subspace or a subspace isomorphic to c_0 or l_1? In fact, question 5.2 is the "uniform" or "local" version of this problem. That is, question 5.2 is equivalent to:

Question 5.2'. *Does every Banach space contain either a superreflexive subspace or subspaces uniformly isomorphic to* l_1^n *or* l_∞^n, *for every* $n = 1, 2, \ldots$ *?*

In this setting, the "l_∞^n" in the question is redundant.

Recently W. T. Gowers [Go] has constructed a Banach space not containing c_0, l_1 or any reflexive subspace. In fact, Gowers' space has no subspace with a separable dual space. It is possible that refinements of this example will give counterexamples to the above problems.

6. A Construction

What is the "best" subspace we can find inside of every Banach space? The major problem here is not just to answer the question, but to formulate the question. We now know that almost every Banach space contains a subspace which fails the approximation property (see [LT2], Chapter 1g). Thanks to W. B. Johnson [J2] we also know of non–Hilbert spaces for which every subspace of every quotient space has a basis. We also know that every Banach space contains a basic sequence but may not contain an unconditional basic sequence [GoM]. The best subspace we could hope to find in a Banach space is a subspace isomorphic to c_0, or l_p, $1 \leq p < \infty$. Not just because we understand these spaces better than any others, but because they have the property that they embed (complementably) into every infinite dimensional subspace of themselves. That is, we can recover all of the properties of the whole space inside of every subspace. But in 1972, B. S. Tsirelson [T] (see also [CS]) showed that there are Banach spaces which do not contain copies of c_0 or l_p, $1 \leq p < \infty$. This example quickly blossomed into an "industry" [CS] and even today has its place in the recent exciting solutions to the unconditional basic sequence problem [GoM], the c_0 l_1, reflexive space counterexample [Go], and the distortion problem [S1], [OS], [MiTJ]. H. P. Rosenthal then raised the question whether every Banach space X might contain a subspace Y which embeds into every one of its infinite dimensional subspaces? Such a space Y is called minimal. This was the "best" subspace we could hope for at the time. In 1982, P. G. Casazza and E. Odell [C0] (see also [CS]) showed that Tsirelson's space contains no minimal subspaces. As of this writing, we know of only two new classes of minimal Banach spaces (besides subspaces of c_0, l_p, $1 \leq p < \infty$) [CJT], [S2], and only the second is complementably minimal. So, we now know that we cannot find, in every Banach space X, a subspace Y so that infinite dimensional properties of Y are invariant under passing to further subspaces.

Our next approach would be to look for a "locally best" subspace in every Banach space, i.e. a subspace Y of X so that Y is crudely finitely representable in every one of its subspaces. This is not possible either as Tsirelson's space again fails this property by the argument of [OS]. Since Tsirelson's space seems to be blocking all our efforts, let's see what property this space does have. The Tsirelson space T_p enjoys the property that there is a constant K so that for every n there is a subspace H of T_p of condimension n and every n–dimensional subspace of H, K–embeds into every infinite dimensional subspace of T_p. From our earlier examples, this is the "best" we can hope for in an arbitrary Banach space. Our next theorem states that, indeed, every Banach space does contain such a subspace (and with K arbitrarily close to 1).

Recall that a Banach space Y (finite or infinite dimensional) almost isometrically embeds into a Banach space X if for every $\varepsilon > 0$, there is a subspace $Z \subset X$ and an operator $T : Y \longrightarrow Z$ so that $\|T\| \, \|T^{-1}\| \leq 1 + \varepsilon$. We can now state the theorem.

Theorem 6.1. *For every Banach space X, for every $\varepsilon_n \downarrow 0$ and for every f : $N \longrightarrow N$ there is a subspace $Y \subset X$ which the following properties:*

(1) *Y has a normalized basis (y_n) with basis constant $\leq 1 + \varepsilon_0$,*
(2) *Every $E \subset span_{n \leq i < \infty} y_i$, with $\dim E \leq f(n)$, is $1 + \varepsilon_n$–isomorphic to a Banach space F, and F almost isometrically embeds into every infinite*

dimensional subspace of Y,

(3) *Every block basis $(z_i)_{i=1}^{f(n)}$ of $(y_i)_{i=n}^{\infty}$ is $1+\varepsilon_n$-equivalent to a basis $\{w_i\}_{i=1}^{f(n)}$ of a Banach space F, and $\{w_i\}_{i=1}^{f(n)}$ is almost isometrically equivalent to a block basis of every basic sequence in Y.*

Before we prove Theorem 6.1, let us consider some of its consequences. Given a Banach space X, let

$$\begin{aligned} p_X &= \sup\{p \,|\, T_p(X) < \infty\} \\ q_X &= \inf\{q \,|\, C_q(X) < \infty\}. \end{aligned}$$

If we choose $f(n) \geq n^n$ in Theorem 6.1, a simple application of the Maurey–Pisier Theorem [MS] yields:

Corollary 6.2. *If X is an infinite dimensional Banach space, there is an infinite dimensional subspace Y of X so that $p_Y = p_Z$ and $q_Y = q_Z$, for every infinite dimensional subspace Z of Y.*

It follows that Y has some type (respectively, cotype) if and only if Y has a subspace with some type (respectively, cotype). Also, Y is a weak Hilbert space if and only if Y contains a weak Hilbert space.

Corollary 6.3. *For every Banach space X there is a subspace $Y \subset X$ so that every spreading model of Y (built from a weakly null sequence) is finitely representable in every (other) infinite dimensional subspace of Y.*

Proof. This is immediate since spreading models of subspaces of Y are finitely representable in $\text{span}_{n\leq i<\infty} y_i$, for every $n = 1, 2, \ldots$. □

Thus, if Y has c_0 as a spreading model (respectively l_1) then Y has no subspaces with cotype (respectively, type). Recall that a basic sequence (x_n) is block finitely representable in a basic sequence (y_n) if for every n and every $\varepsilon > 0$ there is a block basis $(z_i)_{i=1}^{n}$ of (y_i) which is $1+\varepsilon$-equivalent to $(x_i)_{i=1}^{n}$. In this notation, Krivine's Theorem [Kr] (see also [MS]) says that for every basic sequence (x_n) in a Banach space X there is a $1 \leq p < \infty$ (or for c_0) so that the unit vector basis of l_p (or c_0) is block finitely representable in (x_n). Property (3) of Theorem 6.1 implies that whenever the unit vector basis of l_p is block finitely representable on (y_n) then it is block finitely representable on every basic sequence of Y. This result was first observed by H. P. Rosenthal [R].

Corollary 6.4. *For every Banach space X, there is a subspace Y and a $1 \leq p < \infty$ (or c_0) so that the unit vector basis of l_p (or c_0) is block finitely representable in every basic sequence in Y. Moreover, if l_p (or c_0) is block finitely representable in one basic sequence in Y, then it is block finitely representable in every basic sequence in Y.*

Recall that a Banach space X is K-crudely finitely representable in a Banach space Y if for every finite dimensional subspace $E \subset X$ there is a subspace $F \subset Y$ with $d(E,F) \leq K$. If for every $\varepsilon > 0$ and every finite dimensional subspace $E \subset X$ there is a subspace $F \subset Y$ with $d(E,F) \leq 1+\varepsilon$, we say X is finitely representable in Y.

Corollary 6.5. *Suppose the separable Banach space X is K–crudely finitely representable in every infinite dimensional subspace of X. Then there is an equivalent norm $|\cdot|$ on X so that $(X,|\cdot|)$ is finitely representable in every infinite dimensional subspace of X.*

Proof. Choose the subspace Y of X with basis (y_i) from Theorem 6.1. Choose finite dimensional subspaces $E_1 \subset E_2 \subset \cdots$ whose union is dense in X. For each $i,j = 1,2,\ldots$, choose $F_{ij} \subset \operatorname{span}_{l(j)\le k<\infty} y_k$, where $l(j) \geq \dim F_{ij}$, and $d(E_i, F_{ij}) \leq K$. By switching to a subsequence, we may assume $\lim_{j\to\infty} F_{ij} = F_i$, for every $j = 1,2,\ldots$ (the limit in Banach–Mazur distance) and F_i is finitely representable in every infinite dimensional subspace of X, by Theorem 6.1 (2). For each $i = 1,2,\ldots$ let $T_i : E_i \longrightarrow F_i$ be a K–isomorphism. By switching to a subsequence again, we may assume that for every $x \in X$, $|x| = \lim_{i\to\infty} \|T_i\, x\|$ exists. Then $|\cdot|$ is an equivalent norm on X and clearly $(X,|\cdot|)$ is finitely representable in X (again by Theorem 6.1). □

If the above corollary had an infinite dimensional analogue, it could be quite useful for working in minimal Banach spaces. But our proof is local and we have not found a generalization of it for this case. Also, it would be much better if we could show that $(X,|\cdot|)$ is finitely representable in every infinite dimensional subspace of itself.

Again, we might hope that such a strong property would characterize a Hilbert space. But, it is easily seen that c_0 has the property that every finite dimensional subspace isometrically embeds into every infinite dimensional subspace.

We are now ready for the proof of Theorem 6.1.

Proof of Theorem 6.1. Fix $\varepsilon > 0$ and $n \in N$. Choose n–dimensional Banach spaces $H_1, H_2, \ldots, H_m$ which are $1+\varepsilon$–dense in the set of all n–dimensional Banach spaces in the Banach–Mazur distance. We now ask: (+) Is H_1, $1+\varepsilon$–embeddable into every infinite dimensional subspace of X?

If the answer is "yes", put H_1 in the set A and go to H_2. If the answer is "no", put H_1 in the set B and replace X by an infinite dimensional subspace X_1 of X so that X_1 has no subspace $1+\varepsilon$–isomorphic to H_1. Now go to H_2 and start over. After m–steps, we are left with a partition of the H_i's into two groups, $A = \{H_i\}_{i\in I}$ and $B = \{H_i\}_{i\in J}$, $I \wedge J = \phi$, and an infinite dimensional subspace Z of X so that every H_i, for $i \in I$, is $1+\varepsilon$–embeddable into every infinite dimensional subspace of Z while no H_i, $i \in J$, is $1+\varepsilon$–embeddable into Z.

It follows that for any $E \subset Z$, $\dim E = n$, E is $1+\varepsilon$–isomorphic to H_i, for some $i \in I$, and hence E is $(1+\varepsilon)^2$–embeddable into every infinite dimensional subspace of Z. Observe that this property is maintained if we switch to any infinite dimensional subspace of Z. Hence, given $f : N \longrightarrow N$ and $\varepsilon_n \downarrow 0$, we can inductively carry out this procedure to obtain infinite dimensional subspaces of X, $X \supset Z_1 \supset Z_2 \supset \cdots$ so that every $f(n)$–dimensional subspace of Z_n is $1+\varepsilon_n$–embeddable into every infinite dimensional subspace of Z_n. Now choose $y_n \in Z_n$ so that (y_n) is a $1+\varepsilon_0$–basic sequence in X. Now, if $E \subset \operatorname{span}_{n\le i} y_i$, then there are subspaces $F_k \subset \operatorname{span}_{n+k\le i} y_i$ so that $d(E,F_k) \leq 1+\varepsilon_n$. By switching to a subsequence, we may assume $\lim_{k\to\infty} F_k = F$ (in Banach–Mazur distance). So $d(E,F) \leq 1+\varepsilon_n$ but F_k is $1+\varepsilon_{n+k}$ embeddable into every infinite dimensional subspace of Y. Hence, F is finitely representable in every infinite dimensional subspace of Y. This concludes the construction for part (1) of Theorem 6.1. Again, note that part (1) of the Theorem holds if (y_n) is replaced by any block basis of (y_n).

Part (2) of Theorem 6.1 is proved in a similar manner. Again, fixing n and $\varepsilon > 0$, we list out $(x_{1i})_{i=1}^n, (x_{2i})_{i=1}^n, \ldots, (x_{mi})_{i=1}^n$ with the property: Every normalized basis $(z_i)_{i=1}^n$, with basis constant ≤ 2, for a Banach space X, is $1+\varepsilon$–equivalent to one of the $(x_{ki})_{i=1}^n$. Using our basis (y_i) from part (1), we ask: Is $(x_{1i})_{i=1}^n$, $1+\varepsilon$–equivalent to a block basis of every infinite block basis of Y?

As before, if the answer is "yes", put (x_{1i}) in the set A and go on to (x_{2i}). If the answer is "no", put (x_{1i}) in the set B and replace (y_i) by a block basis (y_i^1) of (y_i) so that (x_{1i}) is not $1+\varepsilon$–equivalent to any block basis of (y_i^1). Now go to (x_{2i}) and start over. After m steps, we arrive at a block basis (z_i) of (y_i) so that every $(x_{ji})_{i=1}^n$ in A is $1+\varepsilon$–equivalent to a block basis of every block basis of (z_i) and $(x_{ji})_{i=1}^n$ in B is not $1+\varepsilon$–equivalent to any block basis of (z_i). As in part (1), we perform this construction inductively to produce successive block bases of block bases $(z_i^1), (z_i^2), \ldots$ with the above property for $1+\varepsilon_n$. Then (z_n^n) is the required block basis. Relabeling (z_n^n) as (y_n), we see that we now have property (3) of Theorem 6.1 while maintaining property (2). This completes the proof of the Theorem. □

Theorem 6.1 can be partially strengthened in several directions. For one, if we let $Y_n = \operatorname{span}_{n<i} y_i$, then for every $E \subset Y_n^*$, $\dim E \leq f(n)$, and for every m, there is an $F \subset Y_m^*$ with $d(E,F) \leq 1+\varepsilon$. It is easily seen using $X = l_1$ that we cannot expect to $1+\varepsilon$ embed E into every infinite dimensional subspace of Y^*. With significantly more effort, we can show that our blocks in Theorem 6.1 are constructable in a very strong way. That is, whenever $(z_i)_{i=1}^n$ is a block basis of $(y_i)_{i=n}^\infty$, then for every m_1, there is a $w_1 \in \operatorname{span}_{m_1 \leq k} y_k$, so that for every m_2, there is a $w_2 \in \operatorname{span}_{m_2 \leq k} y_k, \ldots$, so that for every m_n there is a $w_n \in \operatorname{span}_{m_n \leq k} y_k$ and $(w_i)_{i=1}^n$ is $1+\varepsilon_n$–equivalent to $(z_i)_{i=1}^n$.

Now, a final word about Krivine's Theorem. Until recently, there was the possibility for strengthening this powerful and useful result. First, given $\varepsilon > 0$, and $n \in N$, we might look in every Banach space X for a p and a basic sequence (x_k) in X so that every block basis $(y_i)_{i=1}^n$ of (x_k) is $1+\varepsilon$–equivalent to the unit vector basis of l_p^n. It can be shown that Tsirelson's space $T^{(q)}$, fails this property for every $1 \leq q < \infty$. But Tsirelson's space has this property for $\varepsilon = 1/2$. Unfortunately, Schlumprecht's space [S2] fails this property for every $\varepsilon > 0$. Our final hope for a strengthening of Krivine's Theorem has just fallen to the Gowers'–Maurey spaces [GoM]. That is, in general we cannot get basic sequences in a Banach space X for which small numbers of blocks are even well unconditional. In particular, carrying out the construction of proposition 6.1 in the "unconditional" Gowers' spaces [Go2] we obtain,

Proposition 6.6. *For every $K \geq 1$, there is an n and n–vectors $(x_i)_{i=1}^n$ in a Banach space X and an infinite dimensional Banach space Y with an unconditional basis so that (x_i) is 1–block finitely representable on every basic sequence in Y, yet (x_i) is not K–unconditional.*

References

[B] S. Banach, *Theorie des operations lineaires*, Warszawa, 1932.

[B^*] J. Bourgain, *On finite dimensional homogeneous Banach spaces*, Geometric Aspects of Functional Analysis (J. Lindenstrauss and V. D. Milman, eds.), Springer Lecture Notes, **1317**, pp. 232–239.

[BCLT] J. Bourgain, P. G. Casazza, J. Lindenstrauss, and L. Tzafriri, *Banach spaces with a unique unconditional basis, up to a permutation, Memoirs*, AMS, **322** (1985).

[CJT] P. G. Casazza, W. B. Johnson, and L. Tzafriri, *On Tsirelson's space*, Israel J. Math. **47** (1984), 81–98.

[CKT] P. G. Casazza, N. J. Kalton, and L. Tzafriri, *Decompositions of Banach Lattices into direct sums*, Trans. AMS, no. 2, **304** (1987), 771–800.

[CO] P. G. Casazza and E. Odell, *Tsirelson's space and minimal subspaces*, Longhorn Notes, University of Texas, 1982–83.

[CS] P. G. Casazza and T. Shura, *Tsirelson's space*, Springer Lecture Notes, **1362** (1989).

[EW] I. S. Edelstein and P. Wojtaszczyk, *On projections and unconditional bases in direct sums of Banach spaces*, Studia Math. **56** (1976), 263–276.

[GL] Y, Gordon and D. Lewis, *Absolutely summing operators and local unconditional structures*, Acta Math. **133** (1974), 27–48.

[Go] W. T. Gowers, *A space not containing c_0, l_1 or a reflexive subspace*, (preprint).

[Go2] W. T. Gowers, *On the hyperplane problem*, (preprint).

[GoM] W. T. Gowers and B. Maurey, *A counterexample to the unconditional basic sequence problem*, (preprint).

[Ja] R. C. James, *A nonreflexive Banach space that is uniformly nonoctahedral*, Israel J. Math. **18** (1974), 145–155.

[J1] W. B. Johnson, *Homogeneous Banach spaces*, Geometric Aspects of Functional Analysis (J. Lindenstrauss and V. D. Milman, eds.), Springer Lecture Notes, **1317**, 201–203.

[J2] W. B. Johnson, *Banach spaces all of whose subspaces have the approximation property*, Special Topics of Applied Math., functional Analysis, Numerical Analysis and Opimization, Proceedings Bonn 1979, (J. Frehse, D. Pallaschke, and V. Trottenberg, eds.), North Holland, 1980, pp. 15–26.

[K] Ryszard Komorowski, *Weak Hilbert spaces without unconditional bases*, (preprint).

[Kr] J. L. Krivine, *Sous-espaces de demension finie des espaces de Banach re'ticulés*, Ann. of Math. **104** (1976), 1–29.

[LT1] J. Lindenstrauss and L. Tzafriri, *Classical Banach spaces I*, Berlin, Springer, 1977.

[LT2] ———, *Classical Banach spaces II*, Berlin, Springer, 1979.

[MTJ1] P. Mankiewicz and N. Tomczak-Jaegermann, *Random subspaces and quotients of finite dimensional Banach spaces*, Odense University, 1989, (preprint no. 8).

[MTJ2] ———, *A solution of the finite-dimensional homogeneous Banach space problem*, (preprint).

[MTJ3] ———, *Schauder bases in quotients of subspaces of $l_2(X)$*, (preprint).

[Ma] V. Mascioni, *On Banach spaces isomorphic to their duals*, (preprint).

[MS] V. D. Milman and G. Schechtman, *Asymptotic theory of finite dimensional normed spaces*, Springer Lecture Notes **1200** (1986).

[MiTJ] V. Milman and N. Tomczak–Jaegermann, (preprint).

[NTJ] N. Nielsen and N. Tomczak–Jaegermann, *On subspaces of Banach spaces with property (H) and weak Hilbert spaces*, (preprint).

[OS] E. Odell and T. Schlumprecht, *l_p is distortable*, $1 < p < \infty$, (preprint).

[P] G. Pisier, *The volume of convex bodies and Banach space geometry*, Cambridge University Press, 1989.

[R] H. P. Rosenthal, *On a Theorem of J. L. Krivine concerning block finite — representability of l_p in general Banach spaces*, J. Functional Anal. **28** (1978), 197–225.

[S1] T. Schlumprecht, *An arbitrarily distortable Banach space*, (preprint).

[S2] ———, *A complementably minimal Banach space not containing c_0 or l_p*, (preprint).

[T] B. S. Tsirelson, *Not every Banach space contains l_p or c_0*, Functional Anal. & Appl. **8** (1974), 138–141.

[W] P. Wojtaszczyk, *On projections and unconditional bases in direct sums of Banach spaces II*, Studia math. **47** (1973), 197–206.

DEPARTMENT OF MATHEMATICS, UNIVERSITY OF MISSOURI, COLUMBIA, MISSOURI 65211
E-mail address: mathpc@umcvmb.bitnet

Contemporary Mathematics
Volume **144**, 1993

SOME PROBABILISTIC INEQUALITIES WITH APPLICATIONS TO FUNCTIONAL ANALYSIS

S.J. Dilworth

ABSTRACT. This paper is an exposition of some probabilistic results having applications to functional analysis. These include martingale inequalities and moment inequalities for sums of independent random variables. Applications are given to the Banach space structure of several families of function spaces.

1. Introduction

This paper is an expository account of some probabilistic inequalities which have contributed to our understanding of several families of Banach spaces of interest in functional analysis: namely, the Lebesgue spaces L_p, the Orlicz spaces $L_p + L_q$, the Lorentz spaces $L_{p,q}$, and the Lebesgue-Bochner spaces $L_p(\ell_q)$. Our purpose in writing this paper has been to provide a readable but concise account, with more or less complete proofs, of several connected results which are at present scattered throughout the literature. The notes at the ends of the sections contain additional results and references which complement this introduction.

We turn now to give a detailed description of the results presented herein. The first two sections after this introduction contain the proofs of the probabilistic results; they are intended to be accessible to students and to functional analysts with only a modest knowledge of probability theory. Accordingly, we begin our account with the old and well-known (Theorem 2.1), and end with the comparatively recent (Theorem 3.3). The martingale inequalities presented in the second section may all be viewed as consequences of Doob's maximal inequality (Theorem 2.1). From the standpoint of the later applications to functional analysis, the principal results of this section are Theorem 2.3 and its corollaries. This theorem is a special case of one of the Burkholder-Davis-Gundy inequalities [6]; an early form of the result may also be found in [36]. The elegant proof presented below, which dualizes Doob's maximal inequality, appears in [16], for the special case $q = 1$, and was drawn to our attention by Asmar and Montgomery-Smith [2] in the general case. Besides its obvious appeal, it also has the twin merits of avoiding interpolation and of providing explicit values for the constants which arise. The aim of the third section is to prove the general L_p version of Rosenthal's moment inequality for sums of independent symmetric random variables (Theorem 3.3). Of the many ways of formulating these inequalities, we choose to follow [21] in phrasing them first of all in terms of the norm of a *disjoint sum* of random variables in a certain family of sums

---

1991 *Mathematics Subject Classification.* Primary 46E30, 60E15; Secondary 60G42, 60G50.
This paper is in final form and no version of it will be submitted for publication elsewhere.

and intersections of L_p spaces on $(0, \infty)$. Thus formulated, the original inequalities of [34] for sums of independent symmetric random variables in L_p, $p > 2$, can be dualized to the range $0 < p < 2$. This is accomplished by means of a projection on the Lebesgue-Bochner space $L_2(\ell_p)$, whose boundedness is a consequence of the martingale inequalities of the previous section. This argument appeared in [14], and in a less effective form in [9]. Theorem 3.3 is due to Johnson and Schechtman [24], who proved a very general result for sums of independent random variables in rearrangement-invariant spaces.

The formulation of the general Rosenthal inequality in terms of disjoint sums has a natural and important functional analytic interpretation, which lends itself readily to applications. In the second half of the paper we show how the probabilistic inequalities provide information about the Banach space structure of certain families of function spaces. Accordingly, in the fourth section we explore the relationship between the Orlicz space $L_1 \cap L_2$ and the Lebesgue-Bochner space $L_2(\ell_1)$, with special reference to the non-existence of an unconditional basis for $L_1 \cap L_2$ and to a theorem of Aldous [1] on the existence of unconditional bases for Lebesgue-Bochner spaces. In the last section of the paper we prove the existence of Banach space embeddings of the Orlicz space $L_p + L_q$ and the Lorentz space $L_{p,q}$ into the Lebesgue-Bochner space $L_r(\ell_s)$ for appropriate ranges of the parameters. This leads to a short proof of an interesting (special case of a) theorem of Schütt [35] on the existence of an embedding of $L_{p,q}$ into L_1 in the range $1 \leq q \leq p < 2$. The applications to Lorentz spaces were obtained jointly with N.L. Carothers and are taken from [12].

The rest of this introduction is given over to matters of notation. For a random variable f defined on a probability space $(\Omega, \mathcal{F}, P)$, the decreasing rearrangement, denoted f^*, is the function on $[0,1]$ defined by $f^*(t) = \inf\{s > 0 : P(|f| > s) \leq t\}$. For $0 < p < \infty$, $(L_p(\Omega), \|.\|_p)$ denotes the usual Lebesgue space with its standard norm, and when X is a Banach space, $L_p(\Omega; X)$ denotes the Lebesgue-Bochner space of Bochner-integrable functions with the norm

$$\|f\|_{L_p(\Omega;X)} = \left(\int_\Omega \|f(\omega)\|^p \, dP\right)^{1/p}.$$

When $\Omega = [0,1]$ with Lebesgue measure, we simply write $L_p(X)$. For $1 < p < \infty$, p' denotes the Hölder conjugate index $p/(p-1)$. We write $A \approx B$ to mean that there exists a constant C such that $\frac{1}{C}A \leq B \leq CA$. (Usually, the implied constant C will depend on some parameter, typically $p \in (0, \infty)$, entering into the expressions for A and B.) For any unfamiliar Banach space terminology employed in the second half of the paper it is recommended that the reader consult [28]. All other notation will be introduced as and when it is needed.

2. Martingale Inequalities

First we fix some notation. Let $(\Omega, \mathcal{F}, P)$ be a probability space, and let $(\mathcal{F}_n)_{n\geq 1}$ be an increasing sequence of sub-σ-fields of $\mathcal{F}$. For $f \in L_1(\Omega, \mathcal{F}, P)$, let $f_n = \mathbb{E}(f|\mathcal{F}_n)$ and let $f^{**} = \sup_{1\leq n<\infty} |f_n|$. Then, by Jensen's inequality, $(|f_n|)_{n=1}^\infty$ is a *submartingale*, i.e.

$$\int_A |f_{n+1}| \, dP \geq \int_A |f_n| \, dP$$

for all $n \geq 1$ and for all $A \in \mathcal{F}_n$.

Theorem 2.1. (*Doob's maximal inequality* [15]) *For $\lambda > 0$, we have*

$$P(f^{**} > \lambda) \leq \frac{1}{\lambda} \int_{\{f^{**} > \lambda\}} |f| \, dP. \tag{1}$$

Proof. Let $E = \{f^{**} > \lambda\}$. Then E is the disjoint union of the sets $E_k = \{|f_1| \leq \lambda, \ldots, |f_{k-1}| \leq \lambda, |f_k| > \lambda\}$ for $1 \leq k < \infty$. So

$$\begin{aligned} P(f^{**} > \lambda) &= \sum_{k=1}^{\infty} P(E_k) \\ &\leq \sum_{k=1}^{\infty} \frac{1}{\lambda} \int_{E_k} |f_k| \, dP \\ &\leq \sum_{k=1}^{\infty} \frac{1}{\lambda} \int_{E_k} |f| \, dP \\ &= \frac{1}{\lambda} \int_{\{f^{**} > \lambda\}} |f| \, dP, \end{aligned}$$

where the second inequality follows from the submartingale property since $E_k \in \mathcal{F}_k$. □

Corollary 2.2. *Let $1 < p < \infty$. Then*

$$||f^{**}||_p \leq \frac{p}{p-1} ||f||_p.$$

Proof. First we shall assume that $f^{**} \in L_p(\Omega)$. By (1), for each $\lambda \geq 0$, we have

$$\lambda \int_{\Omega} \chi_{\{f^{**} > \lambda\}} \, dP \leq \int_{\Omega} |f| \chi_{\{f^{**} > \lambda\}} \, dP.$$

Multiplying through by λ^{p-2} and integrating from zero to infinity, we have

$$\int_{\Omega} \left(\int_0^{\infty} \lambda^{p-1} \chi_{\{f^{**} > \lambda\}}(\omega) \, d\lambda \right) \, dPle \int_{\Omega} \left(\int_0^{\infty} \lambda^{p-2} \chi_{\{f^{**} > \lambda\}}(\omega) \, d\lambda \right) |f| \, dP,$$

i.e.,

$$\frac{1}{p} \int_{\Omega} |f^{**}|^p \, dP \leq \frac{1}{p-1} \int_{\Omega} |f^{**}|^{p-1} |f| \, dP.$$

Now, by Hölder's inequality, we have

$$\frac{1}{p} \int_{\Omega} |f^{**}|^p \, dP \leq \frac{1}{p-1} \left(\int_{\Omega} |f^{**}|^p \, dP \right)^{1-(1/p)} \left(\int_{\Omega} |f|^p \, dP \right)^{1/p},$$

which gives the result. For the general case, let $f_n^{**} = \max_{1 \leq j \leq n} |f_j|$. Then $f_n^{**} \in L_p(\Omega)$ and f_n^{**} increases monotonely to f^{**}. So the result follows from the Monotone Convergence Theorem. □

Theorem 2.3. *Let $1 \le q \le p < \infty$. Then, for all sequences $(f_n)_{n=1}^\infty$ of integrable functions, we have*

$$\left\| \left(\sum_{n=1}^\infty |\mathbb{E}(f_n|\mathcal{F}_n)|^q \right)^{1/q} \right\|_p \le \left(\frac{p}{q}\right)^{1/q} \left\| \left(\sum_{n=1}^\infty |f_n|^q \right)^{1/q} \right\|_p .$$

Proof. We first prove the result in the case $q = 1$. For $p = 1$, the result follows at once from Jensen's inequality. So we may assume that $p > 1$. Suppose that $g \in L_{p'}(\Omega)$ and that $||g||_{p'} \le 1$. Then

$$\begin{aligned}
\mathbb{E}\left(g \sum_{n=1}^\infty \mathbb{E}(|f_n||\mathcal{F}_n)\right) &= \mathbb{E}\left(\sum_{n=1}^\infty |f_n|\mathbb{E}(g|\mathcal{F}_n)\right) \\
&\le \mathbb{E}\left(\left(\sum_{n=1}^\infty |f_n|\right) g^{**}\right) \\
&\le ||g^{**}||_{p'} \left\| \sum_{n=1}^\infty |f_n| \right\|_p \\
&\le \frac{p'}{p'-1} \left\| \sum_{n=1}^\infty |f_n| \right\|_p
\end{aligned}$$

(by Corollary 2.2)

$$= p \left\| \sum_{n=1}^\infty |f_n| \right\|_p ,$$

and the result follows from the converse of Hölder's inequality. For the case $1 < q \le p$, we use the fact that $|\mathbb{E}(f_n|\mathcal{F}_n)|^q \le \mathbb{E}(|f_n|^q|\mathcal{F}_n)$ a.e., which follows directly from Jensen's inequality. Hence

$$\begin{aligned}
\left\| \left(\sum_{n=1}^\infty |\mathbb{E}(f_n|\mathcal{F}_n)|^q \right)^{1/q} \right\|_p &\le \left\| \sum_{n=1}^\infty \mathbb{E}(|f_n|^q|\mathcal{F}_n) \right\|_{p/q}^{1/q} \\
&\le \left(\frac{p}{q}\right)^{1/q} \left\| \left(\sum_{n=1}^\infty |f_n|^q \right)^{1/q} \right\|_p . \qquad \square
\end{aligned}$$

Before stating the next corollary, let us agree on some more notation. For $0 < p, q, < \infty$, we identify elements of the Lebesgue-Bochner space $L_p(\Omega; \ell_q)$ with sequences $(f_n)_{n=1}^\infty$ of real-valued random variables on Ω with the norm

$$||f||_{L_p(\Omega;\ell_q)} = \left\| \left(\sum_{n=1}^\infty |f_n|^q \right)^{1/q} \right\|_p .$$

Under the duality $((f_n)_{n=1}^\infty, (g_n)_{n=1}^\infty) = \sum_{n=1}^\infty \mathbb{E} f_n g_n$, the dual space of $L_p(\Omega; \ell_q)$ is isometrically isomorphic to $L_{p'}(\Omega; \ell_{q'})$ in the range $1 < p, q < \infty$.

Corollary 2.4. *Let $1 < p < q < \infty$. Then, for all sequences $(f_n)_{n=1}^\infty$ of integrable functions, we have*

$$\left\| \left(\sum_{n=1}^{\infty} |\mathbb{E}(f_n|\mathcal{F}_n)|^q \right)^{1/q} \right\|_p \le \left(\frac{p'}{q'} \right)^{1/q'} \left\| \left(\sum_{n=1}^{\infty} |f_n|^q \right)^{1/q} \right\|_p .$$

Proof. By Theorem 2.3, the projection $P : L_{p'}(\ell_{q'}) \to L_{p'}(\ell_{q'})$, given by $P((f_n)_{n=1}^\infty) = (\mathbb{E}(f_n|\mathcal{F}_n))_{n=1}^\infty$, is continuous and $\|P\| \le (p'/q')^{1/q'}$. The desired result follows from the duality between $L_{p'}(\ell_{q'})$ and $L_p(\ell_q)$, and from the fact that P is formally self-adjoint. □

Lemma 2.5. *Let $(\Omega, \mathcal{F}, P)$ be a probability space, and let $\mathcal{G}_1$, $\mathcal{G}_2$, and $\mathcal{G}_3$ be independent sub-σ-fields of $\mathcal{F}$. Then, for $f \in L_1(\Omega)$, we have*

$$\mathbb{E}(\mathbb{E}(f|\mathcal{G}_1 \vee \mathcal{G}_2)|\mathcal{G}_1 \vee \mathcal{G}_3) = \mathbb{E}(f|\mathcal{G}_1).$$

Proof. Let $\mathcal{H}$ be the collection of all $\mathcal{G}_1 \vee \mathcal{G}_2$ -measurable sets E such that the conclusion holds for $f = \chi_E$. Clearly, $\mathcal{G}_1 \subset \mathcal{H}$; for $G_2 \in \mathcal{G}_2$, we have by independence

$$\mathbb{E}(\chi_{G_2}|\mathcal{G}_1 \vee \mathcal{G}_3) = \mathbb{E}(\chi_{G_2}|\mathcal{G}_1) = P(G_2)\chi_\Omega,$$

and so $\mathcal{G}_2 \subset \mathcal{H}$. Moreover, if $E_1, E_2 \in \mathcal{H}$, then $E_1 \backslash E_2 \in \mathcal{H}$. It follows that $\mathcal{H}$ contains the algebra generated by $\mathcal{G}_1 \cup \mathcal{G}_2$. Furthermore, $\mathcal{H}$ is easily seen to be a monotone class, whence $\mathcal{H} = \mathcal{G}_1 \vee \mathcal{G}_2$. The desired conclusion now follows easily. □

Corollary 2.6. *Let $(\mathcal{G}_k)_{k=1}^n$ be independent sub-σ-fields of $\mathcal{F}$, and let $1 < p < \infty$, $1 \le q < \infty$. Then, for $f_1, \ldots, f_n \in L_1(\Omega)$, we have*

$$\left\| \left(\sum_{k=1}^{n} |\mathbb{E}(f_k|\mathcal{G}_k)|^q \right)^{1/q} \right\|_p \le C_{p,q} \left\| \left(\sum_{k=1}^{n} |f_k|^q \right)^{1/q} \right\|_p ,$$

where $C_{p,q} = \begin{cases} (p/q)^{2/q} & (q \le p) \\ (p'/q')^{2/q'} & (p < q). \end{cases}$

Proof. Let $\mathcal{F}_k$ and $\tilde{\mathcal{F}}_k$ be the sub-σ-fields of $\mathcal{F}$ generated by $\cup_{i=1}^k \mathcal{G}_i$ and $\cup_{i=n+1-k}^n \mathcal{G}_i$, respectively. Then, by Lemma 2.5, $\mathbb{E}(\mathbb{E}(f_k|\mathcal{F}_k)|\tilde{\mathcal{F}}_k) = \mathbb{E}(f_k|\mathcal{G}_k)$, and so the result follows by applying Theorem 2.3 or Corollary 2.4, first with respect to $(\mathcal{F}_k)_{k=1}^n$ and then with respect to $(\tilde{\mathcal{F}}_k)_{k=1}^n$. □

Remark 2.7. If $p = q = 1$ then, by Jensen's inequality, Corollary 2.6 remains valid with $C_{p,q} = 1$. However, if $p = 1$ and $q > 1$ then the result breaks down. We outline a proof of this surprising fact. Let $1 < r < q < 2$, and let $(g_k)_{k=1}^\infty$ be a sequence of independent symmetric r-stable random variables, normalized so that $\|g_k\|_1 = 1$.

For $n \geq 1$ and $1 \leq i \leq n$, let $f_i^n = \prod_{k=1}^n |g_k|$. Then $\mathbb{E}(f_k^n|g_k) = |g_k|$ $(1 \leq k \leq n)$, and so

$$\left\| \left(\sum_{k=1}^n |\mathbb{E}(f_k^n|g_k)|^q \right)^{1/q} \right\|_1 = \left\| \left(\sum_{k=1}^n |g_k|^q \right)^{1/q} \right\|_1$$
$$\geq \left\| \left(\sum_{k=1}^n |g_k|^2 \right)^{1/2} \right\|_1$$
$$\approx \left\| \sum_{k=1}^n g_k \right\|_1$$

(by Khintchine's inequality)

$$= n^{1/r}.$$

On the other hand, we have

$$\left\| \left(\sum_{k=1}^n |f_k^n|^q \right)^{1/q} \right\|_1 = n^{1/q} \|f_1^n\|_1 = n^{1/q},$$

and so the inequality of Corollary 2.6 must fail when $1 < q \leq 2$. The result must also fail when $q > 2$, for otherwise an interpolation argument would establish the result for $q < 2$, which is false as we have seen.

Notes. Further results about conditional expectations with respect to independent sub-σ-fields may be found in [4] and [9]. The deduction of Corollary 2.6 from Theorem 2.3 follows [9], although the argument belongs to folklore. A recent short proof of Theorem 2.3 (in the case $q = 1$) may be found in [17]. For general convex function versions of Theorem 2.3 see [6] or [16]. There is a version of Theorem 2.3 for rearrangement invariant spaces with finite upper Boyd index due to Johnson and Schechtman [23]. Another proof is given by Hitczenko in [19]. Remark 2.7 is taken from [9].

3. Rosenthal's Inequality

We begin with the inequality proved by Rosenthal in [34].

Proposition 3.1. *Let $1 < p < \infty$, and let $f_1, \dots, f_n$ be independent non-negative random variables in $L_p(\Omega)$. Then*

$$\max \left(\sum_{i=1}^n \|f_i\|_1, \left(\sum_{i=1}^n \|f_i\|_p^p \right)^{1/p} \right) \leq \left\| \sum_{i=1}^n f_i \right\|_p$$
$$\leq 2^p \max \left(\sum_{i=1}^n \|f_i\|_1, \left(\sum_{i=1}^n \|f_i\|_p^p \right)^{1/p} \right).$$

Proof. The left-hand inequality follows at once from Hölder's inequality and from the non-negativity of the f_i's. To prove the right-hand inequality note that

$$\left(\sum_{i=1}^n f_i\right)^p \leq 2^{p-1}\left(\sum_{i=1}^n f_i^p + \sum_{i=1}^n \left(\sum_{j\neq i} f_j\right)^{p-1} f_i\right)$$

pointwise. So by independence of $f_1, \dots, f_n$, we have

$$\begin{aligned}\left\|\sum_{i=1}^n f_i\right\|_p^p &\leq 2^{p-1}\left(\sum_{i=1}^n \|f_i\|_p^p + \sum_{i=1}^n \left(\|\sum_{j\neq i} f_j\|_{p-1}^{p-1}\|f_i\|_1\right)\right)\\ &\leq 2^{p-1}\left(\sum_{i=1}^n \|f_i\|_p^p + \left(\sum_{i=1}^n \|f_i\|_1\right)\left\|\sum_{i=1}^n f_i\right\|_p^{p-1}\right)\end{aligned}$$

(by Hölder's inequality and the non-negativity of the f_i's)

$$\leq 2^p \max\left(\sum_{i=1}^n \|f_i\|_p^p, \left(\sum_{i=1}^n \|f_i\|_1\right)\left\|\sum_{i=1}^n f_i\right\|_p^{p-1}\right).$$

Hence

$$\begin{aligned}\left\|\sum_{i=1}^n f_i\right\|_p &\leq \max\left(2\left(\sum_{i=1}^n \|f_i\|_p^p\right)^{1/p}, 2^p\left(\sum_{i=1}^n \|f_i\|_1\right)\right)\\ &\leq 2^p \max\left(\left(\sum_{i=1}^n \|f_i\|_p^p\right)^{1/p}, \sum_{i=1}^n \|f_i\|_1\right). \qquad \square\end{aligned}$$

Our next goal is to present the recent extensions of Proposition 3.1 to the range $0 < p < 1$. To this end, we introduce some function spaces on $(0, \infty)$. For $0 < p < \infty$, the quasi-norm, $\|.\|_{M_p}$, is defined for a measurable function f thus:

$$\|f\|_{M_p} = \|f^*\chi_{[0,1]}\|_2 + \|f^*\chi_{(1,\infty)}\|_p.$$

For $0 < p \leq 2$, we have

$$\frac{1}{2}\max(\|f\|_2, \|f\|_p) \leq \|f\|_{M_p} \leq 2\max(\|f\|_2, \|f\|_p),$$

and so the class of functions f such that $\|f\|_{M_p} < \infty$ coincides with $L_p(0,\infty) \cap L_2(0,\infty)$. For $2 < p < \infty$, M_p coincides with $L_p(0,\infty) + L_2(0,\infty)$. Under the duality $(f,g) = \int_0^\infty f(t)g(t)\,dt$, the dual space of M_p is $M_{p'}$ $(1 < p < \infty)$ with equivalence of quasi-norms. If $(g_n)_{n=1}^\infty$ is any sequence of measurable functions on $[0,1]$, let $\sum_{n=1}^\infty \oplus g_n$ denote the function g on $(0,\infty)$ defined by $g(n-1+t) = g_n(t)$ for $n \geq 1$ and $0 < t \leq 1$. It has become customary by now to call $\sum_{n=1}^\infty \oplus g_n$ a *disjoint sum.* Finally, let $\tilde{\Omega}$ denote the space $[0,1]^{\mathbb{N}}$ equipped with the completion of the infinite product of Lebesgue measure on $[0,1]$. A typical element of $\tilde{\Omega}$ will be denoted by the sequence $(s_1, s_2, \dots)$. Now we can state the principal result of this section.

Theorem 3.2. *Let $0 < p < \infty$, and let $(f_n)_{n=1}^{\infty}$ be a sequence of independent non-negative random variables defined on a probability space $(\Omega, \mathcal{F}, P)$. Then*

$$\left\| \left(\sum_{n=1}^{\infty} f_n^p \right)^{1/p} \right\|_2 \approx \left\| \sum_{n=1}^{\infty} \oplus f_n^* \right\|_{M_p} .$$

(The implied constant depends only on p.)

Proof. The sequence $(f_n)_{n=1}^{\infty}$ has the same joint distribution on Ω as the sequence $(f_n^*(s_n))_{n=1}^{\infty}$ on $\tilde{\Omega}$, and so it suffices to prove that

$$\left\| \left(\sum_{n=1}^{\infty} g_n(s_n)^p \right)^{1/p} \right\|_{L_2(\tilde{\Omega})} \approx \left\| \sum_{n=1}^{\infty} \oplus g_n \right\|_{M_p}$$

for every sequence $(g_n)_{n=1}^{\infty}$ of non-negative functions on $[0,1]$. Consider the linear mapping $T : M_p \to L_p(\tilde{\Omega}; \ell_2)$ defined by $T\left(\sum_{n=1}^{\infty} \oplus g_n\right) = (g_n(s_n))_{n=1}^{\infty}$. For $0 < p < 2$, we have

$$\begin{aligned}
\left\| T\left(\sum_{n=1}^{\infty} \oplus g_n \right) \right\|_{L_p(\tilde{\Omega};\ell_2)} &= \left\| \left(\sum_{n=1}^{\infty} |g_n(s_n)|^p \right)^{1/p} \right\|_2 \\
&= \left\| \sum_{n=1}^{\infty} |g_n(s_n)|^p \right\|_{2/p}^{1/p} \\
&\approx \max\left(\left(\sum_{n=1}^{\infty} \| |g_n|^p \|_{2/p}^{2/p} \right)^{1/2}, \left(\sum_{n=1}^{\infty} \| |g_n|^p \|_1 \right)^{1/p} \right)
\end{aligned}$$

(by Proposition 3.1)

$$\begin{aligned}
&= \max\left(\left(\sum_{n=1}^{\infty} \|g_n\|_2^2 \right)^{1/2}, \left(\sum_{n=1}^{\infty} \|g_n\|_p^p \right)^{1/p} \right) \\
&\approx \left\| \sum_{n=1}^{\infty} \oplus g_n \right\|_{M_p} .
\end{aligned}$$

Hence, for $0 < p < 2$, T defines an isomorphic embedding from M_p onto a closed subspace of $L_2(\tilde{\Omega}; \ell_p)$. Consider the projection P on $L_2(\tilde{\Omega}; \ell_p)$ defined by $P((g_n)_{n=1}^{\infty}) = (\mathbb{E}(g_n | s_n))_{n=1}^{\infty}$. By Corollary 2.6, P is continuous, with operator norm at most 4, for $1 \le p < \infty$. Moreover, the range of P and the range of T are the same. Since P is formally self-adjoint, and since $M_p^* = M_{p'}$, the result follows for $2 < p < \infty$ by duality. $\square$

Finally, we recast Theorem 3.2 in its more familiar form as a result about sums of independent symmetric random variables. To this end, let Y_p denote the linear space of all measurable functions f on $(0,\infty)$ such that

$$\|f\|_{Y_p} = \|f^*\chi_{[0,1]}\|_p + \|f^*\chi_{[1,\infty)}\|_2 < \infty.$$

Then

$$Y_p = \begin{cases} L_p(0,\infty) + L_2(0,\infty) & (0 < p \le 2) \\ L_p(0,\infty) \cap L_2(0,\infty) & (2 < p < \infty). \end{cases}$$

Observe that M_p and $Y_{4/p}$ are related by the formula $\|f\|_{M_p} \approx \||f|^{p/2}\|_{Y_{4/p}}^{2/p}$.

Theorem 3.3. *Let $0 < p < \infty$, and let $(f_n)_{n=1}^{\infty}$ be a sequence of independent symmetric random variables defined on a probability space $(\Omega, \mathcal{F}, P)$. Then*

$$\left\|\sum_{n=1}^{\infty} f_n\right\|_p \approx \left\|\sum_{n=1}^{\infty} \oplus f_n^*\right\|_{Y_p}.$$

(The implied constant depends only on p.)

Proof. By an old result of Marcinkiewicz and Zygmund [31] (alternatively by the classical Khintchine inequalities), we have

$$\begin{aligned}\left\|\sum_{n=1}^{\infty} f_n\right\|_p &\approx \left\|\left(\sum_{n=1}^{\infty} f_n^2\right)^{1/2}\right\|_p \\ &= \left\|\left(\sum_{n=1}^{\infty}\left(|f_n|^{p/2}\right)^{4/p}\right)^{p/4}\right\|_2^{2/p} \\ &\approx \left\|\sum_{n=1}^{\infty} \oplus (f_n^*)^{p/2}\right\|_{M_{4/p}}^{2/p}\end{aligned}$$

(by Theorem 3.2)

$$\approx \left\|\sum_{n=1}^{\infty} \oplus f_n^*\right\|_{Y_p}. \qquad \square$$

Notes. Proposition 3.1 was proved by Rosenthal in [34] in order to construct new isomorphism types of complemented subspaces of L_p for $p > 2$. Proposition 3.1 may also be deduced from Hoffman-Jørgensen's inequality [20]. Rosenthal's proof is modified in [22] to improve the constant 2^p, and a different argument is given which yields that the best order of the constant as $p \to \infty$ is $p/\log p$. A martingale version of Rosenthal's inequality is given in [5], and in [18] it is shown how to reduce the martingale case to the independent case, which yields the best order of the constants in the martingale case. The case $p = 1$ of Theorem 3.3 is implicit in the proof of [4, Theorem 3]. See [24] for versions of Theorems 3.2 and 3.3 for rearrangement-invariant spaces.

4. THE SPACES $L_1 \cap L_2$ AND $L_2(\ell_1)$

In this section and the next we present some applications of the preceding results to functional analysis. Recall that the space M_1 introduced above coincides with the Orlicz space $L_1(0,\infty) \cap L_2(0,\infty)$ (denoted $L_1 \cap L_2$ for short). We consider here an interesting relationship between the spaces $L_1 \cap L_2$ and $L_2(\ell_1)$. In [1] Aldous showed that if a Banach space X has an unconditional basis then the Lebesgue-Bochner space $L_2(X)$ will be isomorphic to a subspace of a space with unconditional basis only if X has the so-called "unconditional martingale difference" property (see e.g. [7] for this concept). In particular, the space $L_2(\ell_1)$ does not embed into a Banach space with unconditional basis. Now, according to Theorem 3.2 and its proof, $M_1 = L_1 \cap L_2$ is isomorphic to a complemented subspace of $L_2(\tilde{\Omega}; \ell_1)$. Our aim now is to prove that $L_1 \cap L_2$, too, does not embed into a space with unconditional basis, which will imply the same result *a fortiori* for $L_2(\ell_1)$. To this end, let $(h_n)_{n=1}^{\infty}$ be the Haar system on $[0,1]$ defined by $h_1(t) \equiv 1$ and

$$h_{2^k+l}(t) = \begin{cases} 1 & \text{for } t \in [(2l-2)2^{-k-1}, (2l-1)2^{-k-1}], \\ -1 & \text{for } t \in [(2l-1)2^{-k-1}, 2l.2^{-k-1}], \\ 0 & \text{otherwise}, \end{cases}$$

for $k \geq 0$ and $1 \leq \ell \leq 2^k$. For $k \geq 1$, let $(h_n^k)_{n=1}^{\infty}$ be the Haar system translated to the interval $[k-1, k]$ and let $H = (\tilde{h}_n)_{n=1}^{\infty}$ be the diagonal ordering of $(h_n^k)_{k,n=1}^{\infty}$. Then H is a Schauder basis for $L_1 \cap L_2$. Let Z be the subspace of $L_1 \cap L_2$ consisting of all functions $f(t)$ such that $\int_{n-1}^{n} f(t)\,dt = 0$ for $n \geq 1$. The subsequence obtained from H by deleting the terms $h_1^1, h_1^2, h_1^3, \ldots$ forms a Schauder basis for Z.

Proposition 4.1. *H is not an unconditional basis for $L_1 \cap L_2$.*

Proof. Let $f_n^k = h_1^k + h_2^k + 2h_3^k + 4h_5^k + \cdots + 2^{2n+1}h_{2^{2n+1}+1}^k$ and let $g_n^k = h_1^k + 2h_3^k + 8h_9^k + 32h_{33}^k + \cdots + 2^{2n+1}h_{2^{2n+1}+1}^k$. It is readily verified that $\|f_n^k\|_1 = 1$, $\|f_n^k\|_2 = 2^{n+1}$, and that $(n+2)/4 \leq \|g_n^k\|_1 \leq n+2$. For each $k \geq 2$, select n_k such that $2^{n_k} \leq \sqrt{k} < 2^{n_k+1}$. Consider $f = \sum_{k=2}^{\infty} \dfrac{f_{n_k}^k}{k(\ln k)^2}$ and $g = \sum_{k=2}^{\infty} \dfrac{g_{n_k}^k}{k(\ln k)^2}$.

Then

$$\|f\|_1 = \sum_{k=2}^{\infty} \frac{\|f_{n_k}^k\|_1}{k(\ln k)^2} = \sum_{k=2}^{\infty} \frac{1}{k(\ln k)^2} < \infty,$$

and

$$\begin{aligned} \|f\|_2^2 &= \sum_{k=2}^{\infty} \frac{\|f_{n_k}^k\|_2^2}{k^2(\ln k)^4} \\ &= \sum_{k=2}^{\infty} \frac{(2^{n_k+1})^2}{k^2(\ln k)^4} \\ &\leq \sum_{k=2}^{\infty} \frac{4k}{k^2(\ln k)^4} \\ &= \sum_{k=2}^{\infty} \frac{4}{k(\ln k)^4} < \infty. \end{aligned}$$

So $f \in L_1 \cap L_2$. On the other hand, we have

$$\begin{aligned}
\|g\|_1 &= \sum_{k=2}^{\infty} \frac{\|g_{n_k}^k\|_1}{k(\ln k)^2} \\
&\geq \sum_{k=2}^{\infty} \frac{n_k + 2}{4k(\ln k)^2} \\
&\geq \sum_{k=2}^{\infty} \frac{1}{8k(\ln k)^2} \frac{\ln k}{\ln 2}
\end{aligned}$$

which diverges. Since g is obtained from f by multiplying the basis expansion of f with respect to H by a sequence of zeros and ones, it follows that H is not an unconditional basis. □

Proposition 4.2. *Suppose that $L_1 \cap L_2$ is isomorphic to a subspace of a Banach space X having a Schauder basis $(x_n)_{n=1}^{\infty}$. Then G is equivalent to a block basis of $(x_n)_{n=1}^{\infty}$.*

Proof. The analogous fact for the Haar system is well-known. Straightforward changes to the proof given in [27], which uses a convexity theorem of Liapounoff, give the result. □

Theorem 4.3. [13] *$L_1 \cap L_2$ is not isomorphic to a subspace of a Banach space with an unconditional basis.*

Proof. It follows easily from Proposition 4.1 that G is not an unconditional basis for Z. Since a block basis of an unconditional basis is itself unconditional, the result follows from Proposition 4.2. □

To conclude this section, we list without proof some other interesting properties of the space M_1.

(1) M_1 is isomorphic to a subspace of $L_1(0,1)$.
(2) Let X be a closed subspace of M_1. Then X is isomorphic to a Hilbert space or X contains a subspace which is isomorphic to ℓ_1 and complemented in M_1.
(3) M_1 is a separable dual space (and so has the Radon-Nikodym property). The natural predual of M_1 also has the property that it does not embed into a space with an unconditional basis.

Notes. See [13] for the proof of the results stated above. In view of (2) and the results of [25], one can say that M_1 has in a sense the simplest subspace structure of any non-reflexive subspace of L_1. A detailed study of the M_p spaces $(0 < p < \infty)$ is undertaken in [14]. This paper also contains a proof of the following fact: $L_2(\ell_p)$ is isomorphic to a subspace of $\ell_2(\ell_p)$ if and only if $p = 2$. See [32] for further results about $L_p(L_q)$. For applications of Theorem 3.2 to the finite-dimensional structure of L_p see [24].

5. Application to Lorentz spaces

First some notation. Throughout this section L_p denotes the usual Lebesgue space of measurable functions on $(0,\infty)$. For $0<r<s<\infty$ and $0<t<\infty$, $L_r+t.L_s$ denotes the Orlicz space L_r+L_s equipped with the norm $\|f\|_{L_r+t.L_s} = K(t,f;L_r,L_s)$, where

$$K(t,f;L_r,L_s) = \inf\{\|g\|_r + t\|h\|_s : f = g+h\} \tag{2}$$

is the standard K-functional of interpolation theory. The Lorentz space $L_{p,q}$ is the space of measurable functions on $(0,\infty)$ with the quasi-norm

$$\|f\|_{p,q} = \left(\int_0^\infty f^*(t)^q d(t^{q/p})\right)^{1/q}.$$

But this form of the $L_{p,q}$ norm is not particularly adapted to the purpose at hand. Much more useful to us is the following formula derived from the Lions-Peetre K-method of interpolation (see e.g. [3]):

$$\|f\|_{p,q} \approx \left(\sum_{n=-\infty}^{\infty} 2^{-n\theta q} K(2^n, f; L_r, L_s)^q\right)^{1/q} \tag{3}$$

provided $\dfrac{1}{p} = \dfrac{1-\theta}{r} + \dfrac{\theta}{s}$. (In other words $L_{p,q}$ is the interpolation space $[L_r, L_s]_{\theta,q}$.)

Theorem 5.1. *Let $0<r<s<\infty$. Then L_r+L_s is linearly isomorphic to a closed subspace of $L_r(\ell_s)$.*

Proof. First observe that $L_r(\ell_s)$ is linearly isometric to $L_r(\tilde{\Omega};\ell_s)$, since $\tilde{\Omega}$ and $[0,1]$ are measure-equivalent spaces by Caratheodory's Theorem. Hence it suffices to prove the result with $L_r(\ell_s)$ replaced by $L_r(\tilde{\Omega};\ell_s)$. Let $f = \sum_{n=1}^\infty \oplus f_n$ belong to L_r+L_s. Recall that $M_p = L_p + L_2$ for $2<p<\infty$. Thus, by Theorem 3.2, we have

$$\begin{aligned}
\|f\|_{L_r+L_s} &= \left\|\sum_{n=1}^\infty \oplus f_n\right\|_{L_r+L_s} \\
&\approx \left\|\sum_{n=1}^\infty \oplus |f_n|^{r/2}\right\|_{L_2+L_{2s/r}}^{2/r} \\
&\approx \left\|\sum_{n=1}^\infty \oplus |f_n|^{r/2}\right\|_{M_{2s/r}}^{2/r} \\
&\approx \left\|\left(\sum_{n=1}^\infty |f_n(s_n)|^s\right)^{r/2s}\right\|_2^{2/r} \\
&= \left\|\left(\sum_{n=1}^\infty |f_n(s_n)|^s\right)^{1/s}\right\|_r \\
&= \|(f_n(s_n))\|_{L_r(\tilde{\Omega},\ell_s)}.
\end{aligned}$$

Hence the mapping given by $\sum_{n=1}^{\infty} \oplus f_n \to ((f_n(s_n))_{n=1}^{\infty}$ defines a continuous linear isomorphism from $L_r + L_s$ into $L_r(\tilde{\Omega}; \ell_s)$. $\square$

For our results on embeddings of $L_{p,q}$ the following simple proposition proves to be useful.

Proposition 5.2. *Let $0 < r < s < \infty$. Then $L_r + t.L_s$ is linearly isometric to $L_r + L_s$.*

Proof. By dilation of (2) by $\lambda > 0$ we have

$$K(t, f; L_r, L_s) = \inf\{\lambda^{1/r}\|g(\lambda x)\|_r + t\lambda^{1/s}\|h(\lambda x)\|_s : f = g + h\}.$$

Let $\lambda = t^{rs/(s-r)}$, so that $t = \lambda^{(s-r)/rs}$. Then

$$\begin{aligned} K(t, f; L_r, L_s) &= \lambda^{1/r} \inf\{\|g(\lambda x)\|_r + \|h(\lambda x)\|_s : f = g + h\} \\ &= \lambda^{1/r} K(1, f(t^{rs/(s-r)}x); L_r, L_s). \end{aligned}$$

Thus the mapping given by $f(x) \to t^{s/(s-r)} f(t^{rs/(s-r)}x)$ defines a linear isometry from $L_r + t.L_s$ onto $L_r + L_s$. $\square$

Corollary 5.3. *Let $0 < r < p < s < \infty$ and let $0 < q < \infty$. Then $L_{p,q}$ is linearly isomorphic to a closed subspace of $\ell_q(L_r + L_s)$.*

Proof. (3) says that $L_{p,q}$ is isomorphic to a subspace of $\left[\sum_{n=-\infty}^{\infty} \oplus(L_r + 2^n.L_s)\right]_q$. By Proposition 6.2, each $L_r + t.L_s$ is linearly *isometric* to $L_r + L_s$, and so the result follows. $\square$

Theorem 5.4. *Let $0 < q < p < s$. Then $L_{p,q}$ is linearly isomorphic to a subspace of $L_q(\ell_s)$.*

Proof. Applying Corollary 5.3 with $r = q$ we see that $L_{p,q}$ embeds into $\ell_q(L_q + L_s)$. Hence, by Theorem 5.1, $L_{p,q}$ embeds into $\ell_q(L_q(\ell_s))$, which in turn is isomorphic to $L_q(\ell_s)$. $\square$

Remark 5.5. It follows from an example of Pisier (see [29 p.97]) that $L_{2,q}$ does not embed into $L_q(\ell_2)$ if $0 < q < 2$. Thus it is in general not possible to replace "s" by "p" in Theorem 5.4.

Corollary 5.6. *Let $0 < q < p < 2$. Then $L_{p,q}$ is linearly isomorphic to a subspace of L_q.*

Proof. Applying Theorem 5.4 with $s = 2$ we see that $L_{p,q}$ embeds into $L_q(\ell_2)$. Let $(r_n(t))_{n=1}^{\infty}$ be the sequence of Rademacher functions. Then by the classical Khintchine inequalities

$$\left\| \sum_{n=1}^{\infty} f_n(s) r_n(t) \right\|_q \approx \left\| \left(\sum_{n=1}^{\infty} f_n^2 \right)^{1/2} \right\|_q .$$

Hence the mapping given by $(f_n(s))_{n=1}^{\infty} \to \sum_{n=1}^{\infty} f_n(s) r_n(t)$ defines an isomorphic embedding of $L_q(\ell_2)$ into $L_q([0,1] \times [0,1])$ (which is isomorphic to L_q), and the desired conclusion follows. $\square$

Notes. The utilization of Proposition 3.1 to define Banach space embeddings of function spaces goes back to [21] (see also [29]). See [8,10] for information about the subspace structure of $L_{p,q}$. In particular, the "disjoint sum" concept is used there to prove the following fact: ℓ_p is isomorphic to a subspace of $L_{p,q}(0,\infty)$ if and only if $p=2$ or $p=q$. See [11] for further probabilistic applications of disjoint sums in Lorentz spaces. Corollary 5.6 for $q=1$ is a special case of the main result of [35], which classifies the Lorentz spaces which embed into $L_1(0,1)$. See [33] for further results about embeddings into L_1.

REFERENCES

1. D.J. Aldous, *Unconditional bases and martingales in $L_p(F)$*, Math. Proc. Camb. Phil. Soc. **85** (1979), 117–123.
2. N. Asmar and S.J. Montgomery-Smith, *Littlewood-Paley theory on solenoids*, Colloquium Mathematicum, to appear.
3. J. Bergh and J. Löfstrom, *Interpolation Spaces. An Introduction* Grundlehren der mathematischen Wissenschaften, Vol. 223, Springer-Verlag, Berlin-Heidelberg-New York, 1976.
4. W. Bryc and S. Kwapień, *On the conditional expectation with respect to a sequence of independent σ-fields*, Z. Warsch. Verw. Gebiete **46** (1979), 221–225.
5. D.L. Burkholder, *Distribution function inequalities for martingales*, Ann. Probab. **1** (1971), 19–42.
6. D.L. Burkholder, B.J. Davis and R.F. Gundy, *Integral inequalities for convex functions of operators on martingales*, Proc. Sixth Berkeley Symp. Math. Statist. Probab. Vol. 2, Univ. of California Press, Berkeley, Calif., 1972, pp. 223–240.
7. D.L. Burkholder, *A geometric condition that implies the existence of certain singular integrals of Banach-space-valued functions*, in Conference on Harmonic Analysis in Honor of Antoni Zygmund, Wadsworth, Belmont, Calif., 1983.
8. N.L. Carothers and S.J. Dilworth, *Geometry of Lorentz spaces via interpolation*, The University of Texas Functional Analysis Seminar Longhorn Notes, 1985-86, pp. 107–134.
9. N.L. Carothers and S.J. Dilworth, *Inequalities for sums of independent random variables*, Proc. Amer. Math. Soc. **104** (1988), 221–226.
10. N.L. Carothers and S.J. Dilworth, *Subspaces of $L_{p,q}$*, Proc. Amer. Math. Soc. **104** (1988), 537–545.
11. N.L. Carothers and S.J. Dilworth, *Equidistributed random variables in $L_{p,q}$*, J. Funct. Anal. **84** (1989), 146–159.
12. N.L. Carothers and S.J. Dilworth, *Some Banach space embeddings of classical function spaces*, Bull. Austral. Math. Soc. **43** (1991), 73–77.
13. S.J. Dilworth, *Intersection of Lebesgue spaces L_1 and L_2*, Proc. Amer. Math. Soc. **103** (1988), 1185–1188.
14. S.J. Dilworth, *A scale of linear spaces related to the L_p scale*, Illinois J. Math. **34** (1990), 140–158.
15. J.L. Doob, *Stochastic Processes*, Wiley, New York, 1953.
16. A. M. Garsia, *Martingale Inequalities*, W.A. Benjamin, Reading, Mass., 1973.
17. P. Hitczenko, *Best constants in the decoupling inequality for non-negative random variables*, Statistics and Probability Letters **9** (1990), 327–329.
18. P. Hitczenko, *Best constants in the martingale version of Rosenthal's inequality*, Ann. Probab. **18** (1990), 1656–1668.
19. P. Hitczenko, *A remark on the paper "Martingale inequalities in rearrangement invariant function spaces" by W.B. Johnson and G. Schechtman*, in The University of Texas at Austin Functional Analysis Seminar Longhorn Notes, 1987-89.
20. J. Hoffman-Jørgensen, *Sums of independent Banach space valued random variables*, Studia Math. **52** (1974), 159–186.
21. W.B. Johnson, B. Maurey, G. Schechtman and L. Tzafriri, *Symmetric structures in Banach spaces*, Mem. Amer. Math. Soc. **217** (1979).

22. W.B. Johnson, G. Schechtman and J. Zinn, *Best constants in moment inequalities for linear combinations of independent and exchangeable random variables*, Ann. Probab. **13** (1985), 234–253.
23. W.B. Johnson and G. Schechtman, *Martingale inequalities in rearrangement invariant function spaces*, Israel J. Math. **64** (1988), 267–275.
24. W.B. Johnson and G. Schechtman, *Sums of independent random variables in rearrangement-invariant function spaces*, Ann. Probab. **17** (1989), 789–808.
25. M.I. Kadeć and A. Pełczyński, *Bases, lacunary sequences and complemented subspaces in the spaces L_p*, Studia Math. **21** (1962), 161–176.
26. J. Khintchine, *Über die diadischen Brüche*, Math. Z. **18** (1923), 109–116.
27. J. Lindenstrauss and A. Pełczyński, *Contributions to the theory of the classical Banach spaces*, J. Funct. Anal. **8** (1971), 225–249.
28. J. Lindenstrauss and L.Tzafriri, *Classical Banach spaces I*, Springer-Verlag, Berlin-Heidelberg-New York, 1977.
29. J. Lindenstrauss and L. Tzafriri, *Classical Banach spaces II*, Springer-Verlag, Berlin-Heidelberg-New York, 1979.
30. M. Loève, *Probability Theory I*, Fourth Edition, Springer-Verlag, Berlin-Heidelberg-New York, 1977.
31. J. Marcinkiewicz and A. Zygmund, *Quelques théoremes sur les fonctions indépendantes*, Studia Math. **7** (1938), 104–120.
32. Y. Raynaud, *Sur les sous-espaces de $L^p(L^q)$*, exposé au Séminaire de Géometrie des Espaces de Banach, Universités Paris 6 et 7, 1984-85.
33. Y. Raynaud and C. Schütt, *Some results on symmetric subspaces of L_1*, Studia Math. **89** (1988), 27–35.
34. H.P. Rosenthal, *On the subspaces of L_p $(p > 2)$ spanned by sequences of independent random variables*, Israel J. Math. **8** (1970), 273–303.
35. C. Schütt, *Lorentz spaces that are isomorphic to subspaces of L_1*, Trans. Amer. Math. Soc. **314** (1989), 583–595.
36. E.M. Stein, *Topics in harmonic analysis related to the Littlewood-Paley theory* Ann. of Math. Studies vol. 63, Princeton University Press, Princeton, New Jersey.

DEPARTMENT OF MATHEMATICS, UNIVERSITY OF SOUTH CAROLINA, COLUMBIA, SOUTH CAROLINA 29208, U.S.A.

Contemporary Mathematics
Volume **144**, 1993

Bochner vs. Pettis norm: examples and results

S.J. DILWORTH AND MARIA GIRARDI

ABSTRACT. Our basic example shows that for an *arbitrary* infinite-dimensional Banach space $\mathfrak{X}$, the Bochner norm and the Pettis norm on $L_1(\mathfrak{X})$ are not equivalent. Refinements of this example are then used to investigate various modes of sequential convergence in $L_1(\mathfrak{X})$.

1. INTRODUCTION

Over the years, the Pettis integral along with the Pettis norm have grabbed the interest of many. In this note, we wish to clarify the differences between the Bochner and the Pettis norms. We begin our investigation by using Dvoretzky's Theorem to construct, for an *arbitrary* infinite-dimensional Banach space, a sequence of Bochner integrable functions whose Bochner norms tend to infinity but whose Pettis norms tend to zero. By refining this example (again working with an arbitrary infinite-dimensional Banach space), we produce a Pettis integrable function that is not Bochner integrable and we show that the space of Pettis integrable functions is not complete. Thus our basic example provides a unified constructive way of seeing several known facts. The third section expresses these results from a vector measure viewpoint. In the last section, with the aid of these examples, we give a fairly thorough survey of the implications going between various modes of convergence for sequences of $L_1(\mathfrak{X})$ functions.

2. EXAMPLES: $L_1(\mathfrak{X})$ vs. $P_1(\mathfrak{X})$

Let $\mathfrak{X}$ be a Banach space with dual $\mathfrak{X}^*$ and let (Ω, Σ, μ) be the usual Lebesgue measure space on $[0,1]$. Let $L_1(\mathfrak{X}) = L_1(\Omega, \Sigma, \mu; \mathfrak{X})$ be the space of $\mathfrak{X}$-valued Bochner integrable functions on Ω with the usual Bochner norm $\|\cdot\|_{\text{Bochner}}$.

1991 *Mathematics Subject Classification.* 46E40, 28A20, 46G10.
This paper is in final form and no version of it will be submitted for publication elsewhere.

A quick review of the Pettis integral. Consider a weakly measurable function $f : \Omega \to \mathfrak{X}$ such that $x^*f \in L_1(\mathbb{R})$ for all $x^* \in \mathfrak{X}^*$. The Closed Graph Theorem gives that for each $E \in \Sigma$ there is an element x_E^{**} of $\mathfrak{X}^{**}$ satisfying $x_E^{**}(x^*) = \int_E x^* f d\mu$ for each $x^* \in \mathfrak{X}^*$. If x_E^{**} is actually in $\mathfrak{X}$ for each $E \in \Sigma$, then we say that f is Pettis integrable with the Pettis integral of f over E being the element $x_E \in \mathfrak{X}$ satisfying

$$x^*(x_E) = \int_E x^* f d\mu$$

for each $x^* \in \mathfrak{X}^*$. In this note we shall restrict our attention to Pettis integrable functions that are strongly measurable. The space $\mathcal{P}_1(\mathfrak{X})$ of (equivalence classes of) all strongly measurable Pettis integrable functions forms a normed linear space under the norm

$$\| f \|_{\text{Pettis}} = \sup_{x^* \in B(\mathfrak{X}^*)} \int_\Omega | x^*(f) | \, d\mu .$$

Clearly if $f \in L_1(\mathfrak{X})$ then $f \in \mathcal{P}_1(\mathfrak{X})$ with

$$||f||_{\text{Pettis}} \leq ||f||_{\text{Bochner}} .$$

If $\mathfrak{X}$ is finite-dimensional, then the Bochner norm and the Pettis norm are equivalent and so $L_1(\mathfrak{X}) = \mathcal{P}_1(\mathfrak{X})$. However, in *any* infinite-dimensional Banach space $\mathfrak{X}$, the next example shows that the two norms are not equivalent on $L_1(\mathfrak{X})$. Specifically, it constructs an *essentially bounded* sequence of Bochner-norm one functions whose Pettis norms tend to 0.

EXAMPLE 1.

Let $\mathfrak{X}$ be an infinite-dimensional Banach space. We now construct a sequence $\{f_n\}$ of $L_1(\mathfrak{X})$ functions which tend to zero in the Pettis norm but not in the Bochner norm.

Let $\{I_k^n : n = 0, 1, \dots ; k = 1, \dots , 2^n\}$ be the dyadic intervals on $[0, 1]$, i.e.

$$I_k^n = \left[\tfrac{k-1}{2^n}, \tfrac{k}{2^n}\right) .$$

Fix $n \in \mathbb{N}$. We define f_n with the help of Dvoretzky's Theorem [D]. Find a 2^n-dimensional subspace E_n of $\mathfrak{X}$ such that the Banach-Mazur distance between E_n and $\ell_2^{2^n}$ is at most 2. So there is an operator $T_n : \ell_2^{2^n} \to E_n$ such that the norm of T_n is at most 2 and the norm of the inverse of T_n is 1. Let $\{e_k^n : k = 1, \dots 2^n\}$ be the image under T_n of the standard unit vectors $\{u_k^n : k = 1, \dots 2^n\}$ of $\ell_2^{2^n}$. Define $f_n : \Omega \to \mathfrak{X}$ by

$$f_n(\omega) = \sum_{k=1}^{2^n} e_k^n \, 1_{I_k^n}(\omega) ,$$

The sequence $\{f_n\}$ has the desired properties.

Since

$$||f_n||_{\text{Bochner}} = \sum_{k=1}^{2^n} \int_{I_k^n} ||e_k^n||_{\mathfrak{X}} \, d\mu = 2^{-n} \sum_{k=1}^{2^n} ||e_k^n||_{\mathfrak{X}}$$

and $1 \leq ||e_k^n||_{\mathfrak{X}} \leq 2$,

$$1 \leq ||f_n||_{\text{Bochner}} \leq 2 .$$

As for the Pettis norm, fix $x^* \in B(X^*)$. Note that the restriction y_n^* of x^* to E_n is an element in E_n^* of norm at most 1. Also, the adjoint T_n^* of T_n has norm at most 2. Thus

$$\begin{aligned} \sum_{k=1}^{2^n} |x^*(e_k^n)| &= \sum_{k=1}^{2^n} |y_n^*(T_n u_k^n)| \\ &= \sum_{k=1}^{2^n} |(T_n^* y_n^*)(u_k^n)| \leq ||T_n^* y_n^*||_{(\ell_2^{2^n})^*} \, ||\sum_{k=1}^{2^n} u_k^n||_{\ell_2^{2^n}} \leq 2\sqrt{2^n} , \end{aligned}$$

and so

$$\int_\Omega |x^*(f_n)| \, d\mu = \sum_{k=1}^{2^n} \int_{I_k^n} |x^*(e_k^n)| \, d\mu = 2^{-n} \sum_{k=1}^{2^n} |x^*(e_k^n)| \leq 2\, 2^{-\frac{n}{2}} ,$$

giving that

$$||f_n||_{\text{Pettis}} \leq 2\, 2^{-\frac{n}{2}} .$$

Thus the Pettis norm of f_n tends to zero. □

This difference between the Bochner and Pettis norms is further emphasized by the following minor variation of Example 1.

EXAMPLE 2.

Fix $0 < \beta < \frac{1}{2}$ and put $\alpha_n \equiv 2^{n\beta}$. Define $\tilde{f}_n \colon \Omega \to \mathfrak{X}$ by $\tilde{f}_n = \alpha_n f_n$ where f_n is as in Example 1. The Bochner norm of $\tilde{f}_n$ tends to infinity at the rate of α_n but the Pettis norm of $\tilde{f}_n$ tends to zero. □

A variation on this example gives a (strongly measurable) Pettis integrable function that is not Bochner integrable – working in *any* infinite-dimensional Banach space. The usual way of constructing such a function uses the Dvoretsky-Rogers Theorem. Recall [DR] that in any infinite-dimensional Banach space $\mathfrak{X}$, there is an unconditionally convergent series $\sum_n x_n$ that is not absolutely convergent. The function $f \colon \Omega \to \mathfrak{X}$ given by

$$f(\cdot) = \sum_{n=1}^{\infty} \frac{x_n}{\mu(E_n)} 1_{E_n}(\cdot),$$

(where $\{E_n\}$ is a partition of Ω into sets of strictly positive measure) is Pettis integrable but not Bochner integrable. Our example is in the same spirit.

EXAMPLE 3.

Let $\mathfrak{X}$ be an infinite-dimensional Banach space. Consider the above construction. Note that the collection $\{I_2^n\}_{n=1}^{\infty}$ partitions $[0,1]$. Let $g_n \colon \Omega \to \mathfrak{X}$ be a normalized f_n (from Example 1) supported on I_2^n, i.e.

$$g_n(\omega) = 2^n f_n(2^n\omega - 1) \, 1_{I_2^n}(\omega) .$$

Define $f\colon \Omega \to \mathfrak{X}$ by

$$f(\omega) = \sum_{n=1}^{\infty} g_n(\omega)\, 1_{I_2^n}(\omega)\,.$$

Clearly f is strongly measurable.

But f is not Bochner integrable since for any $N \in \mathbb{N}$

$$\int_\Omega \|f\|\, d\mu \;\geq\; \sum_{n=1}^{N} \int_{I_2^n} \|g_n\|_{\mathfrak{X}}\, d\mu \;\geq\; \sum_{n=1}^{N} 2^{-n}\, 2^n \;=\; N\,.$$

However f is Pettis integrable. First note that $x^*f \in L_1(\mathbb{R})$ for all $x^* \in \mathfrak{X}^*$ since for a fixed $x^* \in B(X^*)$ we have (using computations from the first example)

$$\int_{I_2^n} |x^* f|\, d\mu \;=\; \int_{I_2^n} |x^* g_n|\, d\mu \;=\; \int_\Omega |x^* f_n|\, d\mu \;\leq\; 2\, 2^{-\frac{n}{2}}\,,$$

and so

$$\int_\Omega |x^* f|\, d\mu \;=\; \sum_{n=1}^{\infty} \int_{I_2^n} |x^* f|\, d\mu \;\leq\; 2\sum_{n=1}^{\infty} 2^{-\frac{n}{2}} \;=\; 2(\sqrt{2}+1)\,.$$

Next fix $E \in \Sigma$. We know there is an element x_E^{**} of $\mathfrak{X}^{**}$ satisfying that $x_E^{**}(x^*) = \int_E x^* f d\mu$ for each $x^* \in \mathfrak{X}^*$. To see that x_E^{**} is actually in $\mathfrak{X}$, consider the sequence $\{s_N\}$ of $L_1(\mathfrak{X})$ functions on Ω where

$$s_N(\omega) = \sum_{n=1}^{N} g_n(\omega)\, 1_{I_2^n}(\omega)\,.$$

Let x_E^N be the Bochner integral of s_N over E. Viewed as a sequence in $\mathfrak{X}^{**}$, $\{x_E^N\}_N$ converges in norm to x_E^{**} since for any $x^* \in B(X^*)$

$$\begin{aligned}
|(x_E^N - x_E^{**})(x^*)| &= \Big|\int_E x^* s_N\, d\mu - \int_E x^* f\, d\mu\Big| \;\leq\; \int_\Omega |x^*(s_N - f)|\, d\mu \\
&= \sum_{n=N+1}^{\infty} \int_{I_2^n} |x^* f|\, d\mu \;\leq\; 2 \sum_{n=N+1}^{\infty} 2^{-\frac{n}{2}}\,.
\end{aligned}$$

Thus x_E^{**} is actually in $\mathfrak{X}$, as needed. □

Janicka and Kalton [JK] showed that if $\mathfrak{X}$ is infinite-dimensional then $\mathcal{P}_1(\mathfrak{X})$ is *never* complete under the Pettis norm. The first example provides a constructive way to see this.

EXAMPLE 4.

Taking some care in our construction in Example 1, we may arrange for $\{E_n\}$ to form a finite dimensional decomposition (FDD) of some subspace of $\mathfrak{X}$. To see this, first choose a basic sequence $\{x_n\}$ in $\mathfrak{X}$. Take a blocking $\{F_n\}$ of the basis so that each subspace F_n is of large enough dimension to find (using the finite-dimensional version of Dvoretzky's Theorem) a 2^n-dimensional subspace E_n of F_n such that the Banach-Mazur distance between E_n and $\ell_2^{2^n}$ is at most 2. Then $\{E_n\}$ forms a FDD.

Keeping with the notation from Example 1, define $h_n : \Omega \to \mathfrak{X}$ by

$$h_n = \sum_{k=1}^{n} f_k .$$

Clearly each h_n is in $\mathcal{P}_1(\mathfrak{X})$. Also, $\{h_n\}$ is Cauchy in the Pettis norm since

$$||h_n - h_{n+j}||_{\text{Pettis}} \leq \sum_{k=n}^{\infty} ||f_k||_{\text{Pettis}} \leq 2\sum_{k=n}^{\infty} 2^{-\frac{k}{2}} .$$

To show that h_n cannot converge to an element in $\mathcal{P}_1(\mathfrak{X})$, we need the following two lemmas whose proofs are given shortly.

LEMMA 1. If the sequence $\{f_n\}$ of functions in $\mathcal{P}_1(\mathfrak{X})$ converges to $f \in \mathcal{P}_1(\mathfrak{X})$ and each f_n is essentially valued in some subspace $\mathcal{Y}$ of $\mathfrak{X}$, then f is also essentially valued in $\mathcal{Y}$. □

LEMMA 2. Let $\{f_n\}$ be a sequence of $\mathcal{P}_1(\mathfrak{X})$ functions that converges in Pettis norm to $f \in \mathcal{P}_1(\mathfrak{X})$. Furthermore, suppose that each f_n is essentially valued in some subspace $\mathcal{Y}$ of $\mathfrak{X}$ and that $T : \mathcal{Y} \to \mathfrak{X}$ is a bounded linear operator. Then Tf_n converges in the Pettis norm to Tf. □

Suppose that h_n converges in the Pettis norm to $h \in \mathcal{P}_1(\mathfrak{X})$. Let $\mathcal{Y} \equiv \sum \oplus E_n$ and let $P_n : \mathcal{Y} \to \mathfrak{X}$ be the natural projection from $\mathcal{Y}$ onto E_n. Note that each h_n, thus also h, are essentially valued in $\mathcal{Y}$. Applying the second lemma to the operator P_n and noting that $P_n h_m \equiv f_n$ for $m \geq n$, we have that $P_n h \equiv f_n$. Thus for each $n \in \mathbb{N}$

$$1 \leq ||P_n h(\omega)||_{\mathfrak{X}} \leq 2$$

for almost all ω. This contradicts the fact that for almost all ω

$$\lim_{N\to\infty} ||h(\omega) - \sum_{n=1}^{N} P_n h(\omega)||_{\mathfrak{X}} = 0 .$$

Thus h_n cannot converge to an element in $\mathcal{P}_1(\mathfrak{X})$. □

Now for the proofs of the lemmas.

PROOF OF LEMMA 1. Let $\{f_n\}$ be a sequence of $\mathcal{P}_1(\mathfrak{X})$ functions that converges in the Pettis norm to $f \in \mathcal{P}_1(\mathfrak{X})$. Furthermore, suppose that each f_n is essentially valued in some subspace $\mathcal{Y}$ of $\mathfrak{X}$ and that f is essentially valued in some subspace $\mathcal{Z}$ with $\mathcal{Y} \subset \mathcal{Z}$. Since f and each f_n are strongly measurable, we may assume that $\mathcal{Z}$ is separable.

Since $\mathcal{Z}$ is separable, there is a sequence $\{z_m^*\}$ of functionals from $\mathcal{Z}^*$ such that an element z from $\mathcal{Z}$ is in $\mathcal{Y}$ if and only if $z_m^*(z) = 0$ for each m.

Since $\{f_n\}$ converges in the Pettis norm to f, for each z_m^* the sequence $\{z_m^* f_n\}_n$ converges to $z_m^* f$ in $L_1(\mathbb{R})$ -norm. But for each m, note that $z_m^* f_n$ is zero almost everywhere and so $z_m^* f$ is also zero almost everywhere. Thus f is also essentially valued in $\mathcal{Y}$. □

PROOF OF LEMMA 2. Let $\{f_n\}$ be a sequence of $\mathcal{P}_1(\mathfrak{X})$ functions which converges in Pettis norm to $f \in \mathcal{P}_1(\mathfrak{X})$. Furthermore, suppose that each f_n is essentially valued in some subspace $\mathfrak{Y}$ of $\mathfrak{X}$. Lemma 1 gives that f is also essentially valued in $\mathfrak{Y}$. Let $T : \mathfrak{Y} \to \mathfrak{X}$ be a bounded linear operator. Note that for a fixed $x^* \in \mathfrak{X}^*$ we may extend T^*x^* to an element in $\mathfrak{X}^*$ without increasing its norm.

Note that Tf (and likewise each Tf_n) is actually in $\mathcal{P}_1(\mathfrak{X})$. To see this, first observe that for a fixed $x^* \in \mathfrak{X}^*$ (viewing T^*x^* as an element in $\mathfrak{X}^*$ of the same norm)

$$\int_\Omega |x^*(Tf)|\,d\mu \;=\; \int_\Omega |(T^*x^*)(f)|\,d\mu \;\leq\; ||T^*x^*||\;||f||_{\text{Pettis}}$$

and so $x^*(Tf) \in L_1$. Next, let x_E be the Pettis integral of f over $E \in \Sigma$. Note that $x_E \in \mathfrak{Y}$ and for each $x^* \in \mathfrak{X}^*$

$$x^*(Tx_E) \;=\; (T^*x^*)(x_E) \;=\; \int_E (T^*x^*)(f)\,d\mu \;=\; \int_E x^*(Tf)\,d\mu\;,$$

and so $T(x_E)$ is the Pettis integral of Tf over $E \in \Sigma$. Thus Tf is indeed in $\mathcal{P}_1(\mathfrak{X})$.

Since for $x^* \in B(X^*)$

$$\int_\Omega |x^*(Tf_n - Tf)|\,d\mu \;=\; \int_\Omega |(T^*x^*)(f_n - f)|\,d\mu \;\leq\; ||T||\;||f_n - f||_{\text{Pettis}}\;,$$

the sequence Tf_n converges in the Pettis norm to Tf. □

3. VECTOR MEASURES

Let's look at our examples from a vector measure viewpoint.

A Pettis integrable function f gives rise to a vector measure $G_f\colon \Sigma \to \mathfrak{X}$ where $G_f(E)$ is the Pettis integral of f over E. The Pettis norm of f is the semivariation of G_f. If $f \in L_1(\mathfrak{X})$, then the Bochner norm of f is the variation of G_f.

Example 2 shows that for *any* infinite-dimensional Banach space $\mathfrak{X}$, there is a sequence of $\mathfrak{X}$-valued measures whose variations tend to infinity but whose semivariations tend to zero. On the other hand, Example 3 gives a $\mathfrak{X}$-valued measure of bounded semivariation but infinite variation.

For $f \in \mathcal{P}(\mathfrak{X})$, the corresponding measure G_f is μ-continuous and has a relatively compact range. In fact [DU], the completion of $\mathcal{P}_1(\mathfrak{X})$ is (isometrically) the space $K(\mathfrak{X})$ of all μ-continuous measures $G : \Sigma \to \mathfrak{X}$ whose range is relatively compact, equipped with the semivariation norm. Thus Example 4 gives a measure in $K(\mathfrak{X})$ which is not representable by a (strongly measurable) Pettis integrable function.

4. CONVERGENCE

We now restrict our attention to sequences of $L_1(\mathfrak{X})$ functions which converge (in the Bochner norm, in the Pettis norm, or weakly in $L_1(\mathfrak{X})$) to functions actually in $L_1(\mathfrak{X})$. We have the following obvious relations:

(1) Bochner-norm convergence implies weak and Pettis-norm convergence;
(2) weak convergence implies neither Bochner-norm nor Pettis-norm convergence (e.g. consider the Rademacher functions in $L_1(\mathbb{R})$);
(3) if $\mathfrak{X}$ is finite dimensional then the Pettis norm and the Bochner norm are equivalent.

Our examples help to clarify the other implications.

If $\mathfrak{X}$ is infinite-dimensional, then there is a sequence of functions which converges in Pettis norm but whose Bochner norms tend to infinity (Example 2). Thus this sequence does not converge in the Bochner norm or weakly.

What if we require the sequence to be bounded ($\sup_n ||f_n||_{L_1(\mathfrak{X})} < \infty$), or uniformly integrable, or even essentially bounded ($\sup_n ||f_n||_{L_\infty(\mathfrak{X})} < \infty$)? Again, if $\mathfrak{X}$ is infinite-dimensional then Example 1 gives a sequence of essentially bounded functions which converges in Pettis norm but not in the Bochner norm.

In certain situations we can pass from Pettis-norm convergence to weak convergence. If $f_n \to f$ in the Pettis norm, then clearly $\int_\Omega g(f_n)\,d\mu \to \int_\Omega g(f)\,d\mu$ for each *simple function* $g \in L_\infty(\mathfrak{X}^*)$. As a first step towards weak convergence, when can we at least conclude that $f_n \to f$ in the $\sigma(L_1(\mathfrak{X}), L_\infty(\mathfrak{X}^*))$-topology? A word of caution, the simple functions need not be dense in $L_\infty(\mathfrak{X}^*)$. For example, the $L_\infty(\ell_2)$ function $g(\cdot) = \sum_{n=1}^\infty e_n 1_{I_2^n}$, where $\{e_n\}$ are the standard unit vectors of ℓ_2, is at least $\frac{\sqrt{2}}{2}$ from any simple function in $L_\infty(\ell_2)$. Our next example illustrates the fact that for any infinite-dimensional Banach space there is a bounded sequence in $L_1(\mathfrak{X})$ that converges in the Pettis norm but not in the $\sigma(L_1(\mathfrak{X}), L_\infty(\mathfrak{X}^*))$-topology.

EXAMPLE 5.

Consider the sequence $\{g_n\}$ from Example 3 where

$$g_n(\omega) \;=\; 2^n\, f_n(2^n\omega - 1)\, 1_{I_2^n}(\omega)\,.$$

Since $||f_n||_{L_1(\mathfrak{X})} \leq 2$, the Bochner norm of each g_n is at most 2.

To examine the Pettis norm, fix $x^* \in B(X^*)$. Computations in Example 1 show that

$$\int_\Omega |x^* g_n|\,d\mu \;=\; \int_{I_2^n} |x^* g_n|\,d\mu \;=\; \int_\Omega |x^* f_n|\,d\mu \;\leq\; 2\,2^{-\frac{n}{2}}$$

and so

$$||g_n||_{\text{Pettis}} \;\leq\; 2\,2^{-\frac{n}{2}}\,.$$

Thus $g_n \to 0$ in the Pettis norm.

As for the $\sigma(L_1(\mathfrak{X}), L_\infty(\mathfrak{X}^*))$-topology, find $y_k^n \in B(\mathfrak{X}^*)$ such that $y_k^n(e_k^n) = \|e_k^n\|_{\mathfrak{X}}$. Note that $\{I_2^n : n = 1, 2, \dots\}$ partitions $[0,1]$ and that $\{I_j^{2n} : j = 2^n + 1, \dots, 2^{n+1}\}$ partitions the interval I_2^n. Define $g \colon \Omega \to \mathfrak{X}^*$ by

$$g(\omega) = \sum_{n=1}^{\infty} \sum_{k=1}^{2^n} y_k^n 1_{I_{2^n+k}^{2n}}(\omega) .$$

Clearly $g \in L_\infty(\mathfrak{X}^*)$. But

$$\begin{aligned} \int_\Omega g(g_n)\, d\mu &= \int_{I_k^n} g(g_n)\, d\mu = \sum_{k=1}^{2^n} \int_{I_{2^n+k}^{2n}} g(g_n)\, d\mu \\ &= \sum_{k=1}^{2^n} \int_{I_{2^n+k}^{2n}} y_k^n(2^n\ e_k^n)\, d\mu \geq \sum_{k=1}^{2^n} 2^n\ 2^{-2n}\ 1 = 1 . \end{aligned}$$

Thus $\{g_n\}$ does not converge in the $\sigma(L_1(\mathfrak{X}), L_\infty(\mathfrak{X}^*))$-topology. □

However, a *uniformly integrable* Pettis-norm convergent sequence does converge in the $\sigma(L_1(\mathfrak{X}), L_\infty(\mathfrak{X}^*))$-topology. To see why, consider a uniformly integrable sequence $\{f_n\}$ such that $\int_\Omega g(f_n)\, d\mu \to 0$ for each simple function $g \in L_\infty(\mathfrak{X}^*)$. For an arbitrary $g \in L_\infty(\mathfrak{X}^*)$, note that there is a sequence of simple functions which converges to g in measure. So we can approximate g (in $L_\infty(\mathfrak{X}^*)$ norm) by a simple function $\tilde{g}$ on a subset $E \in \Sigma$ of full enough measure that $\int_{E^c} \|f_n\|\, d\mu$ is small for each n and the measure of E^c is also small. Note that

$$\left| \int_\Omega g(f_n)\, d\mu \right| \leq \int_{E^c} \|g\|_{\mathfrak{X}^*} \|f_n\|_{\mathfrak{X}}\, d\mu + \int_E \|g - \tilde{g}\|_{\mathfrak{X}^*} \|f_n\|_{\mathfrak{X}}\, d\mu + \left| \int_E \tilde{g}(f_n)\, d\mu \right| .$$

The first two integrals on the left-hand side are small for each n. The last integral converges to zero as $n \to \infty$. Thus $\int_\Omega g(f_n)\, d\mu \to 0$, as needed.

A Pettis-norm convergent sequence need not be uniformly integrable (Example 2). However, if a sequence converges in the $\sigma(L_1(\mathfrak{X}), L_\infty(\mathfrak{X}^*))$-topology then it is *necessarily* uniformly integrable. To see why this is so, recall the usual proof [DU p. 104] that a weakly convergent sequence is uniformly integrable. If a subset K of $L_1(\mathfrak{X})$ is not uniformly integrable, then we can find (with the help of Rosenthal's Lemma) a sequence $\{f_n\}$ in K and a disjoint sequence $\{E_n\}$ in Σ such that

$$\int_{E_n} \|f_n\|\, d\mu > 2\delta \qquad \text{and} \qquad \int_{\cup_{j \neq n} E_j} \|f_n\|\, d\mu < \delta$$

for some $\delta > 0$. For each n we can find $l_n \in L_\infty(\mathfrak{X}^*) \subset [L_1(\mathfrak{X})]^*$ of norm one and supported on E_n such that $\int_{E_n} l_n(f_n) > 2\delta$. Define $l \colon \Omega \to \mathfrak{X}^*$ by

$$l(\cdot) = \sum_{n=1}^{\infty} l_n(\cdot)\, 1_{E_n}(\cdot) .$$

Clearly, l is a norm one element of $L_\infty(\mathfrak{X}^*) \subset [L_1(\mathfrak{X})]^*$. But $\int_\Omega l(f_n) \nrightarrow 0$ since

$$\int_\Omega l(f_n)\,d\mu = \int_{E_n} l_n(f_n)\,d\mu + \int_{F_n} l(f_n)\,d\mu ,$$

where $F_n = \cup_{m\neq n} E_m$, and so

$$\begin{aligned}\left|\int_\Omega l(f_n)\,d\mu\right| &\geq \left|\int_{E_n} l_n(f_n)\,d\mu\right| - \int_{F_n} \|l\|_{\mathfrak{X}^*}\,\|(f_n)\|_{\mathfrak{X}}\,d\mu \\ &\geq 2\delta - \int_{F_n} \|f_n\|\,d\mu \\ &\geq 2\delta - \delta = \delta .\end{aligned}$$

So if $g_n \to g$ in the $\sigma(L_1(\mathfrak{X}), L_\infty(\mathfrak{X}^*))$-topology then the set $\{g_n - g\}$, and thus also the set $\{g_n\}$, are uniformly integrable.

To pass from $\sigma(L_1(\mathfrak{X}), L_\infty(\mathfrak{X}^*))$-convergence to weak convergence, recall [DU] that the dual of $L_1(\mathfrak{X})$ is (isometrically) $L_\infty(\mathfrak{X}^*)$ if and only if $\mathfrak{X}^*$ has the Radon-Nikodým property (RNP). Thus we have:

FACT. *If $\mathfrak{X}^*$ has the RNP, then a uniformly integrable Pettis-norm convergent sequence of $L_1(\mathfrak{X})$ functions also converges weakly.*

As noted above, the uniform integrability condition is necessary. Our next proposition shows that it is also necessary that $\mathfrak{X}^*$ have the RNP. The proof, which uses Stegall's Factorization Theorem, may be pulled out of a result of Ghoussoub and P. Saab [GS] characterizing weakly compact sets in $L_1(\mathfrak{X})$.

PROPOSITION. *If $\mathfrak{X}^*$ fails the RNP, then there is an essentially bounded sequence of $L_1(\mathfrak{X})$ functions that is Pettis-norm convergent but not weakly convergent.*

PROOF.

Since $\mathfrak{X}^*$ fails the RNP, there is a separable subspace $\mathfrak{Y}$ of $\mathfrak{X}$ such that $\mathfrak{Y}^*$ is not separable. We shall construct an essentially bounded sequence $\{g_n\}$ of $L_1(\mathfrak{Y})$ functions such that $g_n \to 0$ in the Pettis-norm on $L_1(\mathfrak{Y})$ but $g_n \nrightarrow 0$ weakly in $L_1(\mathfrak{Y})$. This is sufficient.

Let $\Delta = \{-1,1\}^{\mathbb{N}}$ be the Cantor group with Haar measure ν. Let $\{\Delta_k^n : k = 1, \ldots, 2^n\}$ be the standard n-th partition of Δ. Thus $\Delta_1^0 = \Delta$ and $\Delta_k^n = \Delta_{2k-1}^{n+1} \cup \Delta_{2k}^{n+1}$ and $\nu(\Delta_k^n) = 2^{-n}$. We will work with our underlying measure space being the Cantor group endowed with its Haar measure instead of our usual Lebesgue measure space on $[0,1]$.

Consider the set $C(\Delta)$ of real-valued continuous function on Δ as a subspace of $L_\infty(\mathbb{R})$. Let $\{1_\Delta\} \cup \{h_k^n : n = 0,1,2,\ldots \text{ and } k = 1,\ldots,2^n\}$ be the usual Haar basis of $C(\Delta)$, where $h_k^n : \Delta \to \mathbb{R}$ is given by

$$h_k^n = 1_{\Delta_{2k-1}^{n+1}} - 1_{\Delta_{2k}^{n+1}} .$$

Let $\{e_k^n : n = 0,1,2,\dots \text{ and } k = 1,\dots,2^n\}$ be an enumeration (lexicographically) of the usual ℓ_1 basis and $H: \ell_1 \to L_\infty$ be the Haar operator that takes e_k^n to h_k^n.

By Stegall's Factorization Theorem [S, Theorem 4], H factors through $\mathcal{Y}$, i.e. there are bounded linear operators $R: \ell_1 \to \mathcal{Y}$ and $S: \mathcal{Y} \to L_\infty$ such that $H = SR$. Let $\tilde{R}$ be the natural extension of R to a bounded linear operator from $L_1(\ell_1)$ to $L_1(\mathcal{Y})$.

Consider the sequence $\{f_m\}$ of $L_1(\ell_1)$ functions given by

$$f_m(\cdot) = \frac{1}{m} \sum_{n=1}^{m} \sum_{k=1}^{2^n} h_k^n(\cdot) e_k^n .$$

Let $g_m = \tilde{R}(f_m)$. The sequence $\{g_m\}$ of $L_1(\mathcal{Y})$ functions does the job.

Since $\sup_m \|f_m(\omega)\|_{\ell_1} = 1$ for ν-a.e. ω, we have that $\{g_m\}$ is essentially bounded.

To examine the Pettis norm of f_m, consider the continuous convex function $\psi: \ell_\infty \to \mathbb{R}$ where the image of $x^* = (\alpha_k^n) \in \ell_\infty \simeq \ell_1^*$ (again, lexicographically ordered) is

$$\psi(x^*) = \int_\Delta |x^*(f_m)| \, d\nu = \frac{1}{m} \int_\Delta \left| \sum_{n=1}^{m} \sum_{k=1}^{2^n} h_k^n(\cdot) x^*(e_k^n) \right| d\nu .$$

Since the image of f_m is supported on $\{e_k^n : n = 1,\dots,2^m \text{ and } k = 1,\dots,2^n\}$, we consider the natural restriction map $r_m : \ell_\infty \to \ell_\infty^{2(2^{2^m}-1)} \subset \ell_\infty$ given by

$$\begin{aligned} r_m((\alpha_k^n : n = 0,1,2,\dots \text{ and } k = 1,\dots,2^n)) &= \\ (\alpha_k^n : n = 1,\dots,2^m \text{ and } k = 1,\dots,2^n) . \end{aligned}$$

Let K be the unit ball of $r_m(\ell_\infty)$. The maximum of ψ over K is attained at some extreme point x_0^* of K. Being an extreme point, $x_0^* = (\epsilon_k^n)$ satisfies $|\epsilon_k^n| = 1$ for each admissible n and k. So

$$\begin{aligned} \|f_m\|_{\text{Pettis}} &= \sup_{x^* \in B(\ell_\infty)} \psi(x^*) = \sup_{x^* \in K} \psi(x^*) \\ &= \psi(x_0^*) = \frac{1}{m} \int_\Delta \left| \sum_{n=1}^{m} \sum_{k=1}^{2^n} \epsilon_k^n \, h_k^n(\cdot) \right| d\nu . \end{aligned}$$

Since $\sum_{k=1}^{2^n} \epsilon_k^n \, h_k^n$ behaves like the nth-Rademacher function, Khintchine's inequality gives that $\|f_m\|_{\text{Pettis}}$ decays like $m^{-\frac{1}{2}}$.

Now for the Pettis norm of g_m, note that if $y^* \in \mathcal{Y}^*$ then

$$\begin{aligned} \int_\Delta |y^* g_m| \, d\nu &= \int_\Delta |y^*(Rf_m)| \, d\nu \\ &= \int_\Delta |(R^* \, y^*)(f_m)| \, d\nu \le \|R^*\| \, \|y^*\| \, \|f_m\|_{\text{Pettis}} . \end{aligned}$$

Thus $g_m \to 0$ in the Pettis norm on $L_1(\mathcal{Y})$.

To see that $g_m \not\to 0$ weakly in $L_1(\mathfrak{Y})$, it suffices to show that $\tilde{H}f_m \not\to 0$ weakly in $L_1(L_\infty)$ where $\tilde{H}$ is the natural extension of H to an operator from $L_1(\ell_1)$ to $L_1(L_\infty)$. But since

$$\tilde{H}f_m(\cdot) = \frac{1}{m}\sum_{n=1}^{m}\sum_{k=1}^{2^n} h_k^n(\cdot)h_k^n ,$$

we may view $\tilde{H}f_m$ as a function from Δ into $C(\Delta)$ and thus we only need to show that $\tilde{H}f_m \not\to 0$ weakly in $L_1(C(\Delta))$. To see this, just consider the linear functional $l \in [L_1(C(\Delta))]^*$ given by

$$l(f) = \int_\Delta [f(\omega)](\omega)\, d\nu \qquad \text{where} \quad f \in L_1(C(\Delta))$$

and note that

$$l(\tilde{H}f_m) = \frac{1}{m}\int_\Delta \sum_{n=1}^{m}\sum_{k=1}^{2^n} h_k^n(\omega)h_k^n(\omega)\, d\nu = 1 .$$

Thus the sequence $\{g_m\}$ does all it is supposed to do. □

We close with one last example illustrating the above proposition in the case that $\mathfrak{X}$ is ℓ_1.

EXAMPLE 6. Once again working on the Lebesgue measure space on $[0,1]$, consider the sequence $\{g_n\}$ of $L_1(\ell_1)$ functions where

$$g_n(\cdot) = \frac{1}{n}\sum_{k=1}^{n} r_k(\cdot)e_k$$

where $\{e_k\}$ are the standard unit vectors of ℓ_1 and $\{r_k\}$ are the Rademacher functions.

The sequence is uniformly integrable, indeed it is even essentially bounded by 1. As for the Pettis norm of a g_n, fix $x^* \in (\ell_1)^*$, say $x^* = (\alpha_j) \in \ell_\infty$. Since

$$\int_\Omega |x^* g_n|\, d\mu = \frac{1}{n}\int_\Omega |\sum_{k=1}^{n} \alpha_k r_k|\, d\mu ,$$

Khintchine's inequality shows us that the Pettis norm of g_n behaves like $\frac{1}{\sqrt{n}}$. Thus $g_n \to 0$ in the Pettis norm and so also in the $\sigma(L_1(\mathfrak{X}), L_\infty(\mathfrak{X}^*))$-topology.

However, $\{g_n\}$ does not converge weakly. For consider the vector measure $G\colon \Sigma \to \ell_\infty$ given by

$$G(E) = (\mu(E \cap [r_j = 1])) .$$

Clearly G is a μ-continuous vector measure of bounded variation. Thus G gives rise to a functional $l \in [L_1(\ell_1)]^*$ given by

$$l(f) = \int_\Omega f\, dG \qquad \text{for} \quad f \in L_1(\ell_1) ,$$

where we view G as taking values in ℓ_1^*. Since

$$l(g_n) = \frac{1}{n}\sum_{k=1}^{n} l(e_k r_k) = \frac{1}{n}\sum_{k=1}^{n} l(e_k 1_{[r_k=1]} - e_k 1_{[r_k=-1]}) = \frac{1}{n}\sum_{k=1}^{n} \tfrac{1}{2} - 0 = \frac{1}{2},$$

we see that $\{g_n\}$ does not converge weakly.

Note that the above linear functional $l\colon L_1(\ell_1) \to \mathbb{R}$ is continuous for the Bochner norm but not for the Pettis norm. □

References

[DU]. J. Diestel and J. J. Uhl, Jr., *Vector Measures*, Math. Surveys, no. 15, Amer. Math. Soc., Providence, R.I., 1977.

[D]. A. Dvoretzky, *Some results on convex bodies and Banach spaces*, Proc. Internat. Symp. on Linear Spaces, Jerusalem, 1961, pp. 123–160.

[DR]. A. Dvoretzky and C. A. Rogers, *Absolute and unconditional convergence in normed spaces*, Proc. Nat. Acad. Sci. USA **36** (1950), 192–197.

[GS]. N. Ghoussoub and P. Saab, *Weak Compactness in Spaces of Bochner Integrable Functions and the Radon-Nikodým property*, Pacific J. Math. **110** (1984), no. 1, 65–70.

[KJ]. L. Janicka and N.J. Kalton, *Vector Measures of Infinite Variation*, Bulletin de L'Académie Polonaise Des Sciences **XXV** (1977), no. 3, 239–241.

[P]. B.J. Pettis, *On integration in vector spaces*, Trans. Amer. Math. Soc. **44** (1938), 277–304.

[Sch]. G. Schlüchtermann, *The Mazur Property and Completeness in the Space of Bochner-Integrable Functions $L_1(\mu, \mathfrak{X})$*, Glasgow Math. J. **34** (1992), 201–206.

[S]. Charles Stegall, *The Radon-Nikodým property in conjugate Banach spaces*, Trans. Amer. Math. Soc. **206** (1975), 213–223.

University of South Carolina, Mathematics Department, Columbia, SC 29208

Contemporary Mathematics
Volume **144**, 1993

Extremal Structure of the Unit Ball of $C(K, X)$

PATRICK N. DOWLING
ZHIBAO HU
MARK A. SMITH

ABSTRACT. Strongly extreme points of the unit ball of $C(K, X)$ are characterized very naturally as the norm-one functions that are strongly extreme point valued at each point of K. Known results concerning extreme points of the unit ball of $C(K, X)$ are recalled, and it is shown that a variety of stronger types of extreme points (e.g. denting points) never exist in the unit ball of $C(K, X)$.

During the last fifty years several types of extreme points of convex sets have been invented and studied in general Banach spaces. For a specific Banach space, it is always interesting to discover whether the closed unit ball has any extreme points and, if so, to exactly describe them; it is of further interest to determine which, if any, of these extreme points are in fact one of the many stronger types of extreme points in the literature.

Let K be an infinite compact Hausdorff topological space and let X be a Banach space. Let $C(K, X)$ denote the Banach space of all continuous functions from K into X equipped with the supremum norm.

In this note, the extremal structure of the closed unit ball of $C(K, X)$ relative to the extremal structure of the closed unit ball of X is investigated. First, the rather peculiar but known extreme point structure of $C(K, X)$ is recalled, then strongly extreme points of the unit ball of $C(K, X)$ are very naturally characterized, and finally it is shown that $C(K, X)$ never has many of the stronger type of extreme points (for example, denting points).

Let B_E denote the closed unit ball of Banach space E. A point x in B_E is said to be an *extreme point* of B_E provided that x is not the midpoint of any nontrivial line segment lying in B_E. The unit ball B_E is said to be *stable*

1991 *Mathematics Subject Classification.* Primary: 46B20, 46E40.
Key words and phrases. extreme points, strongly extreme points, denting points.

provided the mapping from $B_E \times B_E$ into B_E that sends (x, y) to $\frac{1}{2}(x+y)$ is an open mapping. The following theorem lists some of the known results (see [**G**]) concerning extreme points of the unit ball of $C(K, X)$; assertions (1) and (2) are straightforward and well known , and assertion (5), which is stated in [**G**], can be proved using (1), (2) and Michael's selection theorem [**M**].

THEOREM 1. *Let f be an element of $C(K, X)$.*

(1) *If f is an extreme point of $B_{C(K,X)}$, then $\|f(t)\| = 1$ for all t in K.*
(2) *If the set of all t in K for which $f(t)$ is an extreme point of B_X is dense in K, then f is an extreme point of $B_{C(K,X)}$.*
(3) [**BLP**] *If f is an extreme point of $B_{C(K,X)}$ and $dimX \leq 3$, then the set of all t in K for which $f(t)$ is an extreme point of B_X is dense in K.*
(4) [**BLP**] *There is a Banach space X with $dimX = 4$ and an extreme point f of $B_{C([0,1],X)}$ such that there is no t in K for which $f(t)$ is an extreme point of B_X.*
(5) *If B_X is stable, then f is an extreme point of $B_{C(K,X)}$ if and only if $f(t)$ is an extreme point of B_X for all t in K.*

For a Banach space E , a point x in B_E is said to be a *strongly extreme point* of B_E provided that for every $\epsilon > 0$ there exists $\delta > 0$ such that if $\|x \pm z\| \leq 1+\delta$ for z in E, then $\|z\| \leq \epsilon$. The following theorem characterizes strongly extreme points of the unit ball of $C(K, X)$.

THEOREM 2. *Let f be an element of $C(K, X)$. Then f is a strongly extreme point of $B_{C(K,X)}$ if and only if $f(t)$ is a strongly extreme point of B_X for all t in K.*

PROOF. Suppose f is in $B_{C(K,X)}$ and there exists t_0 in K such that $f(t_0)$ is not a strongly extreme point of B_X. Then there exist a sequence (z_n) in X and $\epsilon > 0$ such that, for all n in $\mathbb{N}$,

$$\|z_n\| > \epsilon \quad \text{and} \quad \|f(t_0) \pm z_n\| \leq 1 + \frac{1}{n}.$$

Since f is continuous at t_0 , for each n in $\mathbb{N}$, there exists an open neighborhood U_n of t_0 such that

$$\|f(t) - f(t_0)\| < \frac{1}{n} \text{ for all } t \text{ in } U_n.$$

Since K is normal, for each n in $\mathbb{N}$, there exists a continuous function $g_n : K \to [0,1]$ such that

$$supp(g_n) \subseteq U_n \text{ and } g_n(t_0) = 1,$$

where $supp(g_n)$ denotes the support of g_n. For each n in $\mathbb{N}$, define $h_n : K \to X$ by

$$h_n(t) = g_n(t)z_n \quad \text{for } t \text{ in } K.$$

Note that h_n is in $C(K,X)$, that $h_n(t_0) = z_n$, and that $\|h_n\| = \|z_n\|$ for each n in $\mathbb{N}$. Now, on one hand, if t is in U_n, then

$$\begin{aligned} \|f(t) \pm h_n(t)\| &= \|f(t) \pm g_n(t)z_n\| \\ &\leq \|f(t_0) \pm g_n(t)z_n\| + \|f(t) - f(t_0)\| \\ &\leq 1 + \frac{1}{n} + \frac{1}{n}, \end{aligned}$$

and, on the other hand, if t is in $K \backslash U_n$, then

$$\begin{aligned} \|f(t) \pm h_n(t)\| &= \|f(t) \pm g_n(t)z_n\| \\ &= \|f(t)\| \\ &\leq 1. \end{aligned}$$

Thus $\|f \pm h_n\| \leq 1 + \frac{2}{n}$ and $\|h_n\| \geq \epsilon$ for each n in $\mathbb{N}$. This shows that f is not a strongly extreme point of $B_{C(K,X)}$.

Conversely, suppose $f(t)$ is a strongly extreme point of B_X for all t in K. To obtain a contradiction, assume that f is not a strongly extreme point of $B_{C(K,X)}$. Then there exist a sequence (h_n) in $C(K,X)$ and $\epsilon > 0$ such that, for all n in $\mathbb{N}$,

$$\|h_n\| > \epsilon \quad \text{and} \quad \|f \pm h_n\| \leq 1 + \frac{1}{n}.$$

For each n in $\mathbb{N}$, choose t_n in K such that $\|h_n(t_n)\| > \epsilon$. Since K is compact, there exist a subnet (t_{n_λ}) of (t_n) and t_0 in K such that $lim_\lambda t_{n_\lambda} = t_0$. Since $f(t_0)$ is a strongly extreme point of B_X, there exists n_0 in $\mathbb{N}$ such that if $\|f(t_0) \pm z\| \leq 1 + \frac{1}{n_0}$ for z in X, then $\|z\| \leq \epsilon$. Using the continuity of f at t_0, choose λ_0 such that

$$\left\|f(t_{n_{\lambda_0}}) - f(t_0)\right\| < \frac{1}{2n_0} \quad \text{and } n_{\lambda_0} > 2n_0.$$

Then

$$\begin{aligned} \left\|f(t_0) \pm h_{n_{\lambda_0}}(t_{n_{\lambda_0}})\right\| &\leq \left\|f(t_{n_{\lambda_0}}) \pm h_{n_{\lambda_0}}(t_{n_{\lambda_0}})\right\| + \left\|f(t_{n_{\lambda_0}}) - f(t_0)\right\| \\ &< 1 + \frac{1}{n_{\lambda_0}} + \frac{1}{2n_0} \\ &< 1 + \frac{1}{n_0} \end{aligned}$$

yet $\left\|h_{n_{\lambda_0}}(t_{n_{\lambda_0}})\right\| > \epsilon$, contrary to the choice of n_0. This completes the proof.

For a Banach space E, a point x in B_E is said to be a *point of sequential continuity* of B_E provided that whenever (x_n) is a sequence in B_E such that

$x_n \to x$ weakly, it follows that $\|x_n - x\| \to 0$. The following theorem says that the unit ball of $C(K,X)$ never has a point of sequential continuity. As an immediate consequence, the unit ball of $C(K,X)$ can never have any type of extreme point that is stronger than a point of sequential continuity, for example, a point of continuity, a denting point, a strongly exposed point, or a locally uniformly rotund point.

THEOREM 3. *The unit ball of $C(K,X)$ contains no point of sequential continuity.*

PROOF. Let f be in $B_{C(K,X)}$ and set

$$U = \left\{t \in K : \|f(t)\| < \frac{2}{3}\right\} \text{ and } V = \left\{t \in K : \|f(t)\| > \frac{1}{3}\right\}.$$

Then U and V are open sets in K such that $K = U \cup V$. Hence either U or V is infinite.

In the case U is infinite, there exists a pairwise disjoint sequence (U_n) of non-empty, open subsets of U. Choose x_0 in X such that $\|x_0\| = \frac{1}{3}$. Since K is normal, for each n in $\mathbb{N}$, there exists a continuous function $g_n : K \to [0,1]$ such that $supp(g_n) \subseteq U_n$ and $g_n(t_n) = 1$ for some point t_n in U_n. Now, for n in $\mathbb{N}$, define $f_n : K \to X$ by

$$f_n(t) = f(t) + g_n(t)x_0 \quad \text{for } t \text{ in } K.$$

Then each f_n is in $B_{C(K,X)}$ and $\|f_n - f\| = \frac{1}{3}$.

CLAIM. *$f_n \to f$ weakly in $C(K,X)$.*

To establish this claim, it suffices to show that $g_n(\cdot)x_0 \to 0$ weakly in $C(K,X)$. Let F be an element of $C(K,X)^*$. Then (see [**S**]) there exists a countably additive X^*–valued Borel measure G of bounded variation on K such that

$$F(g) = \int_K g(t)\, dG(t) \quad \text{for all } g \text{ in } C(K,X).$$

Thus, since $supp(g_n) \subseteq U_n$,

$$\begin{aligned}
|F(g_n(\cdot)x_0)| &= \left|\int_K g_n(t)x_0 \; dG(t)\right| \\
&= \left|\int_{U_n} g_n(t)x_0 \; dG(t)\right| \\
&\leq \int_{U_n} |g_n(t)| \; \|x_0\| \; d\,|G|\,(t) \\
&\leq |G|\,(U_n)
\end{aligned}$$

for each n in $\mathbb{N}$. But the sequence (U_n) is pairwise disjoint, and hence $|G|(U_n) \to 0$. This establishes the claim which in turn shows that f is not a point of sequential continuity of $B_{C(K,X)}$ since $\|f_n - f\| = \frac{1}{3}$ for each n in $\mathbb{N}$.

In the case V is infinite, there exists a pairwise disjoint sequence (V_n) of non-empty open subsets of V. As above, for each n in $\mathbb{N}$, there exists a continuous function $g_n : K \to [0,1]$ such that $supp(g_n) \subseteq V_n$ and $g_n(t_n) = 1$ for some point t_n in V_n. Now, for n in $\mathbb{N}$, define $f_n : K \to X$ by

$$f_n(t) = (1 - g_n(t))f(t) \quad \text{for } t \text{ in } K.$$

Then each f_n is in $B_{C(K,X)}$ and, since t_n is in V,

$$\begin{aligned}\|f_n - f\| &\geq \|f_n(t_n) - f(t_n)\| \\ &= \|f(t_n)\| \\ &> \frac{1}{3}.\end{aligned}$$

In the same manner as in the previous case, it can be shown that $f_n \to f$ weakly in $C(K, X)$ which in turn shows that f is not a point of sequential continuity of $B_{C(K,X)}$. This completes the proof.

References

[BLP] R.M.Blumenthal, J. Lindenstrauss, and R.R. Phelps, *Extreme operators into $C(K)$*, Pac. J. Math. **15** (1965), 747-756.

[G] R. Grzaslewicz, *Extreme operator-valued continuous maps*, Ark. Mat. **29** (1991), 73-81.

[M] E. Michael, *Continuous selections I*, Annals of Math. **63** (1956), 361-382.

[S] I. Singer, *Linear functionals on the space of continuous mappings of a compact Hausdorff space into a Banach space*, Rev. Math. Pures Appl. **2** (1957), 301-315.

P.N. Dowling, Department of Mathematics and Statistics, Miami University, Oxford, OH 45056

E-mail address: pndowling@miavxl.acs.muohio.edu

Z. Hu, Department of Mathematics and Statistics, Miami University, Oxford, OH 45056

E-mail address: zhu@miavxl.acs.muohio.edu

M. A. Smith, Department of Mathematics and Statistics, Miami University, Oxford, OH 45056

E-mail address: masmith@miamiu.acs.muohio.edu

Contemporary Mathematics
Volume **144**, 1993

RANDOM PRODUCTS OF NONEXPANSIVE MAPPINGS IN SPACES WITH THE OPIAL PROPERTY

JOHN M. DYE, TADEUSZ KUCZUMOW,
PEI-KEE LIN AND SIMEON REICH

ABSTRACT. Let C be a bounded closed convex subset of a reflexive Banach space with Opial's property. Let $\{T_1, T_2, \ldots, T_N\}$ be N nonexpansive self-mappings of C satisfying condition (W) with respect to a common fixed point. A special case of our main result shows that any quasi-periodic random product drawn from these mappings converges weakly to a common fixed point. We also obtain weak convergence results for products of strongly nonexpansive mappings.

Let (C, ρ) be a metric space and let τ be a Hausdorff topology on C such that (C, τ) is sequentially compact. An operator $T\colon C \to C$ is called *nonexpansive* if $\rho(Tx, Ty) \leq \rho(x, y)$ for all x and y in C. We say that T satisfies condition (W) with respect to a fixed point p of T if whenever a ρ-bounded sequence $\{x_n\}$ is τ-convergent to $x \in C$ and $(\rho(x_n, p) - \rho(Tx_n, p)) \to 0$ it follows that $\{Tx_n\}$ is also τ-convergent to x. This condition encompasses a large class of maps, and if T_1 and T_2 satisfy condition (W) with respect to a common fixed point, then so does their composition T_1T_2. Since all strong nonexpansive mappings in the sense of [5] satisfy condition (W), we see that in any uniformly convex Banach space all averaged and firmly nonexpansive mappings (including all linear contractive projections and all resolvents of accretive operator) satisfy condition (W) with respect to any topology τ weaker than the strong topology. An example of a (linear) (W) nonexpansive map which is not strongly nonexpansive can be found in [9, p.161]. Also, every self-mapping of the Hilbert ball (equipped with the hyperbolic metric and the weak topology) that is either an averaged mapping [19] or firmly nonexpansive (of the first or second kind [12]) satisfies condition (W) with respect to each one of its fixed points.

We say that the space (C, ρ, τ) has *Opial's property* if whenever a ρ-bounded sequence $\{x_n\}$ is τ-convergent to $x \in C$ it follows that

$$\limsup_{n\to\infty} \rho(x_n, x) < \limsup_{n\to\infty} \rho(x_n, y)$$

for all $y \neq x$.

1991 *Mathematics Subject Classification.* 46B20, 47H09, 65J15.

Key words and phrases. Nonexpansive mapping, Opial's property, random product.

This research was partially supported by the NSF under Grant DMS-9006855

The paper was written during a visit of the second author to the Department of Mathematics of the University of Southern California. He is grateful to his colleagues at USC for their hospitality.

All the authors thank the referee for many helpful suggestions.

Opial's property in Banach spaces was introduced by Opial [17] to characterize weak limits by means of the norm. He showed that Hilbert spaces and the l^p spaces with $1 < p < \infty$ (more generally all spaces with weakly sequentially continuous duality mappings– see [13]) have this property (with respect to the weak topology). In [7] van Dulst proved that every separable Banach space X admits an equivalent norm with the property that X with this new norm has Opial's property. But on the other hand, he has also proved that every reflexive Banach space admits an equivalent norm such that X with the new norm does not have Opial's property.

In [17] Opial also observed that his property is independent of uniform convexity. For example, $L^p[0,1]$ with $1 \leq p < +\infty$ and $p \neq 2$ does not have Opial's property with respect to the weak topology. Nevertheless, if (Ω, Σ, μ) is a positive σ-finite measure space, then in $L^p(\mu)$ one can introduce the metrizable topology of convergence locally in measure. This topology is induced by

$$d_0(f,g) = \sum_{n=1}^{\infty} \frac{1}{2^n} \frac{1}{\mu(E_n)} \int_{E_n} \frac{|f-g|}{1+|f-g|} d\mu$$

for all f, $g \in L_0$ (the set of all scalar-valued Σ-measurable functions on Ω), where $\{E_n\}$ is a partition of Ω such that $0 < \mu(E_n) < \infty$ for $n = 1, 2, \cdots$ [16]. The space $L^p(\mu), 1 \leq p < \infty$ does have Opial's property with respect to the topology of convergence locally in measure.

There are many examples of spaces with Opial's property with respect to the weak* topology. The space $l^1 = c_0^*$ is a standard example.

We also recall that the Hilbert ball with the hyperbolic metric has Opial's property with respect to the weak topology.

We shall say that the space (C, τ) satisfies condition (M) if (A, τ) is metrizable for all countable subsets A of C. For example, every weakly (weak*) compact subset of a (dual to a separable) Banach space satisfies condition (M).

Now let $\{T_1, T_2, \cdots, T_N\}$ be N (W) nonexpansive operators on C with a common fixed point, and let r be a mapping from the set of natural numbers $\mathbb{N}$ onto $\{1, 2, \cdots, N\}$, which assumes each value infinitely often. An unrestricted (or random) product of these operators is the sequence $\{S_n : n = 1, 2, \dots\}$ defined by

$$S_n = T_{r(n)} T_{r(n-1)} \cdots T_{r(1)}.$$

Such products find application in many areas. See [6] and [10] for examples and references.

Recall that the mapping r is said to be quasi-periodic if there exits a natural number L such that, for each i, $\{r(i), r(i+1) \dots, r(i+L-1)\}$ contains all members of $\{1, 2, \cdots, N\}$

Quasi-periodic products were previously studied by Browder and Kakutani [3,4], Prager [18], Apostol [1] and Dye [8]. More recently it was shown in [10] that if $\{T_1, T_2, \dots, T_N\}$ are N (W) nonexpansive maps on a Hilbert space H sharing a common fixed point, and r is quasi- periodic, then $\{S_n x\}$ converges weakly (to a common fixed point) for each x in H. As mentioned in [10] it would be of interest to find out if this result continues to hold outside Hilbert space. A partial answer to this question was obtained in [11] where the result was shown to hold in those reflexive Banach spaces with a weakly sequentially continuous duality map. In the present paper we prove that this result is true for a much wider class of Banach spaces, namely those which have Opial's property. Moreover, we show that the result is also true for those spaces (C, ρ, τ) which have Opial's property and satisfy condition (M). This class of spaces includes, for example, the Hilbert ball equipped with the hyperbolic metric and the weak topology as well as $L^p(\mu)$, $1 \leq p < \infty$, with the topology of convergence locally in measure.

Our argument is based on the following lemma.

Lemma 1. *Let (C, ρ) be a metric space and let τ be a Hausdorff topology on C. Assume that (C, ρ, τ) has Opial's property and that (C, τ) satisfies condition (M). Let $\{u_1, u_2, \ldots, u_N\}$ be N points in C and let $q : \mathbb{N} \to \{1, 2, \ldots, N\}$ assume each value infinitely often. If $\{\{x_i^m : i \in \mathbb{N}\} : m \in \mathbb{N}\}$ is a sequence of ρ-bounded sequences in C such that for each $m \in \mathbb{N}$ every sequence $\{x_i^m : i \in \mathbb{N}\}$ τ-converges to $u \in C$, and*

$$\limsup_{i\to\infty} \rho(x_i^m, u) \geq \limsup_{i\to\infty} \rho(x_i^{m+1}, u_{q(m)}),$$

then $u_k = u$ for all $1 \leq k \leq N$.

Proof. Fix $1 \leq k \leq N$ and let $\{m_j : j \geq 1\}$ be a sequence tending to ∞ such that $q(m_j - 1) = k$. By condition (M) the subspace (A, τ) of (C, τ) that consists of the sequences $\{x_i^m : i \in \mathbb{N}\}, m = 1, 2, \ldots$, the point u, and the points $\{u_1, u_2, \ldots, u_N\}$ is metrizable by a metric d. Let

$$r_m = \limsup_{i\to\infty} \rho(x_i^m, u).$$

Since

$$\limsup_{i\to\infty} \rho(x_i^{m+1}, u) \leq \limsup_{i\to\infty} \rho(x_i^{m+1}, u_{q(m)})$$

by Opial's property, we see that $\{r_m\}$ decreases to a limit $r \geq 0$. For each $j = 1, 2, \ldots$ we can find an $i_j \in \mathbb{N}$ such that

$$d(x_{i_j}^{m_j}, u) < \frac{1}{j}$$

$$\rho(x_{i_j}^{m_j}, u) \geq r_{m_j} - \frac{1}{j}$$

$$\rho(x_{i_j}^{m_j}, u_{q(m_j-1)}) \leq r_{m_j-1} + \frac{1}{j}.$$

Hence

$$\{x_{i_j}^{m_j}\} \text{ is } \tau\text{-convergent to } u,$$

$$\limsup_{j\to\infty} \rho(x_{i_j}^{m_j}, u) \geq r,$$

and

$$\limsup_{j\to\infty} \rho(x_{i_j}^{m_j}, u_k) \leq r.$$

Opial's condition now implies that $u = u_k$. □

We also need the following lemma which is established by a diagonal argument.

Lemma 2. *Let $\{S_{n_i}x\}$ be a subsequence of a random product of $\{S_nx\}$. Then there is a subsequence $\{S_{p_i}x\}$ of $\{S_{n_i}x\}$ and a map $q : \mathbb{N} \to \{1, 2, \dots, N\}$ such that $r(p_i + j) = q(j)$ for all $1 \le j \le i$.*

Theorem 1. *Let (C, ρ) be a metric space and let τ be a Hausdorff topology on C such that (C, τ) is sequentially compact. Assume that (C, ρ, τ) has Opial's property and that (C, τ) satisfies condition (M). Let $\{T_1, T_2, \dots, T_N\}$ be N nonexpansive self-mappings of C satisfying condition (W) with respect to a common fixed point. Let r be a mapping from $\mathbb{N}$ onto $\{1, 2, \dots, N\}$ and let $S_n = T_{r(n)}T_{r(n-1)} \cdots T_{r(1)}$. If r is quasi-periodic, then $\{S_nx\}$ τ-converges (to a common fixed point) for each x in C.*

Proof. . Let a subsequence $\{S_{n_i}x\}$ of $\{S_nx\}$ τ-converge to u. Let $\{S_{p_i}x\}$ be the subsequence of $\{S_{n_i}x\}$ obtained in Lemma 2. Let $x_i^1 = S_{p_i}x$, and for each $m \ge 2$

$$x_i^m = T_{r(p_i+m-1)}T_{r(p_i+m-2)} \cdots T_{r(p_i+1)}S_{p_i}x.$$

Note that

$$x_i^m = T_{q(m-1)}T_{q(m-2)} \cdots T_{q(1)}S_{p_i}x$$

for $i \ge m - 1$. Let p be the common fixed point mentioned in the statement of the theorem. Since each T_j satisfies condition (W) and

$$\rho(x_i^m, p) - \rho(S_{p_i}x, p) \underset{i\to\infty}{\longrightarrow} 0$$

for each $m \ge 2$, $\{x_i^m\}$ τ-converges to u. Also,

$$\rho(x_i^m, u) \ge \rho(x_i^{m+1}, T_{q(m)}u), \quad i \ge m,$$

and q is quasi-periodic (because r is). Consequently, Lemma 1 shows that u is a fixed point for all the operators $\{T_j : 1 \le j \le N\}$. Opial's property can now be used to show that there is only one subsequential limit. The proof is complete.

Corollary 1. *Let $\{T_1, T_2, \dots, T_N\}$ be N (W) nonexpansive self-mappings of a weakly (weak*) compact subset K of a (dual to a separable) Banach space X sharing a common fixed point. Let r be a quasi-periodic mapping from $\mathbb{N}$ onto $\{1, 2, \dots, N\}$ and let $S_n = T_{r(n)}T_{r(n-1)} \cdots T_{r(1)}$. If X has Opial's property with respect to the weak (weak*) topology then $\{S_nx\}$ converges weakly (weakly*) (to a common fixed point) for each x in K.*

Corollary 2. *Let $\{T_1, T_2, \dots, T_N\}$ be N (W) nonexpansive self-mappings of a ρ-closed, ρ-convex subset K of the Hilbert ball sharing a common fixed point. Let r be a quasi-periodic mapping from $\mathbb{N}$ onto $\{1, 2, \dots, N\}$ and let $S_n = T_{r(n)}T_{r(n-1)} \cdots T_{r(1)}$. Then $\{S_nx\}$ converges weakly (to a common fixed point) for each x in K.*

A similar result holds for $L^p(\mu)$, $1 \le p < \infty$, with τ the metrizable topology of convergence locally in measure mentioned earlier. However, in this case we obtain a sharper result. As a matter of fact, the analogues of Theorems 1 - 4 of [11] are valid. As an example, we mention the following result. (Recall that a point w in a closed subset C of a Banach space is said to be *a weak internal point* (WIP) for C if given any $z \in C$, there exists a positive $r = r(z)$ such that $w + s(w - z) \in C$ for all $0 \le s \le r$.)

Theorem 2. *Let $X = L^p(\mu)$, $1 < p < \infty$, and let τ be the topology of convergence locally in measure. Let $\{T_1, T_2, \ldots, T_N\}$ be N (W) nonexpansive self-mappings of a τ-compact subset K of X whose fixed point sets share a common WIP. Let r be a mapping from $\mathbb{N}$ onto $\{1, 2, \ldots, N\}$ which assumes each value infinitely often and let $S_n = T_{r(n)}T_{r(n-1)} \cdots T_{r(1)}$. Then $\{S_n x\}$ converges locally in measure (to a common fixed point) for each x in K.*

Proof. Using a result in [2] it is not difficult to show that if $\{z_n\}$ is a sequence in $L^p(\mu)$, $1 < p < \infty$, such that $|z_n| \to s$ and $z_n \to 0$ locally in measure, then $\lim_{n\to\infty} |u + z_n| = \delta(|u|, s)$ for all u in X, where $\delta(r, s) = (r^p + s^p)^{\frac{1}{p}}$. The result now follows from the discussion following theorem 4 in [11]. □

Remark 1. In the settings of Corollary 1, Corollary 2 (cf. [15, 20]) and Theorem 2, the limit of $\{S_n x\}$ defines a nonexpansive retraction onto the common fixed point set of $\{T_1, T_2, \ldots, T_N\}$.

Remark 2. The arguments leading to Theorem 1 also yield a generalization of Theorem 1 in [14] to the setting of (C, ρ, τ).

We say that a nonexpansive map $T : C \to C$ has condition (S) with respect to a fixed point p of T if whenever a ρ-bounded sequence $\{x_n\}$ satisfies $(\rho(x_n, p) - \rho(Tx_n, p)) \to 0$ it follows that $\rho(x_n, Tx_n) \to 0$. We don't know if Theorem 1 continues to hold when r is not quasi-periodic or when (C, ρ, τ) lacks Opial's property. However, when we replace condition (W) by condition (S) which is stronger in our setting we are able to establish τ-convergence for all random products drawn from two maps. To this end, we need the following lemma.

Lemma 3. *Let (C, ρ) be a metric space and let τ be a Hausdorff topology on C. Assume that (C, ρ, τ) has Opial's property and that the nonexpansive map $T : C \to C$ has condition (S) with respect to a fixed point p. Suppose that $\{v_n\}$ is a ρ-bounded sequence that τ-converges to u. If $(\rho(v_n, p) - \rho(Tv_n, p)) \to 0$, then $u = Tu$.*

Proof. We have

$$\rho(v_n, Tv_n) \to 0$$

because T satisfies condition (S). Hence

$$\limsup_{n\to\infty} \rho(v_n, Tu) = \limsup_{n\to\infty} \rho(Tv_n, Tu) \le \limsup_{n\to\infty} \rho(v_n, u)$$

and $u = Tu$ because (C, ρ, τ) has Opial's property. □

Theorem 3. *Let (C, ρ) be a metric space and let τ be a Hausdorff topology on C such that (C, τ) is sequentially compact. Assume that (C, ρ, τ) has Opial's property, and let T_1 and T_2 be two nonexpansive mappings on C, whose fixed point sets have a nonempty intersection. If T_1 and T_2 satisfy condition (S) with respect to a common point, then any random product $\{S_n x\}$ drawn from T_1 and T_2 τ-converges (to a common fixed point) for each x in C.*

Proof. Let a subsequence $\{S_{n_k} x\}$ of $\{S_n x\}$ τ-converge to u. By possibly reindexing and/or dropping to a further subsequence we may assume that there exist (homogeneous) words $T_1^{p(k)}$ with $p(k) \ge 1$, in the original sequence, and a point w such that

$$S_{n_{k+1}} = (\ldots)T_2 T_1^{p(k)} S_{n_k}$$

and

$$T_1^{p(k)-1} S_{n_k} x \tau\text{-converges to } w.$$

Multiple applications of Lemma 3 show that w is a common fixed point for T_1 and T_2 and that u is a fixed point of T_1. Hence,

$$\begin{aligned} &\lim_{k\to\infty} \sup \rho(S_{n_k} x, u) \\ &\geq \lim_{k\to\infty} \sup \rho(T_1^{p(k)-1} S_{n_k} x, u) \\ &\geq \lim_{k\to\infty} \rho(T_1^{p(k)-1} S_{n_k} x, w) \end{aligned} \tag{1}$$

where the first inequality follows by the nonexpansivity of T_1 and the second is a consequence of Opial's property.

Finally, we note that

$$\begin{aligned} &\lim_{k\to\infty} \rho(T_1^{p(k)-1} S_{n_k} x, w) \\ &\geq \lim_{k\to\infty} \rho\left((\ldots)T_2 T_1^{p(k)} S_{n_k} x, w\right) \\ &\geq \lim_{k\to\infty} \sup \rho(S_{n_k} x, u) \end{aligned} \tag{2}$$

where the fact that w is a common fixed point is used in the first inequality and Opial's property is used again in the second inequality. A comparison of (1) and (2) shows, in light of Opial's property, that $w = u$. In particular, u is a common fixed point for T_1 and T_2, and the result follows as in Theorem 1. □

If the maps in Theorem 3 are further assumed to be retractions (i.e., $P^2 = P$), then a modification of the proof of Theorem 3 shows that the result can be extended from products drawn from two maps to products drawn from three maps.

Theorem 4. *Let (C, ρ) be a metric and let τ be a Hausdorff topology on C such that (C, τ) is sequentially compact. Assume that (C, ρ, τ) has Opial's property, and let P, Q and R be three nonexpansive retractions onto closed sets with a nonempty intersection. If each one of these retractions satisfies condition (S) with respect to a certain point in this intersection, then any random product $\{S_n x\}$ drawn from P, Q and R τ-converges (to a common fixed point) for each x in C.*

References

1. C. Apostol, *Products of contractions in Hilbert space*, Acta Sci. Math. (Szeged) **33** (1972), 91–94.
2. H. Brezis and E. Lieb, *A relation between pointwise convergence of functions and convergence of functionals*, Proc. Amer. Math. Soc. **88** (1983), 486–490.
3. F.E. Browder, *On some approximation methods for solutions of the Dirichlet problem for linear elliptic equations of arbitrary order*, J. Math. Mech. **7** (1958), 69–80.
4. F.E. Browder, *Convergence theorems for sequences of nonlinear operators in Banach spaces*, Math. Z. **100** (1967), 201–225.
5. R.E. Bruck and S. Reich, *Nonexpansive projections and resolvents of accretive operators in Banach spaces*, Houston J. Math. **3** (1977), 459–470.

6. F. Deutsch, Math. Rev. (1985), no. Review #85m:41047.
7. D. van Dulst, *Equivalent norms and the fixed point property for nonexpansive mappings*, J. London Math. Soc. **25** (1982), 139–144.
8. J. Dye, *Convergence of random products of compact contractions in Hilbert space*, J. Integral Equations and Operator Theory **12** (1989), 12–22.
9. ———, *A generalization of a theorem of Amemiya and Ando on the convergence of random products of contractions in Hilbert space*, J. Integral Equations and Operator Theory **12** (1989), 155–162.
10. J. M. Dye and S. Reich, *Unrestricted iterations of nonexpansive mappings in Hilbert spaces*, Nonlinear Analysis **18** (1992), 199–207.
11. ———, *Unrestricted iterations of nonexpansive mappings in Banach spaces*, Nonlinear Analysis (to appear).
12. K. Goebel and S. Reich, *Uniform Convexity, Hyperbolic Geometry and Nonexpansive Mappings*, Marcel Dekker, New York and Basel, 1984.
13. J.P. Gossez and E. Lami-Dozo, *Some geometric properties related to the fixed point theory for nonexpansive mappings*, Pacific J. Math. **40** (1972), 565–573.
14. T. Kuczumow, *Weak convergence theorems for nonexpansive mappings and semigroups in Banach spaces with Opial's property*, Proc. Amer. Math. Soc. **93** (1985), 430–432.
15. ———, *Nonexpansive retracts and fixed points of nonexpansive mappings in the Cartesian product of n Hilbert balls*, Nonlinear Analysis **9** (1985), 601–604.
16. C. Lennard, *A new convexity property that implies a fixed point property for L^1*, Studia Math. **100** (1991), 95–108.
17. Z. Opial, *Weak convergence of the sequence of successive approximations for nonexpansive mappings*, Bull. Amer. Math. Soc. **73** (1967), 591–597.
18. M. Prager, *On a principle of convergence in Hilbert space*, Czech. Math. J. **10** (1960), 271–282.
19. S. Reich, *Averaged mappings in the Hilbert ball*, J. Math. Anal. Appl. **109** (1985), 199–206.
20. S. Reich, *The alternating algorithm of von Neumann in the Hilbert ball*, Dynamic Systems and Applications (to appear).

Department of Mathematics, California State University, Northridge, CA 91330 U.S.A.
E-mail address: jdye@vax.csun.edu

Department of Mathematics, Memphis University, Memphis, TN 38152 U.S.A.
E-mail address: linpk@hermes.msci.memst.edu

Department of Mathematics, Lublin Technical University, 20-618 Lublin, Poland
E-mail address: cieslak@plumcs11.bitnet

Department of Mathematics, University of Southern California, Los Angeles, CA 90089 U.S.A.
E-mail address: sreich@mtha.usc.edu

Contemporary Mathematics
Volume **144**, 1993

POLYNOMIAL SCHUR AND POLYNOMIAL DUNFORD-PETTIS PROPERTIES

JEFF FARMER AND WILLIAM B. JOHNSON

ABSTRACT. A Banach space is *polynomially Schur* if sequential convergence against analytic polynomials implies norm convergence. Carne, Cole and Gamelin show that a space has this property and the Dunford-Pettis property if and only if it is Schur. Herein is defined a reasonable generalization of the Dunford–Pettis property using polynomials of a fixed homogeneity. It is shown, for example, that a Banach space will has the P_N Dunford–Pettis property if and only if every weakly compact $N-$homogeneous polynomial (in the sense of Ryan) on the space is completely continuous. A certain geometric condition, involving estimates on spreading models and implied by nontrivial type, is shown to be sufficient to imply that a space is polynomially Schur.

1. INTRODUCTION

The relationship between holomorphic functions defined on an infinite dimensional Banach space and (geometric or topological) properties of the space has been of recent interest (see, for example, [AAD], [ACG], [CCG], [CGJ], [F], [R 1]). As in the one–dimensional case, holomorphic functions are defined in terms of Taylor series, which in the infinite–dimensional case have terms consisting of homogeneous analytic polynomials. Just as in the case of linear functionals (1–homogeneous polynomials), one can consider properties of the topologies induced by the polynomials on the space. In this paper we consider the properties which are analagous to the Schur property and the Dunford–Pettis property; i.e., those obtained by replacing weak sequential convergence with sequential convergence against an arbitrary N–homogeneous analytic polynomial. We relate these properties to one another and to the geometric property of type and the existence of certain spreading models.

X will be a complex infinite–dimensional Banach space. An *$N-$homogeneous analytic polynomial* on X is the restriction to the diagonal of an $N-$linear form on the $N-$fold Cartesian product of X with itself, or equivalently, a linear functional on the $N-$fold projective tensor product of X with itself. Indeed, given an

---

1991 *Mathematics Subject Classification.* Primary 46B05, 46B20 Secondary 46G20.

This paper forms a portion of the first author's doctoral dissertation written under the supervision of the second author.

The first author was supported in part by NSF Grant #DMS-9021369.

The second author was supported in part by NSF Grant #DMS-9003550.

This paper is in final form and no version of it will be submitted for publication elsewhere.

$N-$homogeneous analytic function P on X, one obtains an N-linear form on X by taking the Nth derivative and dividing by $N!$; the form is related to the polynomial by the polarization formula:

$$A_P(x_1, \ldots, x_n) = \text{Avg}\{\epsilon_i = \pm 1\}(\Pi\epsilon_i)P(\sum_{i=1}^{n} \epsilon_i x_i).$$

The form A_P is clearly symmetric (invariant under permutations of the coordinates). Likewise any bounded symmetric $N-$linear form will give rise to an $N-$homogeneous analytic polynomial. Such a form can be linearized by taking the projective tensor product of X with itself N times and extending the form to a linear functional on this tensor product. The subspace of symmetric linear functionals is the dual of the symmetric $N-$fold projective tensor product, which is a complemented subspace of the $N-$fold projective tensor product. The projection is given by extending the following map linearly:

$$(x_1 \otimes x_2 \otimes \cdots \otimes x_n) \to \frac{1}{n!} \sum_{\pi \in S_n} (x_{\pi_1} \otimes x_{\pi_2} \otimes \cdots \otimes x_{\pi_n}).$$

We denote the symmetric projective tensor product by $\widehat{\otimes}_s^N X$. The N-linear form A_P associated with P can now be considered a linear functional on $\widehat{\otimes}_s^N X$. The supremum norm of the polynomial is related to that of the linear functional as follows:

$$||P|| \leq ||A_P|| \leq \frac{N^N}{N!}||P||$$

If we call the space of polynomials $\mathcal{P}_N$ the above simply says that $\mathcal{P}_N$ is isomorphic to $(\widehat{\otimes}_s^N X)^*$. Since for our purposes the index N will be fixed, we will suppress reference to this isomorphism and use the same label for a polynomial and its associated symmetric linear functional. More details about the above relationships may be obtained from [M] or [R 1]. We will study the topologies generated by these polynomials, especially with respect to sequential convergence.

We define the $\mathcal{P}_N-$weak topology on X to be the topology generated by the all of the homogeneous analytic polynomials of degree less than or equal to N; that is, a net $\{x_\alpha\}$ converges to x in the $\mathcal{P}_N-$weak topology if, for every $M \leq N$, for every $M-$homogeneous analytic polynomial P, $P(x_\alpha) \to P(x)$. Note that for $N = 1$ this is the usual weak topology, and that for $M > N$ the $\mathcal{P}_M-$weak topology is finer than the $\mathcal{P}_N-$weak one. We call the weak polynomial topology the topology which is generated by the union of $\mathcal{P}_N$ for all $N \in Z_+$. In analogy to the Schur property, we say a space is $\mathcal{P}_N-$*Schur* if whenever $P(x_n) \to 0$ for all $P \in \mathcal{P}_N$ then x_n is norm null. If $P(x_n) \to 0$ for all $P \in \mathcal{P}_N$ for all N implies that x_n is norm null, then we say X is $\mathcal{P}-$*Schur*. It is evident (multiply linear functionals) that every $\mathcal{P}_N-$Schur space is $\mathcal{P}-$Schur and that every Schur space is $\mathcal{P}_N-$Schur for every N (and $\mathcal{P}-$Schur).

These topologies were introduced in [R 1], and the weak polynomial topology also appeared in [CCG] which considered relations between the Dunford-Pettis property, the Schur property and the $\mathcal{P}-$Schur property (in the terminology of [CCG], "X is $\mathcal{P}-$Schur "= "X is a $\Lambda-$space").

Let $\theta : X \to \widehat{\otimes}_s^N X$ by $\theta(x) = x \otimes \cdots \otimes x$ (N times) and define $\theta(X) = \Delta_N(X)$.

This set is a non-convex, norm-closed subset of $\widehat{\otimes}_s^N X$ with the property that $\lambda x \in \Delta_N(X)$ whenever $x \in \Delta_N(X)$.

Now θ is a continuous $N-$homogeneous function, which is the (nonlinear) preadjoint of the isomorphism between $\mathcal{P}_N$ and $(\widehat{\otimes}_s^N X)^*$. Notice that θ is an $N-$to$-$one map; we have

$$\theta^{-1}(x \otimes \cdots \otimes x) = \{e^{\frac{2\pi n i}{N}} x | n \in Z\}$$

(Use separating functionals to the Nth power to show equality.) We reserve the symbol θ_N for this function.

If $P(x_\alpha) \to P(x)$ for all $P \in \mathcal{P}_N(X)$ then the net need not converge against polynomials in $\mathcal{P}_M$ for all $M < N$, but since $P(x) = P(y)$ for all $P \in \mathcal{P}_N$ implies that $\frac{x}{y}$ is a complex Nth root of unity, if also $x_\alpha \to x$ weakly, then x_α $\mathcal{P}_N-$weakly converges to x. Thus in practice it is easy to pass from convergence against all $N-$homogeneous polynomials to $\mathcal{P}_N-$weak convergence.

Although for any one polynomial P, $P(x - x_\alpha) \to 0$ and $P(x_\alpha) \to P(x)$ are not in general equivalent, the following known fact is useful.

Lemma 1.1. *$x_\alpha \to x$ in the $\mathcal{P}_N-$weak topology, if and only if $x - x_\alpha \to 0$ in the $\mathcal{P}_N-$weak topology.*

Proof (sketch). Let P be an N-homogeneous polynomial and let $x_\alpha \to x$ in the $\mathcal{P}_N-$weak topology. Then, letting A_P be the N-linear form associated with P we have

$$P(x - x_\alpha) = \sum_{i=1}^{N} (-1)^i \binom{N}{i} A_P(x, x, \ldots, x, x_\alpha, \ldots, x_\alpha)$$

where in each term x appears i times and x_α appears $N-i$ times. Since convergence in the $\mathcal{P}_N-$weak topology implies convergence against any polynomial of lesser homogeneity, we consider each term as an $(N-i)-$homogeneous polynomial (x being fixed), to see that the sum indeed converges to zero. The converse is obtained, using the same expansion, by induction on N.

2. Polynomial Dunford–Pettis Spaces

One result of [CCG] is that a Banach space is Schur if and only if it is polynomially Schur and has the Dunford–Pettis property. We can obtain an analagous result for polynomials of fixed homogeneity by defining an appropriately analagous Dunford-Pettis property. Our first task is to adapt Lemma 7.3 of [CCG] for our purposes.

Propopsition 2.1. *The following are equivalent for any Banach space X, and any fixed positive integer N.*

(i) *Any polynomial on X is $\mathcal{P}_N-$weakly sequentially continuous.*

(ii) *If $\{x_n\}_{n=1}^\infty$ is a $\mathcal{P}_N-$weakly null sequence in X (i.e. if $\{x_k \otimes \cdots \otimes x_k\}$ (N times) is weakly null), then $\{x_k \otimes \cdots \otimes x_k\}$ (m times) is weakly null in $\widehat{\otimes}^m X$ for $m > N$.*

(iii) *For $m > N$ the function $\theta = \theta(N,m)$ which takes $\theta : \Delta_N(X) \to \Delta_m(X)$ by*

$$x \otimes \cdots \otimes x \ (N \ times) \mapsto x \otimes \cdots \otimes x \ (m \ times)$$

is weak to weak sequentially continuous.

Proof. The proof of these equivalences is an exercise, and can be adapted easily from the proof of Lemma 7.3 in [CCG].

For $n = 1$, the equivalent properties of 2.1 were shown to be implied by the Dunford–Pettis property. We will now define a polynomial Dunford–Pettis property which will imply the conditions of 2.1 for each positive integer.

We say that a space X has the $\mathcal{P}_N$ Dunford–Pettis property provided that it satisfies any of the equivalent conditions of Proposition 2.2.

Proposition 2.2. *Let X be a Banach space. For fixed N, the following are equivalent :*

(i) *Whenever $\{P_n\}_{n=1}^{\infty}$ is a weakly null sequence of $N-$homogeneous polynomials (or equivalently, symmetric bounded $N-$linear forms on X) and $\{x_n\}_{n=1}^{\infty}$ converges $\mathcal{P}_N-$weakly to x in X, then $P_n(x_n) \to 0$.*

(ii) *Every weakly compact operator on $\widehat{\otimes}_s^N X$ is completely continuous when restricted to $\Delta_N(X)$.*

(iii) *If K is a weakly compact set in any Banach space Y, and J is $\mathcal{P}_N-$weakly compact in X, then $\theta_N(J) \otimes K$ is a weakly compact set in $(\widehat{\otimes}_s^N X)\widehat{\otimes} Y$*

where by $\theta_N(J) \otimes K$ we mean simply the set $\theta_N(J) \times K$, that is, the set of all $\theta_N(j) \otimes k$ with $j \in J$ and $k \in K$.

In the case $N = 1$ these conditions reduce to known equivalent statements of the classical Dunford–Pettis property; this proposition justifies
the definition of the $\mathcal{P}_N$ Dunford–Pettis property. Before giving the proof, we make the following remark.

R. Ryan in [R 2] considers $N-$homogeneous polynomials from $X \to Y$; as in the scalar case, we can equivalently consider linear operators from $\widehat{\otimes}_s^N X$ to Y; such a polynomial is weakly compact if it maps bounded sets to weakly compact ones (i.e. if the associated linear operator is weakly compact). Ryan investigates some conditions which are equivalent to weak compactness of such polynomials. Using this definition it is easy to see that (ii) above is equivalent to

(ii)′ Every weakly compact N-homogeneous polynomial from X to any Banach space Y is completely continuous (on X).

Proof.

(i)⇒(iii) We want to show that $\theta_N(J) \times K$ is weakly compact in $\widehat{\otimes}_s^N X \otimes Y$. Take a sequence $\theta_N(x_n) \otimes k_n$ in $\theta_N(J) \otimes K$; by hypothesis assume we have passed to a subsequence such that $\theta_N(x_n)$ and k_n are weakly convergent to $\theta_N(x)$ and k in $\widehat{\otimes}_s^N X$ and Y, respectively. That is, $\theta_N(x_n) - \theta_N(x)$ and $k_n - k$ are weakly null. If

$$\phi \in (\widehat{\otimes}_s^N X \otimes Y)^* \equiv \mathcal{B}\big(Y, (\widehat{\otimes}_s^N X)^*\big)$$

then $\phi(k_n - k)$ is weakly null by continuity and (i) applies. We then have

$$\Big\langle \phi(k_n - k), \theta_N(x_n) - \theta_N(x) \Big\rangle =$$

$$\langle \phi(k_n), \theta_N(x_n) \rangle - \langle \phi(k), \theta_N(x_n) \rangle - \langle \phi(k_n), \theta_N(x) \rangle + \langle \phi(k), \theta_N(x) \rangle$$

Taking limits as $n \to \infty$ we see

$$0 = \lim_{n \to \infty} \langle \phi(k_n), \theta_N(x_n) \rangle - 2\langle \phi(k), \theta_N(x) \rangle + \langle \phi(k), \theta_N(x) \rangle$$

$$= \lim_{n \to \infty} \langle \phi(k_n), \theta_N(x_n) \rangle - \langle \phi(k), \theta_N(x) \rangle$$

This says exactly that $\theta_N(x_n) \otimes k_n$ is weakly convergent to $\theta_N(x) \otimes k$.

(iii)⇒(ii) Let $T : \widehat{\otimes}_s^N X \to Y$ be weakly compact and $\theta_N(x_n) \to \theta_N(x)$ weakly in the symmetric tensor product. Choose ϕ_n to be norming functionals for $T(\theta_N(x_n) - \theta_N(x))$ in the sphere of Y^*. Since T^* is also weakly compact, assume by passing to a subsequence that $T^*(\phi_n)$ converges weakly, say to ψ. By passing to a subsequence we can also assume that $\{T^*(\phi_n) - \psi\}_{n=1}^{\infty}$ is basic (or norm null, in which case the argument is simpler). Apply (iii) to the sets

$$K = \{T^*(\phi_n) - \psi\}_{n=1}^{\infty} \cup \{0\} \quad \textit{and} \quad J = \{x_n\}_{n=1}^{\infty} \cup \{x\}$$

to see that $\theta_N(J) \otimes K$ is weakly compact in $(\widehat{\otimes}_s^N X)\widehat{\otimes}(\widehat{\otimes}_s^N X)^*$. The sequence $\theta_N(x_n) \otimes \{T^*(\phi_n) - \psi\}$ thus has a convergent subsequence (we pass to that). First we claim that this subsequence must go weakly to zero; indeed, it goes to zero in a weaker Hausdorff topology, namely that generated by considering the weak topology on the second co-ordinate. Since it is clear that $\theta_N(x) \otimes \{T^*(\phi_n) - \psi\}$ is weakly null, we can conclude that

$$w - \lim\Big([\theta_N(x_n) - \theta_N(x)] \otimes \{T^*(\phi_n) - \psi\}\Big) = 0$$

Now consider the functional associated with the identity operator; call it Γ. We have

$$\begin{aligned}
0 &= \lim_{n \to \infty} \Gamma\big(\{T^*(\phi_n) - \psi\} \otimes [\theta_N(x_n) - \theta_N(x)]\big) \\
&= \lim_{n \to \infty} \langle \{T^*(\phi_n) - \psi\}, \theta_N(x_n) - \theta_N(x) \rangle \\
&= \lim_{n \to \infty} \Big(\langle T^*(\phi_n), \theta_N(x_n) - \theta_N(x) \rangle - \langle \psi, \theta_N(x_n) - \theta_N(x) \rangle \Big) \\
&= \lim_{n \to \infty} \langle \phi_n, T(\theta_N(x_n)) - T(\theta_N(x)) \rangle \\
&= \lim_{n \to \infty} \|T(\theta_N(x_n)) - T(\theta_N(x))\|
\end{aligned}$$

This gives (ii).

(ii)⇒(i) Let $\{P_n\}_{n=1}^{\infty}$ be the weakly null sequence and $\{x_n\}_{n=1}^{\infty}$ go $\mathcal{P}_N$−weakly to x and define a map T from $\Delta_N(X)$ to c_0 by

$$T(\theta_N(z)) = \big(P_n(z)\big)_n \quad \forall z \in X$$

The map T extends linearly (via the polarization formula) to all of $\widehat{\otimes}_s^N X$. Since $T^*(e_n) = P_n$ goes weakly to 0 we see that the map T is weakly compact. Applying (ii) we get $T(\theta_N(x_n))$ going in the norm on c_0 to $T(\theta_N(x))$. But since the norm on c_0 is the sup norm, this gives (i) and completes the proof.

We note that the condition (ii) is sharply stated with the following example.

Example. We will see momentarily that l_2 is $\mathcal{P}_2$–Schur and therefore is $\mathcal{P}_2$–Dunford Pettis. Consider the operator

$$Id\otimes Q_1 : l_2\widehat{\otimes}l_2 \to c_0$$

where Q_1 is the projection onto the first basis vector. Consider the symmetrized version, that is, restrict the operator to the symmetric tensor product, which is a complemented subspace. This operator is weakly compact and therefore completely continuous on $\theta_2(l_2)$ by Proposition 2.2 but is clearly not completely continuous on the entire symmetric tensor product; consider the image of $e_n \otimes e_1 + e_1 \otimes e_n$, for example, which is weakly null but whose image is the unit vector basis of c_0.

Proposition 2.3. *If a Banach space is $\mathcal{P}_N$ Dunford–Pettis then it satisfies the equivalent conditions in Proposition 2.1.*

Proof. We prove 2.2(i) implies 2.1(ii).

Let $\{x_n\}_{n=1}^{\infty}$ be a $\mathcal{P}_N$−weakly null sequence in X, i.e. $\theta_N(x_n)_{n=1}^{\infty}$ is weakly null in $\widehat{\otimes}^N X$ and $\theta_M(x_n)_{n=1}^{\infty}$ is also weakly null in $\widehat{\otimes}^M X$ whenever $1 \leq M < N$. Let $m = N+1$, $\phi \in (\widehat{\otimes}^{N+1}X)^*$ and consider ϕ as a linear operator from X to $(\widehat{\otimes}^N X)^*$. Since $\{x_n\}_{n=1}^{\infty}$ is weakly null in X, so is its image in $(\widehat{\otimes}^N X)^*$ under ϕ. Thus

$$\big\langle \phi, \theta_{N+1}(x_n)\big\rangle = \big\langle \phi(x_n), \theta_N(x_n)\big\rangle \to 0$$

by the first formulation of the $\mathcal{P}_N$ Dunford–Pettis property (notice that for this application it matters not whether ϕ is symmetric). This proves the proposition for $m = N+1$ and by induction (in an obvious way) for $m = qN+1$ for $q = 1, 2, \ldots$. But we can also write an analogous proof for $m = N+k$ for $2 \leq k < N$ and extend it by induction as well.

Proposition 2.4. *Let X be a Banach space. For fixed N, the following are equivalent :*

(i) *X is $\mathcal{P}_N$−Schur.*
(ii) *X has the $\mathcal{P}_N$ Dunford–Pettis property and is $\mathcal{P}$−Schur.*
(iii) *X satisfies (i)–(iii) of proposition 1.1 and is $\mathcal{P}$−Schur.*

Proof. (i)⇒(ii) requires only Lemma 1.1 and (ii)⇒(iii) is Proposition 2.3, so (iii)⇒(i) remains. Let $\theta_N(x_n)$ be weakly null. Then $\theta_M(x_n)$ is weakly null for all M by 2.1. But since X is $\mathcal{P}_N-$Schur, x_n must go to 0 in norm.

It is of interest to note that if we are not in the context of the $\mathcal{P}-$Schur property, the conditions of 2.1 are weaker than the $\mathcal{P}_N$ Dunford–Pettis property; T^*, the original Tsirelson space (or, in fact any space having the approximation property with $\mathcal{P}_N(X)$ reflexive for all N; see [F] for further discussion of such spaces) will satisfy 2.1 for all N but fail to be $\mathcal{P}_N$ Dunford–Pettis.

Examples. It is clear from the classical work of Pitt [P] that l_p spaces for $(1 \leq p < \infty)$ are $\mathcal{P}_N-$Schur for $N \geq p$, and it is proved in [CCG] that L_p spaces $(2 \leq p < \infty)$ are $\mathcal{P}-$Schur (in fact $\mathcal{P}_N-$Schur for $N \geq p$); we can thus conclude that they are $\mathcal{P}_N$ Dunford–Pettis. The space c_0 is Dunford–Pettis and therefore $\mathcal{P}_N$ Dunford–Pettis for every N. This implies (for example) that $l_3 \oplus c_0$ is $\mathcal{P}_3$–Dunford–Pettis but not $\mathcal{P}_N-$Schur for any N. In the next section we discuss further exactly which spaces may be $\mathcal{P}_N-$Schur.

3. Spaces with Type are Polynomially Schur

In this section we will give some sufficient criteria for spaces to be Polynomially Schur. We will show, for example, that any space having non-trivial type is $\mathcal{P}-$Schur and indeed is $\mathcal{P}_N-$Schur for some N.

(Jaramillo and Prieto [JP] have independently shown that every superreflexive space is polynomially Schur). In particular, L_p spaces

are Polynomially Schur for all $1 < p < \infty$.

It is convenient to use the concept of a spreading model, the construction of which is due to Brunel and Sucheston [BS 1].

Finite versions of Ramsey's Theorem allow that given any property of n-tuples of elements from a sequence, one can pass to a subsequence with the property that all n-tuples formed from the subsequence share the property or else all fail it. By using the size of the norm of a sum of n elements as the property one can, by repeatedly applyling the theorem, approximately stabilize the norm (to within any desired ϵ_n) of any finite combination as long as many of the beginning terms are thrown away. More preciesely we have the following fact (see [B] or [BS 1]):

Proposition 3.1. *Let (f_n) be a bounded sequence with no norm-Cauchy subsequence in a Banach space X. Then there exists a subsequence (e_n) of (x_n) and a norm L on the vector space S of finite sequences of scalars such that*

$$\forall \epsilon > 0 \quad \forall a \in S \quad \exists k \in N \ s.t. \ \forall k < k_1 < k_2 < \cdots < k_M$$

we have

$$\left| \left\| \sum a_i e_{k_i} \right\| - L(a) \right| < \epsilon.$$

The completion of $[e_i]$ (call it F) under the norm L is called a spreading model for the sequence (e_n). The reason for the terminology is that the sequence (e_n)

is invariant under spreading with respect to the norm F, that is to say, for every finite sequence of scalars (a_i) and every subsequence σ of the natural numbers

$$\left\|\sum_{i=1}^{M} a_i e_i\right\|_F = \left\|\sum_{i=1}^{M} a_i e_{\sigma(i)}\right\|_F .$$

Thus any norm estimate satisfied by sequences in the spreading model will be approximately satisfied for sequences of finite length to any desired degree provided we go out far enough in the sequence (x_n). If the original sequence was weakly null then the resulting sequence will be unconditional; that is to say, we have the following (Lemma 2 of [B], or see [BS 2]).

Proposition 3.2. *If (x_n) is weakly null, then the sequence (e_n) is unconditional in F with unconditional constant at most 2.*

Now we are ready to state the criterion.

Theorem 3.3. *Suppose a Banach space X has the property that for every normalized weakly null sequence $\{y_n\}$ in X there exists a subsequence and a sequence $\{f_n\}$ in X^* biorthogonal to it which has an (unconditional) spreading model with an upper p-estimate for some $p > 1$. Then X is $\mathcal{P}-$Schur. If the same p works for every such sequence and $N > p'$, then the space X is $\mathcal{P}_N-$Schur.*

We know of no space which is $\mathcal{P}-$Schur but which fails the above property. The space $(l_3 \oplus l_4 \oplus l_5 \oplus \cdots)_2$ is easily seen to be $\mathcal{P}-$Schur although it fails cotype (and hence type and superreflexivity), yet is reflexive; it does satisfy the hypothesis of 3.3. A Schur space satisfies the hypothesis vacuously.

Proof. We will pass to subsequences and relabel without mercy. Start with any bounded sequence in X which is not norm null; we need to find a polynomial which is bounded away from zero on a subsequence. By Rosenthal's theorem [D, Chapter XI] there is either a weakly Cauchy subsequence or a subsequence equivalent to the unit vector basis of l_1. Since the unit vector basis of l_1 is not weakly null, we are done in this case. For the same reason (linear functionals are polynomials) we are finished if our weakly Cauchy sequence is not weakly null. So we have reduced to the case of a bounded weakly null sequence which is not norm null and can apply the
hypothesis to that sequence
(it is purely formal that the "normalized"
condition can be replaced by "bounded").
Let $\{y_n, f_n\}$ be the biorthogonal system obtained, and assume with no loss of generality that the f_n are normalized. By the
definition of a spreading model we know that for any $c > 0$ we can find a constant C so that for every M we have

$$\left\|\sum_{i=c\log M}^{M} f_i\right\| \le C\Big(\sum_{i=c\log M}^{M} \|f_i\|^p\Big)^{\frac{1}{p}} .$$

This is because once we have a spreading model we may always improve the stability estimates by passing to a subsequence. Choosing c small enough so that $m = c\log M \le M^{\frac{1}{p}}$ for all $M \in \mathcal{N}$, and letting (a_i) be scalars of modulus less than or equal to 1, we obtain

$$\left\|\sum_{i=1}^{M} a_i f_i\right\| \le \sum_{i=1}^{m-1} ||f_i|| + \left\|\sum_{i=m}^{M} a_i f_i\right\| \le m + C\Big(\sum_{i=m}^{M} |a_i|^p\Big)^{\frac{1}{p}} \le C_1 M^{\frac{1}{p}}.$$

Now by general Banach lattice techniques (see Lemma 3.4 below), we know that for any $r < p$ (we choose r so that $p' < r' < N$) we can get a different C so that for any sequence (a_i) we will have

$$\left\|\sum_{i=1}^{M} a_i f_i\right\| \le C\Big(\sum_{i=1}^{M} |a_i|^r\Big)^{\frac{1}{r}}.$$

Now, given $y \in B$ choose $(a_i) \in l_r$ of l_r norm one to norm the sequence $(f_i(y))_{i=1}^{M}$ in $l_{r'}$. We then have that

$$\left|\Big(\sum_{i=1}^{M} f_i(y)^N\Big)\right|^{\frac{1}{N}} \le \Big(\sum_{i=1}^{M} |a_i|^r\Big)^{\frac{1}{r}} \Big(\sum_{i=1}^{M} |f_i(y)|^{r'}\Big)^{\frac{1}{r'}}. = \sum_{i=1}^{M} a_i f_i(y)$$

$$\le \left\|\sum_{i=1}^{M} a_i f_i\right\| \le C\Big(\sum_{i=1}^{M} |a_i|^r\Big)^{\frac{1}{r}} = C$$

which says that

$$P(y) = \sum_{i=1}^{\infty} f_i(y)^N \le C^N \quad \text{hence} \quad ||P|| \le C^N.$$

But we know that $P(y_n) = f_n(y_n)^N$ is bounded away from zero because the sequence (y_n, f_n) was biorthogonal. Therefore no sequence bounded away from 0 in X can be polynomially null, and so X is $\mathcal{P}-$Schur. If there is a uniform value for p then X is $\mathcal{P}_N-$Schur for $N > p'$.

It remains for us to prove the following standard fact.

Lemma 3.4. *Suppose that $\{x_i\}$ is a normalized sequence in a Banach space satisfying*

$$\left\|\sum_{i\in B} a_i x_i\right\| \le C|B|^{\frac{1}{p}} \max_{i\in B} |a_i|$$

for all scalars (a_i), and all finite subsets B of the natural numbers. Then for any $1 < r < p$ there exists a constant D so that for all (a_i) and all M

$$\left\|\sum_{i=1}^{M} a_i x_i\right\| \le D\Big(\sum_{i=1}^{M} |a_i|^r\Big)^{\frac{1}{r}}.$$

Proof. We use the convention that $\frac{1}{p}+\frac{1}{q}=1$ and $\frac{1}{r}+\frac{1}{s}=1$. Given the scalars, assume by homogeneity that $\sum_{i=1}^{M}|a_i|^r=1$. Define

$$B_n=\{i \mid 2^{-n-1}<|a_i|\leq 2^{-n}\}$$

and write

$$\Big\|\sum_{i=1}^{M} a_i x_i\Big\| \leq \sum_n \Big\|\sum_{i\in B_n} a_i x_i\Big\| \leq C\sum_n 2^{-n}|B_n|^{\frac{1}{p}}r.$$

Now just compute:

$$\sum_n 2^{-n}|B_n|^{\frac{1}{p}} = \sum_n 2^{-\frac{nr}{p}}2^{n(\frac{r}{p}-1)}|B_n|^{\frac{1}{p}} \leq \Big(\sum_n 2^{-nr}|B_n|\Big)^{\frac{1}{p}}\Big(\sum_n 2^{nq(\frac{r}{p}-1)}\Big)^{\frac{1}{q}}$$

$$= 2^{\frac{r}{p}}\Big(\sum_n 2^{-(n+1)r}|B_n|\Big)^{\frac{1}{p}} D(p,r) \leq 2^{\frac{r}{p}}\Big(\sum_{i=1}^{M}|a_i|^r\Big)^{\frac{1}{p}} D(p,r)$$

$$= 2^{\frac{r}{p}} D(p,r)\Big(\sum_{i=1}^{M}|a_i|^r\Big)^{\frac{1}{r}}.$$

The condition in Theorem 3.3 is rather hard to check. However, a much simpler criterion is sufficient.

Theorem 3.5. *Suppose the dual space of X, X^*, has type $p>1$. Then for every normalized weakly null sequence $\{y_n\}$ in X there exists a subsequence and a sequence $\{x_n\}$ in X^* biorthogonal to it which has an upper p-estimate. In particular, X satisfies the hypothesis of 3.3, and thus is $\mathcal{P}_N-$Schur for all $N>p$.*

In view of the fact that every space with non-trivial type also has a dual with some non-trivial type, we can state the following corollary.

Corollary 3.6. *Suppose X has type. Then for some N, X is $\mathcal{P}_N-$Schur.*

Proof (of 3.5). Recall that the fact that X^* has type p means that there is a constant T_p so that

$$\int_0^1 \Big\|\sum_{i=1}^{M} r_i(t)x_i\Big\| dt \leq T_p\Big(\sum_{i=1}^{M}\|x_i\|^p\Big)^{\frac{1}{p}} \qquad \forall M\in\mathcal{N}$$

for any finite number of elements $x_1,\dots x_n$ (where r_i are the Rademacher functions). Now suppose that (y_n) is a normalized weakly null sequence, which we can assume is basic by passing to a subsequence. We can find a bounded sequence (x_n) in X^* of functionals biorthogonal to (y_n). Since l_1 i s not embeddable in X^* we know by Rosenthal's theorem that we can find a weakly Cauchy subsequence of (x_n). Pass to the odd terms of (y_n), relabel and replace (x_n) with $(x_{2n+1}-x_{2n})$. Then we have a biorthogonal system (x_n,y_n) with $(x_n)\to 0$ weakly. By proposition 3.1 and the remark following it we know that (x_n) has an unconditional spreading model F.

Now F is finitely representable in X^* (this means that given any finite dimensional subspace of F we can find a $1+\epsilon$–isomorphic copy of that subspace in X^*, see again [B] or [BS 1]). Since the definition of type is local, F will also have type p with constant $\leq T_p$. Since the basis e_n of F is unconditional with constant at most 2, it has an upper p–estimate with constant less than or equal to $2T_p$. Thus we have satisfied the hypothesis of Theorem 3.3.

REFERENCES

[AAD] R. Alencar, R. Aron and S. Dineen, *A Reflexive space of holomorphic functions in infinitely many variables*, Proc. Amer. Math. Soc. **90** (1984), 407-411.

[ACG] R. Aron, B. Cole and T. Gamelin, *Spectra of algebras of analytic functions on a Banach space* (to appear).

[B] B. Beauzamy, *Banach–Saks properties and spreading models*, Math. Scand. **44** (1979), 357–384.

[BS 1] A. Brunel and L. Sucheston, *On B−convex Banach spaces*, Math. Sytems Thy. **7** (1973), 294–299.

[BS 2] A. Brunel and L. Sucheston, *On J−convexity and some ergodic super-properties of Banach spaces*, Trans. Amer. Math. Soc. **204** (1975), 79–90.

[CCG] T. Carne, B. Cole and T. Gamelin., *A uniform algebra of analytic functions on a Banach space*, Proc. Amer. Math. Soc. **314** (1989), 639-659.

[CGJ] B. Cole, T. Gamelin, and W. B. Johnson, *Analytic disks in fibers over the unit ball of a Banach space* (to appear).

[D] J. Diestel, *Sequences and series in Banach spaces*, Springer–Verlag, New York, 1984.

[F] J. Farmer, *Polynomial reflexivity in Banach spaces* (to appear).

[JP] J. Jaramillo and Prieto, *Weak-polynomial convergence on a Banach space* (to appear).

[M] J. Mujica, *Complex analysis in Banach spaces*, Notas de Mathematica, vol. 120, North-Holland, Amsterdam, 1986.

[P] H. R. Pitt, *A note on bilinear forms*, J. London Math. Soc. **11** (1936), 174–180.

[R 1] R. Ryan, *Applications of topological tensor products to infinite dimensional holomorphy*, Thesis, Trinity College, Dublin, 1980.

[R 2] R. Ryan, *Weakly compact holomorphic mappings on Banach spaces*, Pac. J. Math. **131** (1988), 179-190.

Contemporary Mathematics
Volume **144**, 1993

ON EMBEDDINGS OF FUNCTION SPACES INTO $L^p + L^q$

A. GARCÍA DEL AMO AND FRANCISCO L. HERNÁNDEZ

ABSTRACT. The isomorphic embedding of Orlicz function spaces into the spaces $L^p + L^q$ is studied, complementing some recent results given in [H–R].

It has been shown recently in [H–R] that the spaces $L^p + L^q$, defined as usual in Interpolation Theory, are universal spaces for the class of all Orlicz function spaces $L^F(0, \infty)$ with Boyd indices between p and q:

Theorem 1. *Let $0 < p < q < \infty$. Every Orlicz space $L^F(0, \infty)$ with Boyd indices $p < p_F \leq q_F < q$ is lattice-isomorphic to a sublattice of $L^p + L^q$.*

Recall that the Boyd indices p_F and q_F associated to an Orlicz space $L^F(0, \infty)$ can be computed as

$$p_F = \lim_{s \to \infty} \frac{\log \left(\inf_{t>0} \left\{ \frac{F(st)}{F(t)} \right\} \right)}{\log s}$$

and

$$q_F = \lim_{s \to \infty} \frac{\log \left(\sup_{t>0} \left\{ \frac{F(st)}{F(t)} \right\} \right)}{\log s} .$$

In the extreme case of the lower Boyd index p_F of $L^F(0, \infty)$ be equal to p we have also some embeddings into $L^P + L^q$:

Theorem 2. *If $0 < p < q < \infty$ and F is an Orlicz function with $F(x) = x^p$ at ∞, $F(x)x^{-p}$ non-decreasing on $\mathbf{R}^+$ and $F(x)x^{-q}$ non-increasing on $\mathbf{R}^+$, then $L^F(0, \infty)$ is lattice-isomorphic to a sublattice of $L^p + L^q$.*

Applying the above results to the special class of Orlicz spaces $L^{x^r \wedge x^s}(0, \infty) = L^r + L^s$, we get the following:

1991 *Mathematics Subject Classification.* Primary 46E30.
Research supported in part by DGICYT grant PB 88/0141.
This paper is in final form and no version of it will be submitted for publication elsewhere.

0271-4132/93 $1.00 + $.25 per page

Proposition 3. *Let $0 < p < q < \infty$. The spaces $L^r + L^s$ are isomorphic to a sublattice of the space $L^p + L^q$ for $p \leq r \leq s < q$.*

In particular it follows that the space $L^p + L^q$ has a sublattice isomorphic to L^r for any $p \leq r < q$.

The embeddings of the spaces $L^2 + L^s$ into $L^2 + L^q$ for $2 < s < q$ was proved earlier by S. Dilworth in ([D] Thm. 5.7), where several Banach structure properties of this scale of spaces $L^2 + L^q$ were studied (see also [C–D]).

The purpose of this note is to present some results which complement the above in the sense of precising the appropiate ranges of the indices for the embeddings into the spaces $L^p + L^q$. For example, it is shown that L^q cannot be isomorphically embedded into $L^p + L^q$ for $2 < p < q$. Also it holds that the spaces $L^r + L^q$ are isomorphic to a sublattice of $L^p + L^q$ for $0 < p \leq r < q < \infty$. In the proofs we use some crucial results given in the memoir [J–M–S–T].

First, let us give some notations and definitions. By F we denote an Orlicz function, i.e. a non–decreasing positive function defined for $x \geq 0$ so that $F(0) = 0$ and $F(1) = 1$. If I stands for the interval $(0,1)$ or $(0,\infty)$, with the Lebesgue measure μ, the Orlicz function space $L^F(I)$ is defined as the space of all equivalence classes of μ–measurable scalar functions on I such that

$$m(\frac{f}{s}) = \int_I F\left(\frac{|f|}{s}\right) d\mu < \infty$$

for some $s > 0$, endowed with the p–norm

$$\|f\|_F = \inf\left\{s > 0 : m\left(\frac{f}{s^{1/p}}\right) \leq 1\right\}$$

when F is a p–convex function $(0 < p \leq 1)$. We will assume that the function F satisfies the Δ_2–condition, hence the space $L^F(I)$ is separable.

Given $L^F(0,\infty)$ and $L^G(0,\infty)$, the sum space $L^F(0,\infty) + L^G(0,\infty)$ is the space of all functions f on $(0,\infty)$ which can be written as $g + h$ such that $g \in L^F(0,\infty)$ and $h \in L^G(0,\infty)$, endowed with the quasi–norm

$$\|f\|_{F+G} = \inf\left\{\|g\|_F + \|h\|_G : f = g + h\right\} .$$

If F and G are Orlicz functions satisfying that $F(x) \leq G(x)$ for $0 \leq x \leq 1$ and $G(x) \leq F(x)$ for $x \geq 1$, then $L^F(0,\infty) + L^G(0,\infty)$ coincides, up to an equivalent renorming, with the Orlicz space $L^{F\wedge G}(0,\infty)$, where $F \wedge G$ denotes the Orlicz function $F \wedge G(x) = \min\{F(x), G(x)\}$. If $F(x) = x^p$ and $G(x) = x^q$ for $0 < p < q < \infty$, we can write in short L^p and $L^p + L^q$ (since $L^p(0,1)$ and $L^p(0,\infty)$ are lattice–isomorphic).

We refer to [L–T] and [J–M–S–T] for other properties of $L^p + L^q$ spaces and Orlicz spaces (see also [K], [H–Rs], [R]).

The following remarkable characterization on the embedding of r.i. function spaces on $(0,1)$ into Orlicz spaces was given in ([J–M–S–T] Thm. 7.7).

Theorem 4. *Let $L^F(0,\infty)$ (resp. $L^F(0,1)$) be p–convex for some $p > 2$. A r.i. function space $X(0,1) \neq L^2(0,1)$ embeds isomorphically into $L^F(0,\infty)$ (resp $L^F(0,1)$) if and only if $X(0,1) = L^G(0,1)$, up to an equivalent renorming, for an Orlicz function G of the form*

$$G(t) = \int_0^\infty \frac{F(st)}{F(s)}\, d\nu(s)\ ,$$

where ν is a probability Borelian measure supported on $(0,\infty)$ (resp. on $[1,\infty)$).

Given a p–convex Orlicz funtion F for $p > 2$, it follows easily from the above result that if $L^G(0,1)$ embeds isomorphically into $L^F(0,1)$ and $L^G(0,1) \neq L^2(0,1)$ then $L^{\sigma_F}(0,1) \subset L^G(0,1) \subset L^F(0,1)$, where σ_F is the function

$$\sigma_F(t) = \sup_{1\le s<\infty} \frac{F(st)}{F(s)} \qquad \text{for } t \ge 1\ .$$

In particular when F is submultiplicative and p–convex at ∞, for $p > 2$, we deduce from $L^G(0,1)$ being isomorphically embedded into $L^F(0,1)$ and $G \neq x^2$ that $L^G(0,1) = L^F(0,1)$. Moreover it happens that this property characterizes the submultiplicative functions at ∞ (see [J–M–S–T] Coroll. 7.9).

Proposition 5. *Let $L^F(0,1)$ be p–convex for some $p > 2$, G be a submultiplicative Orlicz function at ∞ with $\frac{F(x)}{G(x)} \searrow 0$ at ∞. If $L^G(0,1)$ is isomorphic to a subspace of $L^F(0,1)$ and $L^G(0,1) \neq L^2(0,1)$, then $L^G(0,1) = L^F(0,1)$.*

Proof. It follows from Theorem 4 that, up to equivalence at ∞, the function G is of the form

$$G(t) = \int_1^\infty \frac{F(st)}{F(s)}\, d\nu(s)\ ,$$

for a probability measure ν in $(1,\infty)$. For $K \ge 1$ denoting the submultiplicity constant of G, we take $t_0 \ge 1$ such that $\nu((t_0,\infty)) < \frac{1}{2K}$. Then

$$\begin{aligned} G(t) &= \int_1^{t_0} \frac{F(st)}{F(s)}\, d\nu(s) + \int_{t_0}^\infty \frac{F(st)}{F(s)}\, d\nu(s) \\ &\le F(t_0 t) + \int_{t_0}^\infty \frac{G(st)}{G(s)}\, d\nu(s) \\ &\le F(t_0 t) + \frac{G(t)}{2} \end{aligned}$$

which implies that G is mayorated by F at ∞, and hence $L^G(0,1) = L^F(0,1)$. Q.E.D.

Example. The spaces $L^{\frac{x^p}{\log(1+x)}}(0,1)$ and $L^{x^p\log(1+x)}(0,1)$ do not contain any subspace isomorphic to L^p for $p>2$.

As far as we know it is an open question to find a concrete Orlicz function F such that $L^F(0,1)$ contains a sublattice isomorphic to L^p for $p>2$.

We pass now to Orlicz spaces on infinite measures. It is known ([J–M–S–T] p. 169 and 198) that if an Orlicz space $L^F(0,\infty)$ contains a subspace isomorphic to another space $L^G(0,1)$, for $L^G(0,1)\neq L^2(0,1)$, then the Orlicz space $L^G(0,1)$ must be included in the space $L^F(0,1)$ (hence $\limsup_{x\to\infty} F(x)/G(x)<\infty$). We will give below some conditions under which the isomorphic embedding of $L^G(0,1)$ into $L^F(0,\infty)$ implies in fact that $L^G(0,1)$ has to be isomorphically embedded into $L^F(0,1)$.

For an Orlicz funtion F we will consider the associated Orlicz functions F_1 and F_2 defined by

$$F_1(t)=\begin{cases} F(t) & \text{if } 0\le t\le 1\\ 1/F(1/t) & \text{if } 1<t\end{cases}$$

and

$$F_2(t)=\begin{cases} 0 & \text{if } t=0\\ 1/F(1/t) & \text{if } 0<t<1\\ F(t) & \text{if } 1\le t\ .\end{cases}$$

And we will denote by $\overline{F}$ the function on $[1,\infty)$ defined by $\overline{F}(t)=F(t)F(1/t)$.

Theorem 6. *Let $L^F(0,\infty)$ be p–convex for some $p>2$. Suppose that F is super-multiplicative at 0 and at ∞, $\overline{F}$ is non–increasing on $[1,\infty)$ and*

$$\inf_{0<a<1}\limsup_{t\to\infty}\frac{\overline{F}(t)}{\overline{F}(t^a)}=0\ .$$

Then $L^{F_1}(0,1)$ embeds isomorphically into $L^F(0,\infty)$ if and only if $L^{F_1}(0,1)$ embeds isomorphically into $L^F(0,1)$.

Proof. The "only if" assertion is clear. Let us see the "if" assertion. Suppose that $L^{F_1}(0,1)$ is embedded into $L^F(0,\infty)$, then, by Theorem 4, there exists a probability measure ν on $(0,\infty)$ such that the function

$$G(t)=\int_0^\infty \frac{F(st)}{F(s)}\,d\nu(s)$$

is equivalent to F_1 at ∞.

Let us denote by K a positive constant satisfying that $G(t)\ge KF_1(t)$ for $t\ge 1$, $F_1(st)\ge KF_1(s)F_1(t)$ for $0<s,t<1$ and $F_2(st)\ge KF_2(s)F_2(t)$ for $s,t\ge 1$. And we take $0<a<1$ and $t_0\ge 1$ such that $\overline{F}(t)/\overline{F}(t^a)\le K^2/4$ and $\nu((0,t^{-a}])\le K^2/4$

for $t \geq t_0$. From this and the hypothesis we deduce easily the following inequalities:

(1) $F_1(st) \leq K^{-1}F_1(s)F_1(t)$ for $t \geq 1$ and $0 \leq s \leq t^{-1}$

(2) $F_2(st) \leq K^{-1}F_2(s)F_2(t)$ for $t \geq 1$ and $t^{-1} \leq s \leq 1$

(3) F_2/F_1 is non–increasing on $(0,1]$

(4) $\dfrac{F_2(t)F_2(t^{-a})}{F_1(t)F_1(t^{-a})} \leq \dfrac{K^2}{4}$ for $t \geq t_0$.

Now, for $t \geq 1$ we have

$$\begin{aligned} K &\leq \frac{G(t)}{F_1(t)} \\ &= \frac{1}{F_1(t)}\int_0^{1/t} \frac{F_1(st)}{F_1(s)}\,d\nu(s) + \frac{1}{F_1(t)}\int_{1/t}^{1/t^a} \frac{F_2(st)}{F_1(s)}\,d\nu(s) \\ &\quad + \frac{1}{F_1(t)}\int_{1/t^a}^{1} \frac{F_2(st)}{F_1(s)}\,d\nu(s) + \frac{1}{F_1(t)}\int_1^{\infty} \frac{F_2(st)}{F_2(s)}\,d\nu(s) \\ &= I_t^1 + I_t^2 + I_t^3 + I_t^4 \ . \end{aligned}$$

Using (1) we have $I_t^1 \leq K^{-1}\nu((0,1/t])$. And, by (2), (3) and (4) we have

$$I_t^2 \leq \frac{F_2(t)F_2(t^{-1})}{KF_1(t)F_1(t^{-1})}\nu((1/t,1/t^a]) \leq \frac{1}{K}\nu((1/t,1/t^a])$$

and

$$I_t^3 \leq \frac{F_2(t)F_2(t^{-a})}{KF_1(t)F_1(t^{-a})}\nu((1/t^a,1]) \leq \frac{K}{4} \ .$$

Then, we deduce that $I_t^4 \geq K/2$ for t big enough near ∞. This implies that F_1 is equivalent at ∞ to the function

$$\overline{G}(t) = \int_0^{\infty} \frac{F(st)}{F(s)}\,d\overline{\nu}(s) \ ,$$

where $\overline{\nu}(A) = \nu(A \cap [1,\infty))$. Thus, normalizing $\overline{\nu}$ and using Theorem 4, we conclude that $L^{F_1}(0,1)$ is embedded isomorphically into $L^F(0,1)$. Q.E.D.

Reasoning in a similar way as above it can be proved the following:

Proposition 7. *Let $L^F(0,\infty)$ be p–convex for some $p > 2$, $F(x) = x^q$ at 0 for $q \geq p$ and $F(x)x^{-q} \searrow$ in $[1,\infty)$. Then L^q is isomorphic to a subspace of $L^F(0,\infty)$ if and only if L^q is isomorphic to a subspace of $L^F(0,1)$.*

A direct consequence of Theorem 6 (or Proposition 7) is the following:

Corollary 8. *Let $2 < p < q < \infty$. Then L^q is not isomorphic to a subspace of $L^p + L^q$.*

When $0 < p \leq q \leq 2$, the space L^q is isomorphic to a subspace of $L^p + L^q$ (since L^q embeds isomorphically into L^p).

Corollary 9. *Let $L^F(0,\infty)$ be p–convex for some $p > 2$, $F(x) = x^q$ at 0 and different from x^q at ∞ with $F(x)x^{-q} \searrow$ in $[1,\infty)$. Then L^q is not isomorphic to a subspace of $L^F(0,\infty)$.*

Proof. It follows from Propositions 7 and 5. Q.E.D.

Example. (a) If $F(x)$ is x^p at 0 and $x^p/\log(1+x)$ at ∞, for $p > 2$, then L^p is not isomorphic to a subspace of $L^F(0,\infty)$.

Example. (b) If $F(x)$ is $x^p/|\log(e-1+x)|$ at 0 and x^p at ∞, for $p > 2$, then $L^{x^p \log(1+x)}(0,1)$ is not isomorphic to a subspace of $L^F(0,\infty)$.

Applying Proposition 6 in [H–R] we get that, for $1 \leq p < r < q < 2$, the spaces $L^r + L^q$ are isomorphic to a subspace of $L^p + L^q$. This holds in general:

Proposition 10. *Let $0 < p \leq r < q < \infty$. Then $L^r + L^q$ is lattice isomorphic to a sublattice of $L^p + L^q$.*

Proof. Let F and G be the Orlicz functions $x^r \wedge x^q$ and $x^p \wedge x^q$ for $p < r$. In virtue of ([J–M–S–T] p. 188) or Proposition 4 in [H–R] it is enough to show that F is equivalent to the function $H(t) = \int_0^\infty G(st)/G(s)\, d\nu(s)$, where ν is the probability measure on $(0,\infty)$ defined by

$$\nu(E) = \int_{E\cap[0,1]} (q-r)s^{q-r-1}\, ds\ .$$

Indeed, for $0 \leq t \leq 1$, it is clear that $H(t) = t^q$. For $t \geq 1$, we have

$$H(t) = \int_0^{1/t} \frac{G(st)}{G(s)}\, d\nu(s) + \int_{1/t}^1 \frac{G(st)}{G(s)}\, d\nu(s) = I_t^1 + I_t^2\ ,$$

where $I_t^1 = t^q\nu((0,1/t]) = t^r$ and

$$\begin{aligned} I_t^2 &= t^p \int_{1/t}^1 s^{p-q}\, d\nu(s) \\ &= (q-r)t^p \int_{1/t}^1 s^{p-r-1}\, ds = \frac{q-r}{r-p}(t^r - t^p)\ . \end{aligned}$$

Hence $F(t) \leq H(t) \leq (1 + \frac{q-r}{r-p})F(t)$ for $t \geq 0$, which concludes the proof. Q.E.D.

[C–D] N. L. Carothers and S. J. Dilworth, *Some Banach space embeddings of classical function spaces*, Bull. Austral. Math. Soc. **43** (1991), 73–77.

[D] S. J. Dilworth, *A scale of linear spaces related to the L^p-scale*, Illinois J. Math. **34** (1990), 140–158.

[H–Rs] F. L. Hernández and B. Rodríguez–Salinas, *Remarks on the Orlicz function spaces* $L^F(0, \infty)$, Math. Nachr. **156** (1992), 225–232.

[H–R] F. L. Hernández and C. Ruiz, *Universal classes of Orlicz function spaces*, Pacific J. Math. (to appear).

[J–M–S–T] W. B. Johnson, B. Maurey, G. Schechtman and L. Tzafriri, *Symmetric structures in Banach spaces*, Mem. Amer. Math. Soc. **217** (1979).

[K] N. J. Kalton, *Lattice structures on Banach spaces* (to appear).

[L–T] J. Lindenstrauss and L. Tzafriri, *Classical Banach spaces II*, Function spaces, Springer Verlag, Berlin, 1979.

[R] C. Ruiz, *Estructura de espacios de Orlicz de funciones y de sucesiones con pesos. Subespacios distinguidos*, Ph. D. Thesis, Univ. Complutense Madrid, 1990.

FACULTAD MATEMATICAS, UNIVERSIDAD COMPLUTENSE, 28040 MADRID, SPAIN

Contemporary Mathematics
Volume **144**, 1993

ON AUERBACH BASES

B. V. GODUN, BOR-LUH LIN AND S. L. TROYANSKI*

ABSTRACT. Every nonseparable Banach space X with separable norming subspace in X^* admits an equivalent norm $|\cdot|$ such that $(X, |\cdot|)$ does not have Auerbach basis.

A family of vectors $\{x_i\}_{i\in I}$ in a Banach space $(X, \|\cdot\|)$ is called an *Auerbach system* if $dist_{\|\cdot\|}(x_i, [x_j]_{j\neq i}) = \inf_{x\in[x_j]_{j\neq i}} \|x_i - x\| = \|x_i\| = 1$ for all $i \in I$. Equivalently, if there exists $\{x_i^*\}_{i\in I}$ in X^* such that $\{x_i, x_i^*\}_{i\in I}$ is biorthogonal and $\|x_i^*\| = \|x_i\| = 1$, $i \in I$. An *Auerbach basis* of X is an Auerbach system $\{x_i\}_{i\in I}$ which spans X.

It is well-known that every finite dimensional Banach space has Auerbach bases (cf., e.g., [S], p. 118), and every separable infinite dimensional Banach space has an infinite Auerbach system with additional properties (cf., e.g., [Si], p. 168). Furthermore, for every bounded biorthogonal system $\{x_i, x_i^*\}_{i\in I}$ in X, there exists an equivalent norm on X so that $\{x_i\}_{i\in I}$ becomes an Auerbach system under the new norm. Still open is a question of A. Pelczynski ([PB], p. 185): Does every separable Banach space contain an Auerbach basis? In [G1], a nonseparable Banach space with unconditional basis but without Auerbach basis is given. Recently, in [G2], an equivalent norm in $\ell_1([0,1])$ is constructed such that $\ell_1([0,1])$ fails to have Auerbach basis in this norm. Our aim is to extend this result to a class of nonseparable Banach spaces that includes $\ell_1([0,1])$. The main tool is a construction of V. I. Gurarii (cf. [S], p. 96) on interpolation bases with nodes in $C(K)$ where K is a compact metric space.

Recall that a subspace F in X^* is said to be a norming subspace of X if $\|x\| = \sup_{\substack{f\in F\\ \|f\|=1}} f(x)$, $x \in X$. We prove the following result.

Theorem. If X is a nonseparable Banach space with a separable norming subspace in X^*, then X admits an equivalent norm $|\cdot|$ such that $(X, |\cdot|)$ does not have Auerbach basis.

The Theorem applies to the class of Banach spaces including $\ell_1([0,1])$, $C(K)$ where K is a nonmetrizable, separable compact space and nonseparable dual Banach spaces with separable predual.

1991 *Mathematics Subject Classification.* 46B15, 46B20, 46B26.

Key words and phrases. Auerbach bases, nonseparable Banach spaces.

*S. L. Troyanski was partially supported by Bulgarian Ministry of Education and Science, contract # MM-3/91.

This paper is in final form and no version of it will be submitted for publication elsewhere.

Let x be an element in X and let Y be a subspace of X. Since $dist_{\|\cdot\|}(x,Y) = \|\hat{x}\|_{X/Y}$, we get the following result.

Lemma 1. Let Y be a subspace of X. Then $d(x,Y) = \sup\{|y^*(x)| : y^* \in Y^\perp, \|y^*\| = 1\}$ for all x in X.

Lemma 2. If K is a compact metrizable space, then for every dense sequence $\{t_n\}_{n\in\mathbb{N}}$ in K, there is an equivalent norm $|\cdot|$ on $C(K)$ with the property: If $\{f_n\}_{n\in\mathbb{N}}$ is a sequence in $C(K)$, $|f_n| = 1$, $n \in \mathbb{N}$ and $\lim_n f_n(t_m) = 0$, $m \in \mathbb{N}$ then $\lim_n \; dist_{|\cdot|}(f_n, G) = 1$ for all finite dimensional subspace G in $C(K)$.

Proof. Put $F_m = \{f \in C(K) : f(t_k) = 0,\ k = 1,2,\dots,m\}$, $m \in \mathbb{N}$. According to Gurarii's theorem ([S], p. 96), there exists in $C(K)$, an interpolating basis $\{e_n, e_n^*\}_{n\in\mathbb{N}}$ with nodes $\{t_n\}_{n\in\mathbb{N}}$, i.e., $R_n(C(K)) = F_n$, $n \in \mathbb{N}$ where $R_n(f) = \sum_{k>n} e_k^*(f)e_k$, $n \in \mathbb{N}$. For $f \in C(K)$, define

$$|f| = \sup_n \|R_n f\| \; .$$

It is clear that $|\cdot|$ is an equivalent norm on $C(K)$ and $|R_n| = 1$, $n \in \mathbb{N}$.

Suppose $\{f_n\}_{n\in\mathbb{N}}$ is a sequence in $C(K)$, $|f_n| = 1$, $n \in \mathbb{N}$ and $\lim_n f_n(t_m) = 0$, $m \in \mathbb{N}$. We first show that $\lim_n dist_{|\cdot|}(f_n, F_m) = 0$, $m \in \mathbb{N}$.

Indeed, embedds K canonically in $C(K)^*$ and let $T_m = [t_k]_{k\le m}$, $m \in \mathbb{N}$ in $C(K)^*$. It is easy to see that $F_m^\perp = T_m$, $m \in \mathbb{N}$. Since T_m is finite dimensional and $\lim_n f_n(t_m) = 0$, $m \in \mathbb{N}$, it follows from Lemma 1 that for all $m \in \mathbb{N}$,

$$\lim_n \; dist_{|\cdot|} \; (f_n, F_m) = \lim_n \; \sup\{x^*(f_m) : x^* \in T_m,\ |x^*| = 1\} = 0 \; .$$

Let G be any finite dimensional subspace in $C(K)$. It suffices to show that for any subsequence $\{f_{n_k}\}$ of $\{f_n\}$, $\lim_k dist(f_{n_k} G) = 1$. Since G is finite dimensional, without loss of generality, we may assume that there exist g_n, $n \in \mathbb{N}$ and g in G such that $dist_{|\cdot|}(f_n, G) = |f_n - g_n|$, $n \in \mathbb{N}$ and $\lim_n |g_n - g| = 0$.

Given any $\epsilon > 0$, choose m such that $|R_m(g)| < \epsilon$. Fix m, choose N such that for all $n \ge N$, $dist_{|\cdot|}(f_n, F_m) < \epsilon$ and $dist_{|\cdot|}(f_n, G) > |f_n - g| - \epsilon$. Let $h_n \in F_m$ such that $|f_n - h_n| < \epsilon$. Since $R_m(h_n) = h_n$, we have for all $n \ge N$,

$$\begin{aligned} dist_{|\cdot|}(f_n, G) &> |f_n - g| - \epsilon > |R_m(f_n - g)| - \epsilon \ge |R_m(f_n)| - |R_m(g)| - \epsilon \\ &\geqq |R_m(h_n)| - |R_m(f_n - h_n)| - 2\epsilon > |h_n| - 3\epsilon \\ &\geqq |f_n| - |h_n - f_n| - 3\epsilon > 1 - 4\epsilon \; . \end{aligned}$$

□

Corollary 3. Let X be a separable Banach space and let F be a separable norming subspace of X in X^*. Then X admits an equivalent norm $|\cdot|$ such that if $\{x_n\}_{n\in\mathbb{N}}$ is a sequence in X, $|x_n| = 1$, $n \in \mathbb{N}$ and $\sigma(X,F) - \lim_n x_n = 0$ then $\lim_n dist_{|\cdot|}(x_n, Y) = 1$ for all finite dimensional subspace Y in X.

Proof. Let $\{t_n\}_{n\in\mathbb{N}}$ be a norm dense subset of the unit sphere of F and let K be the weak* closure of $\{t_n\}$ in X^*. Since X is separable, K is compact metrizable. We now embedd X canonically in $C(K)$ and apply Lemma 2. □

Lemma 4. Let X be a separable Banach space and let F be a separable norming subspace of X in X^*. Then X admits an equivalent norm $|\cdot|$ such that $dist_{|\cdot|}(x^*, F) < |x^*|$ for all $x^* \neq 0$ in X^*.

Proof. Let $\|\cdot\|$ be a norm in X with the property in Corollary 3. Let $\{X_n\}_{n\in\mathbb{N}}$ be an increasing sequence of finite dimensional subspaces such that $\cup_n X_n$ is norm dense in X. For $x^* \in X^*$, define

$$|||x^*||| = \sum_n 2^{-n} dist_{\|\cdot\|}(x^*, X_n^{\perp})$$

and

$$|x^*| = \|x^*\| + |||x^*||| \ .$$

It is clearly that $|\cdot|$ is an equivalent norm on X^*, $\|x^*\| \leq |x^*| \leq 2\|x^*\|$, $x^* \in X^*$. It is easy to see that $|||\cdot|||$ is weak* lower semi-continuous, hence $|\cdot|$ is a dual norm in X^*. Let $|\cdot|$ be the equivalent norm in X with $(X, |\cdot|)^* = (X^*, |\cdot|)$.

It is clear that $|x^{**}| \leq \|x^{**}\|$ for all $x^{**} \in X^{**}$. We shall show that $|x^{**}| \geq \|x^{**}\|$ for all $x^{**} \in F^{\perp}$ and so $|x^{**}| = \|x^{**}\|$, $x^{**} \in F^{\perp}$.

Fix $x^{**} \in F^{\perp}$, $|x^{**}| = 1$. Let $\{x_\alpha\}$ be a net in X, $|x_\alpha| \leq 1$ and $w^* - \lim_\alpha x_\alpha = x^{**}$. Then $\liminf_\alpha \|x_\alpha\| \geq \|x^{**}\|$ and since $x^{**} \in F^{\perp}$, we have $\sigma(X, F) - \lim_\alpha x_\alpha = 0$. Since the $\sigma(X, F)$-topology is metrizable on bounded sets, we may apply Corollary 3 not only for sequences but also for nets. Thus for $n \in \mathbb{N}$, $\lim_\alpha dist_{\|\cdot\|}(\frac{x_\alpha}{\|x_\alpha\|}, X_n) = 1$. By Lemma 1, $\lim_\alpha \sup_{x^* \in X_n^{\perp}} \frac{|x^*(x_\alpha)|}{\|x^*\| \ \|x_\alpha\|} = 1$, $n \in \mathbb{N}$. Therefore, for all $n \in \mathbb{N}$, there is a net $\{x_\alpha^*\}$ in $X_n^{\perp}$ such that $\|x_\alpha^*\| = 1$ and $\liminf_\alpha x_\alpha^*(x_\alpha) = \liminf_\alpha \|x_\alpha\|$.

Given any $\epsilon > 0$, choose $n \in \mathbb{N}$, such that $2^{-n+1} < \epsilon$. Then for every $x^* \in X_n^{\perp}$, $|||x^*||| \leq \epsilon \|x^*\|$. Hence

$$\begin{aligned} |x^{**}| &= \lim_\alpha |x_\alpha| \geq \liminf_\alpha \frac{x_\alpha^*(x_\alpha)}{|x_\alpha^*|} = \liminf_\alpha \frac{x_\alpha^*(x_\alpha)}{\|x_\alpha^*\| + |||x_\alpha^*|||} \\ &\geq (1+\epsilon)^{-1} \liminf_\alpha \|x_\alpha\| \geq (1+\epsilon)^{-1} \|x^{**}\| \ . \end{aligned}$$

Thus $|x^{**}| \geq \|x^{**}\|$.

Let $x^* \neq 0$ in X^*. Since $\cup_n X_n$ is dense in X, there exists n such that $dist_{\|\cdot\|}(x^*, X_n^{\perp}) > 0$. This implies that $\|x^*\| < |x^*|$. By Lemma 1,

$$dist_{|\cdot|}(x^*, F) = \sup_{\substack{x^{**} \in F^{\perp} \\ |x^{**}|=1}} |x^{**}(x^*)| = \sup_{\substack{x^{**} \in F^{\perp} \\ \|x^{**}\|=1}} |x^{**}(x^*)| =$$
$$dist_{\|\cdot\|}(x^*, F) \leq \|x^*\| < |x^*| \ .$$

□

Proof of Theorem. Let F be a separable norming subspace of X in X^*. Choose a separable subspace Y in X such that $\|x^*\| = \sup_{\substack{y \in Y \\ \|y\|=1}} |x^*(y)|$, $x^* \in F$. Embedd Y canonically in F^*. Then Y is a separable norming subspace of F in F^*. According to Lemma 4, F admits an equivalent norm $|\cdot|$ such that $dist_{|\cdot|}(e^*, Y) < |e^*|$, for all $e^* \neq 0$ in F^*. The norm $|\cdot|$ on F^* induces an equivalent norm on X with the property that $dist_{|\cdot|}(x, Y) < |x|$, $x \neq 0$ in X. Suppose X has an Auerbach basis $\{x_i\}_{i\in I}$. Then I is uncountable. Since Y is separable, there is a countable subset J of I such that $Y \subset [x_j]_{j\in J}$. then for any $i \in I\backslash J$,

$$dist_{|\cdot|}(x_i, [x_j]_{j\neq i}) \leq dist_{|\cdot|}(x_i, [x_j]_{j\in J}) \leq dist(x_i, Y) < |x_i| \; .$$

This is a contradiction that $\{x_i\}_{i\in I}$ is an Auerbach basis. □

References

[G1] B. V. Godun, *A Special Class of Banach Spaces*, Math. Notes **37** (1985), 220–223.

[G2] B. V. Godun, *Bases of Auerbach in Banach Spaces Isomorphic to* $\ell_1([0,1])$, C. R. Bulg. Acad. Sci. **43** (1990), 19–21. (Russian)

[PB] A. Pelczynski and C. Bessaga, *Some Aspects of the Present Theory of Banach Spaces*, in the new edition of Theory of Linear Operations by S. Banach; English translated by F. Jellett, North Holland, 1987, pp. 163–237.

[S] Z. Semedeni, *Schauder Bases in Banach Spaces of Continuous Functions*, in Lecture Notes in Math, vol. 918, Springer-Verlag, 1982.

[Si] I. Singer, *Bases in Banach Spaces*, vol. II, Springer-Verlag, 1981.

Department of Mathematics, University of Iowa, Iowa City, Iowa 52242 U.S.A.

Current address: B. V. Godun, Department of Mathematics, Kharkov Institute of Civil Engineers, 12 Revolucii Str., 31002 Kharkov, Ukraine

S. L. Troyanski, Department of Mathematics, Sofia University, 5 James Bouchier Blvd, 1126 Sofia, Bulgaria

Contemporary Mathematics
Volume **144**, 1993

RENORMING BANACH SPACES WITH FUNDAMENTAL BIORTHOGONAL SYSTEM

B. V. GODUN AND S. L. TROYANSKI

ABSTRACT. Let X be a Banach space with fundamental biorthogonal system. Then X has an equivalent norm that shares many properties of $l_1(I)$. For example, the unit ball is the closed convex hull of its strongly exposed points (with a uniformity in the strong exposition), but has no point of locally uniformly rotundity similarly to the ball of $l_1(I)$. If in addition $\operatorname{dens} X = \operatorname{dens} X^*$, then X has an equivalent norm such that the dual norm of X^* shares many properties of $l_1(I)$.

1. INTRODUCTION

In [S] it is proved that if X is weakly compactly generated then for each $\epsilon > 0$ there exists in X an equivalent norm with respect to which X has (α, ϵ)-property.

Recently in [FS] it was proved that if X has separable dual then for each $\epsilon > 0$ there exists in X an equivalent norm with respect to which X has (β, ϵ)-property in the strong sense. This result was extended in [F] for Banach spaces X with separable projectional resolution of the identity and $\operatorname{dens} X = \operatorname{dens} X^*$.

The main aim of this paper is to extend the results from [S] and [F] over the wide class of Banach spaces with suitable minimal system of cardinality $\operatorname{dens} X$. We automatically get corresponding geometrical applications analogous to those from [FS] and [F]. We give also some new applications. The results of this paper are announced in [GT].

2. NOTATIONS, DEFINITIONS AND MAIN RESULTS

As usual N denotes naturals. The unit ball of a Banach space X is denoted by U_X, the unit sphere by S_X. For $\{x_i\}_{i\in I} \subset X$ we denote $U_l(\{x_i\}_{i\in I}) = TU_{l_1(I)}$, where T is bounded linear operator from $l_1(I)$ into X such that $Te_i = x_i$, $i \in I$, and $\{e_i\}_{i\in I}$ is the unit vector basis in $l_1(I)$.

The system $\{x_i, x_i^*\}_{i\in I} \subset X \times X^*$ is called biorthogonal if $x_i^*(x_j) = \delta_{i,j}$. If, in addition, $\sup_i \|x_i\| \cdot \|x_i^*\| \le c$, the system $\{x_i, x_i^*\}_{i\in I}$ is called a c-bounded biorthogonal system. A biorthogonal system $\{x_i, x_i^*\}_{i\in I}$ is called fundamental if $[x_i]_{i\in I} = X$. The Banach space X has FBS if there exists in $X \times X^*$ a fundamental biorthogonal system.

1991 *Mathematics Subject Classification.* Primary 46B20.

Partially supported by Ministry of Science and Education of Bulgaria, Contract MM-3/91.
This paper is in final form and no version of it will be submitted for publication elsewhere

A sequence $\{x_n^*\}_{n\in N} \subset X^*$ is called w^*- basic [JR] provided that there exists a sequence $\{x_n\}_{n\in N} \subset X$ so that $\{x_n, x_n^*\}_{n\in N}$ is biorthogonal and for each x^* in the w^*-closure of the space $[x_n^*]_{n\in N}$

$$\sum_{n=1}^{k} x^*(x_n)x_n^* \xrightarrow{w^*} x^*.$$

Definition 1.

(a) *The Banach Space X has (α, ϵ)-property if there exists a system $\{x_i, x_i^*\}_{i\in I} \subset X \times X^*$ such that*

(1) $$x_i^*(x_i) = \|x_i\| = \|x_i^*\| = 1, \ |x_i^*(x_j)| < \epsilon, \quad i \neq j,$$

(2) $$U_X = \overline{conv}(\{\pm x_i\}_{i\in I}).$$

(b) *X has (β, ϵ)- property if there exists a system $\{x_i, x_i^*\}_{i\in I} \subset X \times X^*$ with (1) and $\|x\| = \sup_i |x_i^*(x)|$.*

The above properties have been used in the study of norm attaining operators ([L], [P], [S]).

Definition 2.

(a) *The Banach space X has strictly (α, ϵ)-property if X has (α, ϵ)-property and in addition*

(3) $$\inf_{j\in I} \max\{|x_j^*(x_i)|, |x_i^*(x_j)|\} = 0 \quad \text{for all } i \in I.$$

(b) *X has (β, ϵ) - property in the strong sense if X has (β, ϵ)-property, and in addition*

(4) $$U_{X^*} = \overline{conv}(\{\pm x_i^*\}_{i\in I}).$$

Theorem 1. *If X has biorthogonal system with cardinality equal to $dens\, X$, then for each $\epsilon \in (0,1)$ X admits an equivalent norm $|\cdot|$ such that $(X, |\cdot|)$ has (α, ϵ)-property. If X in addition has FBS then $(X, |\cdot|)$ has strictly (α, ϵ)-property.*

Theorem 2. *If X has FBS and $dens\, X = dens\, X^*$, then for each $\epsilon \in (0,1)$ X admits an equivalent norm $|\cdot|$ such that $(X, |\cdot|))$ has (β, ϵ)-property in the strong sense.*

The conditions of this theorem are satisfied for example from the dual to the James tree space which has no Markushevich basis and has no separable projectional resolution [FG]. Other examples of this kind can be found in [E].

3. Proofs of theorems 1 and 2

By $B(X)$ we denote the set of all sequences of elements of X equivalent to the unit vector basis in l_1.

Lemma 1. *Let E be a separable Banach space and $\eta > 0$. There exists an $(2+\eta)$-bounded biorthogonal system $\{v_n, v_n^*\}_{n\in N}$ such that $\{v_n^*\}_{n\in N} \notin B(E^*)$ and the set $U_l(\{v_n^*\}_{n\in N})$ is w^*-compact.*

Proof. According to [JR] from every normalized w^*-convergent to 0 sequence in E^* we can find subsequence which is w^*-basic asymptotically monotone sequence. Let $\{v_n, v_n^*\}_{n\in N} \subset E\times E^*$ be $a(2+\eta)$-bounded biorthogonal system such that $\{v_n^*\}_{n\in N}$ is normalized w^*-basic sequence. Since $\{v_n^*\}_{n\in N}$ is w^*-basic then it is easy to see that the set $U_l(\{v_n^*\}_{n\in N})$ is w^*-compact.

If $\{v_n^*\}_{n\in N} \notin B(E^*)$ Lemma 1 is proved.

If $\{v_n^*\}_{n\in N} \in B(E^*)$ then according to the James theorem (cf. [**LT**, p. 97]) we can find $\{w_n^*\}_{n\in N}$ which is normalized block-basic sequence of $\{v_n^*\}_{n\in N}$ and $\{w_n^*\}_{n\in N}$ is $(1+\eta)$ - equivalent to the unit vector basic in l_1. Since $\{v_n^*\}_{n\in N}$ is w^*-convergent to 0, then it is easy to see that $w_n^* \xrightarrow{w^*} 0$. According to [JR] without loss of generality we can assume that $\{w_n^*\}$ is w^*-basic sequence. The subspace $F = [w_n^*] \subset X^*$ is w^*-closed. Indeed, let $f_k \in F$ and $f_k \xrightarrow{w^*} f$. Since $\{w_n^*\}$ is an w^*-basic sequence then

$$g_k = \sum_{n=1}^{k} f(w_n)w_n^* \xrightarrow{w^*} f,$$

where $\{w_n\}$ is such that $\{w_n, w_n^*\}$ is biorthogonal. Since $\{w_n^*\}$ is $(1+\eta)$-equivalent to the unit vector basis of l_1 then $U_F \subseteq (1+\eta)W$, where $W = U_l(\{w_n^*\})$. Since W is w^*-compact and $g_n \in (1+\eta)cW$, where $c = \sup \|g_n\|$, we get $f \in (1+\eta)cW \subset$ F. Put $H = E/F_\perp$ .Let T be the canonical map from E onto H. Then T^* is isometry between H^* and F. Set $x_n = Tw_n, x_n^* = T^{*-1}w_n^*$. Obviously, $\{x_n\}$ is basis in $H, (1+\eta)$-equivalent to the unit vector basis in c_o. Therefore,

$$y_k = \sum_{n=1}^{k} x_n,\ y_n^* = x_n^* - x_{n+1}^*$$

is an $(2+2\eta)$-bounded biorthogonal system and $\{y_n^*\} \notin B(H^*)$. We can find $z_k \in E$ such that $Tz_k = y_k$ and $\{z_k, T^*y_k^*\}$ is an $(2+3\eta)$-bounded biorthogonal system. Since T^* is isometry $\{T^*y_n^*\}_{n\in N} \notin B(E^*)$.

Proposition 1. *If X has FBS for each $\eta > 0$, then there exist two $(4+\eta)$-bounded biorthogonal systems*

$$\{u_{i,n}, u_{i,n}^*\}_{(i,n)\in I\times N},\ \{v_{i,n}, v_{i,n}^*\}_{(i,n)\in I\times N} \subset X \times X^*,$$

$|I| =$ *dens* X, *with the following properties:*

$$\{u_{i,n}\}_{n\in N} \notin B(X),\ \{v_{i,n}^*\}_{n\in N} \notin B(X^*)$$

for any $i \in I$, the sets

$$\{0\} \cup \{u^*_{i,n}\}_{(i,n)\in I\times N}, \quad \text{and } U_l(\{v^*_{i,n}\}_{(i,n)\in I\times N})$$

are w^*-compact.

Proof. Since X has FBS then X has a quotient space $E = X/Y$ such that $\operatorname{dens} E = \operatorname{dens} X$ and E has a separable projectional decomposition [Pl], i.e. there exists a transfinite set of projections $P_\alpha \colon E \to E$, $\alpha < \alpha_0$, such that

(i) $\|P_\alpha\| = 1$;
(ii) $P_\alpha P_\beta = P_{\min(\alpha,\beta)}$;
(iii) the space $E_\alpha = Q_\alpha E$ is infinite dimensional and separable for any $\alpha < \alpha_0$, where $Q_\alpha = P_{\alpha+1} - P_\alpha$;
(iv) $\left[\bigcup_{\alpha<\alpha_0} E_\alpha\right] = E$.

For any $\eta > 0$ in $E_\alpha \times E^*_\alpha$ there exists $(1+\eta)$-bounded fundamental biorthogonal system $\{e_{\alpha,n}, e^*_{\alpha,n}\}_{n\in N}$ such that the system $\{e^*_{\alpha,n}\}_{n\in N}$ is not total in E_α [DJ]. Therefore, $\{e_{\alpha,n}\}_{n\in N}$ is not a basis and we get $\{e_{\alpha,n}\}_{n\in N} \notin B(E_\alpha)$. Extend $e^*_{\alpha,n}$ to E by taking $Q^*_\alpha e^*_{\alpha,n}$ and denote them by $f_{\alpha,n}$. Obviously, $\{e_{\alpha,n}, f_{\alpha,n}\}_{\alpha<\alpha_0, n\in N}$ is an $(2+2\eta)$-bounded fundamental biorthogonal system in E.

Let $T \colon X \to E$ be the corresponding quotient map. According to the lifting theorem [G1] there exist a system $\{u_{\alpha,n}\}_{\alpha<\alpha_0, n\in N}$ in X such that $Tu_{\alpha,n} = e_{\alpha,n}$, $\{u_{\alpha,n}\}_{n\in N} \notin B(X)$ for each $\alpha < \alpha_0$ and $\{u_{\alpha,n}, T^* f_{\alpha,n}\}_{\alpha<\alpha_0, n\in N}$ is an $(4 + 5\eta)$-bounded fundamental biorthogonal system. Since $\{y_{\alpha,n}\}$ is fundamental then $\{0\} \cup \{T^* f_{\alpha,n}\}_{\alpha<\alpha_0, n\in N}$ is w^*-compact.

According to Lemma 1 we can find an $(2 + \eta)$-bounded biorthogonal system $\{w_{\alpha,n}, w^*_{\alpha,n}\}_{n\in N} \subset E_\alpha \times E^*_\alpha$ such that $\{w^*_{\alpha,n}\}_{n\in N} \notin B(E^*_\alpha)$ and the set $U_l(\{w^*_{\alpha,n}\}_{n\in N})$ is w^*-compact.

Extend $w^*_{\alpha,n}$ to E by taking $Q^*_\alpha w^*_{\alpha,n}$ and denote them by $g_{\alpha,n}$. Obviously, $\{w_{\alpha,n}, g_{\alpha,n}\}_{\alpha<\alpha_0, n\in N}$ is an $(4 + 2\eta)$-bounded biorthogonal system in E, and $\{g_{\alpha,n}\}_{n\in N} \notin B(E^*)$ for each $\alpha < \alpha_0$. We can find $v_{\alpha,n} \in X$, such that $Tv_{\alpha,n} = w_{\alpha,n}$ and $\{v_{\alpha,n}, T^* g_{\alpha,n}\}_{\alpha<\alpha_0, n\in N}$ is an $(4+3\eta)$-bounded biorthogonal system in X. Since T^* is isometry between E^* and $Y^\perp$ we have $\{T^* g_{\alpha,n}\}_{n\in N} \notin B(X^*)$. The set $U_l(\{g_{\alpha,n}\}_{\alpha<\alpha_0, n\in N})$ is w^*-compact since $U_l(\{g_{\alpha,n}\}_{n\in N})$ are w^*-compact for each $\alpha < \alpha_0$ and $\left[\bigcup_{\alpha<\alpha_0} E_\alpha\right] = E$.

Proposition 2. *Let X have biorthogonal system $\{y_{i,n}, y^*_{i,n}\}_{(i,n)\in I\times N}$, $|I| = \operatorname{dens} X$, such that*

$$\|y_{i,n}\| = 1, \ \|y^*_{i,n}\| \le c, \ (i,n) \in I \times N, \tag{5}$$

$$\{y_{i,n}\}_{n\in N} \notin B(X), \ i \in I. \tag{6}$$

Then for each $\epsilon \in (0,1)$, X admits an equivalent norm $|\cdot|$ such that $(X, |\cdot|)$ has (α,ϵ)-property.

Moreover

(a) *if* $\{0\} \cup \{y^*_{i,n}\}_{(i,n)\in I\times N}$ *is* w^**-compact then* $(X, |\cdot|)$ *has strictly* (α,ϵ)*-property;*

(b) *if* $U_l(\{y_{i,n}\}_{(i,n)\in I\times N})$ *and* U_X *are* $\sigma(X,E)$*-compact for some subspace* $E \subset X^*$ *containing* $\{y^*_{i,n}\}_{(i,n)\in I\times N}$, *then* $U_{(X,|\cdot|)}$ *is* $\sigma(X,E)$*-closed.*

Proof. From (6) it follows that furnishing $y_{i,n}$ with a proper sign we can assume that for any $i \in I$, there exists a sequence of numbers $\{c_{k,i,n}\}$ such that

$$c_{k,i,n} \geq 0, \ \sum_n c_{k,i,n} = 1, \ \lim_{k\to\infty} \|\sum_n c_{k,i,n} y_{i,n}\| = 0. \tag{7}$$

Let $\epsilon > 0$. Pick $\delta = \epsilon/c(1+\epsilon)$ and denote $x_{i,n} = \delta z_i + y_{i,n}$, $(i,n) \in I \times N$,

$$V = \overline{\text{conv}}(\{\pm x_{i,n}\}_{(i,n)\in I\times N}),$$

where $\{z_i\}_{i\in I}$ is a dense subset of U_X. Clearly, $V \subset (1+\delta)U_X$ and from (7) $\delta U_X \subset V$, so that the Minkowski functional of V is an equivalent norm $|\cdot|$ on X. We note that

$$y^*_{i,n}(x_{i,n}) \geq 1 - c\delta > 0, \ (i,n) \in I \times N. \tag{8}$$

Therefore the functionals

$$x^*_{i,n} = y^*_{i,n}/y^*_{i,n}(x_{i,n}), \ (i,n) \in I \times N$$

are well defined and

$$x^*_{i,n}(x_{i,n}) = 1, \ (i,n) \in I \times N. \tag{9}$$

If $(i,n) \neq (j,m)$, then (8) implies

$$|x^*_{i,n}(x_{j,n})| = |y^*_{i,n}(\delta z_j + y_{j,n})|/y^*_{i,n}(x_{i,n}) \leq \epsilon. \tag{10}$$

Obviously $|\ x_{i,n}\ | \leq 1, (i,n) \in I \times N$, and using (9) we obtain $|\ x^*_{i,n}\ | \geq 1, (i,n) \in I\times$N. Since $U_{(X,|\cdot|)}$ is closed absolutely convex hull of the system $\{\pm x_{i,n}\}_{(i,n)\in I\times N}$ we have for every $f \in X^*$

$$|f| = \sup_{i,n} |f(x_{i,n})|.$$

But now, from (9), (10) it follows $|x^*_{i,n}| \leq 1$ which combined with (10) implies that $|x_{i,n}| \geq 1, (i,n) \in I \times N$. Therefore $|x_{i,n}| = |x^*_{i,n}| = 1, (i,n) \in I \times N$, and the conditions (9), (10) give that $(X, |\cdot|)$ has (α,ϵ)-property.

(a) Fix $(i,n) \in I \times N$. From (8) we have

$$|x^*_{i,n}(x_{j,m})| = \delta |y^*_{i,n}(z_j)|(1 + \delta y^*_{i,n}(z_j)), \tag{11}$$

$$|x^*_{j,m}(x_{i,n})| = \delta |y^*_{j,m}(z_j)|(1 + \delta y^*_{j,m}(z_j)). \tag{12}$$

Since the system $\{z_j\}_{j\in I}$ is dense in $U_{(X,\|\cdot\|)}$ and its elements can be approximated arbitrarily close by elements of the hyperplane $[y^*_{i,n}]_\perp$ then there exists a sequence $\{j_k\}$ such that $\lim_{k\to\infty} y^*_{i,n}(z_{j_k}) = 0$.

On the other hand, since the set $\{0\} \cup \{u^*_{i,n}\}_{(i,n)\in I\times N}$ is w^*-compact

$$\lim_{k\to\infty} y^*_{j_k,m}(z_i) = 0.$$

Now passing to the limit on $\{j_k\}$ in (11) and (12) we obtain that $(X, |\cdot|)$ has strictly (α, ϵ)-property.

(b) The set V is $\sigma(X,E)$-closed. Indeed, let $v_\beta = \sum_{i,n} \lambda_{\beta,i,n} x_{i,n}, \beta \in B$ be a net of elements of V converging in the $\sigma(X,E)$-topology to some $v \in$ X. We have to show that $v \in V$. Without loss of generality we may assume that $\lambda_{\beta,i,n} \geq 0$ and $\lim_\beta \lambda_{\beta,i,n} = \lambda_{i,n}$. Let $\lambda = \sum_{i,n} \lambda_{i,n}$. Clearly $0 \leq \lambda \leq 1$, and given $\eta > 0$ we can find a finite subset $A \subset I \times N$ such that

$$\lambda - \sum_{(i,n)\in A} \lambda_{i,n} < \eta.$$

For $\beta \in B$ let

$$s_\beta = \sum_{(i,n)\in A} \lambda_{\beta,i,n} x_{i,n}, \quad t_\beta = \delta \sum_{(i,n)\in C} \lambda_{\beta,i,n} z_i, \quad u_\beta = \sum_{(i,n)\in C} \lambda_{\beta,i,n} y_{i,n},$$

where $C = (I \times N)\backslash A$. One has $v_\beta = s_\beta + t_\beta + u_\beta$. We may assume that $\{s_\beta\}_{\beta\in B}$, $\{t_\beta\}_{\beta\in B}$, $\{u_\beta\}_{\beta\in B}$ converge in the $\sigma(X,E)$ topology to s, t and u, respectively. Clearly

$$s = \sum_{(i,n)\in A} \lambda_{i,n} x_{i,n} \in \lambda V.$$

On the other hand there is an $\beta_0 \in B$ such that for $\beta \geq \beta_0$

$$\sum_{(i,n)\in C} \lambda_{\beta,i,n} < 1 - \lambda + \eta,$$

hence $t_\beta \in \delta(1-\lambda+\eta)U_X$. Since δU_X is $\sigma(X,E)$-compact and is contained in V we have $t \in (1-\lambda+\eta)V$, and hence $s + t \in (1+\eta)V$.

Since $U_l(\{y_{i,n}\}_{(i,n)\in I\times N})$ is $\sigma(X,E)$-compact, $u = \sum_{(i,n)\in C} \lambda_{i,n} y_{i,n}$ and therefore $\|u\| \leq \eta$. So $\operatorname{dist}(v, (1+\eta)V) \leq \|u\| \leq \eta$. This shows that V is $\sigma(X,E)$-closed.

The proofs of Theorems 1 and 2 follow immediately from Propositions 1 and 2.

Remark. If the continuum hypothesis is assumed, in [KV] a nonseparable Banach space X is constructed which possesses the following property: in every uncountable family $\{x_i\} \subset X$ there is an element x_j such that $x_j \in \overline{\operatorname{conv}}(\{x_i\}_{i\neq j})$. Clearly, in such a space there is no uncountable biorthogonal systems. Since the geometrical meaning of the condition (1) is done by

$$\operatorname{dist}(x_j, \overline{\operatorname{conv}}(\{x_i\}_{i\neq j})) > 1 - \epsilon, \text{ for any } j \in I,$$

this space (as well as every nonseparable subspace of it) does not have the (α,ϵ)-property for any equivalent norm and any $\epsilon \in (0,1)$.

4. Geometrical applications

As we mentioned we get all applications given in [FS]. We give some new applications also. Let us recall some definitions.

Definition 3. *An element $x \in S_X$ is called*

- *strongly exposed point ($x \in strexp\, U_X$) if for some functional $f_x \in S_{X^*}$, the conditions $\{x_n\} \subset U_X, f_x(x_n) \to 1$ imply $\|x_n - x\| \to 0$;*
- *locally uniformly rotund point ($x \in lur\, U_X$), if the conditions $\{x_n\} \subset U_X, \|x_n + x\| \to 2$ imply $\|x_n - x\| \to 0$.*

Theorem 3. *If X has (α, ϵ)-property for some $\epsilon \in (0,1)$, then $strexp\, U_X = \{\pm x_i\}_{i \in I}$, where $\{x_i\}_{i \in I}$ is a system with properties (1) and (2). In addition if X has strictly (α, ϵ)-property for some $\epsilon \in (0, 1/2]$ then*

$$lur\, U_X = \emptyset.$$

Proof. In [FS] it is essentially proved that $\{\pm x_i\}_{i \in I} \subseteq \operatorname{strexp} U_X$.

Now we show that strexp $U_X \subseteq \{\pm x_i\}_{i \in I}$. Indeed, as we noted before, for $f \in X^*$

$$\|f\| = \sup_i |f(x_i)|. \tag{13}$$

Then from (13) it follows that for arbitrary $i, j \in I (i \neq j)$,

$$\|x_i^* \pm x_j^*\| \leq \max\{1 + |x_i^*(x_j)|, 1 + |x_j^*(x_i)|, 2\epsilon\} \leq 1 + \epsilon$$

and therefore

$$\begin{aligned} \|x_i \pm x_j\| &\geq (x_i^* \pm x_j^*, x_i \pm x_j)/\|x_i^* \pm x_j^*\| \\ &\geq 2\frac{1 - \max\{|x_i^*(x_j)|, |x_j^*(x_i)|\}}{\max\{1 + |x_i^*(x_j)|, 1 + |x_j^*(x_i)|, 2\epsilon\}} \geq 2\frac{1-\epsilon}{1+\epsilon} > 0. \end{aligned} \tag{14}$$

Let now $x \in$ strexp U_X, and f_x be the corresponding functional supporting x. From (13) it follows that we can find sequence $\{\mu_n x_{i_n}\}, \mu_n = \pm 1$ such that $\|\mu_n x_{i_n} - x\| \to 0$. But according to (14) this is possible only if the sequence $\{\mu_n x_{i_n}\}$ is constant, i.e. $\mu_n x_{i_n} = x$ for every n.

Let now X have strictly (α, ϵ)-property and $0 < \epsilon \leq 1/2$. Then

$$\operatorname{lur} U_X \subseteq \operatorname{strexp} U_X \subseteq \{\pm x_i\}_{i \in I}.$$

But from (2) and (14) there exists for every $i \in I$ a sequence $\{j_n\}_{n \in N}$ such that $\lim_{n \to \infty} \|x_i \pm x_{j_n}\| = 2$, i.e., $\pm x_i \notin \operatorname{lur} U_X$ for all $i \in$ I.

Corollary 1. *The space $l_\infty(I)$ admits an equivalent norm such that the unit ball is the closed convex hull of the set of its strongly exposed points.*

The proof follows simply from Theorem 3 since $l_\infty(I)$ has FBS [G2].

Corollary 2. *Every separable Banach space can be equivalently renormed in such a way that with respect to the new norm the set of locally uniformly rotund points is void.*

References

[DJ] W.J.Davis, and W. B. Johnson, *On the existence of fundamental and total bounded biorthogonal systems in Banach spaces*, Studia Math. **43** (1975), 175–179.

[E] G. Edgar, *Measurability in Banach space II*, Indiana Univ. Math. J. **28** (1979), 559–579.

[F] K. Finet, *Renorming Banach spaces with many projections and smoothness properties*, Math. Ann. **284** (1989), 675–679.

[FG] K. Finet, and G. Godefroy, *Biorthogonal systems and big quotient spaces*, Contemp. Math. **85** (1989), 87–110.

[FS] K. Finet, and W. Schachermayer, *Equivalent norms in separable Asplund spaces*, Studia Math **92** (1989), 275–283.

[G1] B. V. Godun, *Complete biorthogonal systems in Banach spaces,*, Funktsional. An. Applications **17** (1983), 1–5.

[G2] B. V. Godun, *Biorthogonal systems in the spaces of bounded functions*, Dokl. Acad. Nauk Ukr. SSR, ser A **3** (1983), 7–9, (Russian).

[GT] B. V. Godun, and S. L. Troyanski, *Operators which attain their norm and geometry of the unit ball of Banach spaces*, Soviet. Math. Doklady **42** (1991), 532–534.

[JR] W. B. Johnson, and H. P. Rosenthal, *On $w*$-basic sequence and their applications to the study of Banach spaces*, Studia Math. **43** (1972), 74–92.

[KV] K. Kunen, and J. Vaugham, *Handbook of set theoretic topology*, North Holland, 1984.

[L] J. Lindenstrauss, *On operators which attain their norm*, Israel J. Math. **1** (1963), 139–148.

[LT] J. Lindenstrauss, and L. Tzafriri, *Classical Banach Spaces I*, Springer, Berlin-Heidelberg-New York, 1977.

[P] R. Partington, *Norm attaining operators*, Israel J. Math. **43** (1983), 273–276.

[Pl] A. N. Plichko, *Banach space without fundamental biorthogonal system*, Soviet Math. Dokl. **22** (1980), 450–453.

[S] W. Schachermayer, *Norm attaining operators and renorming of Banach spaces*, Israel J. Math. **44** (1983), 201–212.

Department of Mathematics and Informatics, Sofia University, 5, J. Boucher blvd, 1126 Sofia, BULGARIA

Department of Mathematics, Kharkov Institute of Civil Engineers, 12, Revolucii str., 31002 Kharkov, UKRAINE

Contemporary Mathematics
Volume **144**, 1993

QUANTITATIVE ESTIMATES ON RANDOM SUBSPACES WHICH MISS ARBITRARY MEASURABLE SETS ON THE SPHERE

YEHORAM GORDON

ABSTRACT. Let S^{n-1} be the unit sphere of $\mathbb{R}^n$ and $d\mu$ denote the normalized rotation invariant measure on S^{n-1}. Given $1 \le k < n$ and $0 < \delta < 1$, we find $\epsilon = \epsilon(n,k,\delta)$ such that for every closed set $A \subset S^{n-1}$ with $\mu(A) = \delta$, there exists a k-codimensional subspace F such that $F \cap S^{n-1} \subset A_\epsilon$ (= the ϵ-neighborhood, in the geodesic metric, of the set A). The function $\epsilon(n,k,\delta)$ refines the previous known estimates for ϵ.

Let $S^{n-1} = \{x \in \mathbb{R}^n;\ \|x\| = 1\}$ be the unit sphere of $\mathbb{R}^n$, $d\mu$ the normalized Haar measure on S^{n-1}, and by $\rho(x,y)$ we denote the geodesic distance between two points, $x, y \in S^{n-1}$. Given a closed subset $A \subset S^{n-1}$, $A_\epsilon = \{x \in S^{n-1};\ \rho(x,A) \le \epsilon\}$ is the ϵ-neighborhood of A. We are interested in improving known estimates for the following problems:

Problems. Given $1 \le k < n$ and $0 < \mu < 1$, what is the least $\epsilon > 0$ such that for any closed subset A of S^{n-1} with $\mu(A) = \mu$, there exists a subspace E of codimension k such that

(a) $E \cap S^{n-1} \subseteq A_\epsilon$?

(b) $E \cap S^{n-1} \subseteq (\partial A)_\epsilon$?

In [G] we obtained some estimates for ϵ in special situations. These estimates are useful in the proofs of Dvoretzky's theorem for normed spaces, or quasi-normed spaces (see e.g., [MS], [FLM], [P2], [GL]) and in the proof of the Milman quotient-subspace theorem ([M]) as it was developed by [G,GK] and many other important applications. We refer also to [MS], [P] and [TJ] for more details and references therein.

We intend to sharpen here the known estimates for ϵ, so that they will be interesting, we hope, even for "small" dimensions n, k, and arbitrary values n, k, μ as well, irrespective of the shape of the set A. Our starting point is the following theorem:

1991 *Mathematics Subject Classification.* Primary 46B07, 46B20.

Supported by the USA-Israel Binational Science Foundation Grant #86-0074, and the Fund for the Promotion of Research at the Technion #100-834.

This paper is in final form and no version of it will be submitted in publication elsewhere.

THEOREM 1. *Let $1 \le k < n$, and $a_n = \sqrt{2}\Gamma(\frac{n+1}{2}/\Gamma(\frac{n}{2})$. Given any closed subset $A \subseteq S^{n-1}$, define $s(A) = a_n \int_{S^{n-1}} \max_{x\in A}\langle x, u\rangle d\mu(u) = \mathbb{E}(\max_{x\in A}\sum_{i=1}^n x_i g_i)$, where $\{g_i\}_{i=1}^n$ is an orthonormal sequence of Gaussian variables.*

(i) If $a_k > s(A)$, then there is a $k-$codimensional subspace E such that $A \cap S^{n-1} = \phi$.

(ii) If $a_k < s(A)$, then there is a $k-$dimensional subspace F and an orthogonal projection P_F onto F, such that $O \in P_F(\text{int}(\text{co}(A)))$, in particular dim(span(A)) $\ge k$.

Proof. (i) is proved in [G]. We recall some parts of the proof which we shall need in the sequel. Consider the Gaussian operator $G_\omega = \sum_{i=1}^n \sum_{j=1}^k g_{ij}(\omega) e_i \otimes e_j$ from $\mathbb{R}^n$ to $\mathbb{R}^k$, where $\{g_{ij}\}$ are orthonormal Gaussians over some probability space $(\Omega, \sum, P)$. It was proved in [G] that

$$\mathbb{E}(\min_{x\in A}\|G_\omega(x)\|) \ge a_k - s(A)(>0, \text{ by assumption}) ,$$

hence, if $E_\omega = G_\omega^{-1}(0)$, $\dim(E_\omega) = n - k$ a.e., and therefore $\exists E_{\omega_0}$ of dimension $n - k$ such that $E_{\omega_0} \cap A = \phi$.

To prove (ii), the same proof in [G] shows that

$$\mathbb{E}(\min_{\|y\|=1}\max_{x\in A}\langle G_\omega(x), y\rangle) \ge s(A) - a_k(>0, \quad \text{by assumption}) .$$

Let $F_\omega = G_\omega^*(\mathbb{R}^k)$, then $\dim F_\omega = k$ a.e., hence there exists ω_0 such that for every $y \in S^{k-1}$, $\exists x \in A$, such that $\langle x, G_{\omega_0}^*(y)\rangle > 0$. If $P_{F_{\omega_0}}$ is the orthogonal projection onto F_{ω_0}, then this implies $0 \in P_{F_{\omega_0}}(\text{int}(\text{co}(A)))$. □

REMARK. $a_n^2 = (\mathbb{E}(\sum_1^n g_i^2)^{\frac{1}{2}})^2 \le \mathbb{E}(\sum_1^n g_i^2) = n$, and $a_n \ge \frac{a_n a_{n+1}}{\sqrt{n+1}} = \frac{n}{\sqrt{n+1}}$.

LEMMA 2. *For all $0 \le \theta \le \epsilon$, $((A_\epsilon)^c)_\theta \subseteq (A_{\epsilon-\theta})^c$.*

Proof. Let $x \in ((A_\epsilon)^c)_\theta$, then $\exists y \in (A_\epsilon)^c$, s.t. $\rho(x, y) \le \theta$, $y \notin A_\epsilon$, hence $\rho(y, z) > \epsilon$ for all $z \in A$, therefore

$$\rho(x, z) \ge \rho(y, z) - \rho(x, y) > \epsilon - \theta .$$ □

LEMMA 3. *Let A be a closed subset of S^{n-1} with $\mu(A) = \mu$, $0 < \mu < 1$. Let $0 < \theta_0 < \pi$ be the unique angle such that*

$$\mu = \int_0^{\theta_0} \sin^{n-2}\varphi d\varphi / \int_0^\pi \sin^{n-2}\varphi d\varphi = \frac{\Gamma(\frac{n}{2})}{\sqrt{\pi}\Gamma(\frac{n-1}{2})}\int_0^{\theta_0}\sin^{n-2}\varphi d\varphi ,$$

and assume $0 \le \epsilon \le \pi - \theta_0$. Then

(i) $s((A_\epsilon)^c)/a_n \le 1 - \frac{\Gamma(\frac{n}{2})}{\sqrt{\pi}\Gamma(\frac{n-1}{2})}\int_0^\epsilon \sin\theta \int_0^{\epsilon+\theta_0-\theta}\sin^{n-2}\varphi\, d\varphi d\theta$.

(ii) $s(A_\epsilon)/a_n \ge \frac{\Gamma(\frac{n}{2})}{\sqrt{\pi}\Gamma(\frac{n-1}{2})}\int_0^{\pi-\epsilon-\theta_0}\sin\theta\int_0^{\epsilon+\theta_0+\theta}\sin^{n-2}\varphi d\varphi d\theta - \cos(\epsilon+\theta_0)$.

Proof. (i) Denote by α the smallest angle such that $\mu(((A_\epsilon)^c)_\alpha) = 1$. Then $\mu(((A_\epsilon)^c)_\theta)$ is non-decreasing in $0 \le \theta \le \alpha$, and

$$\begin{aligned}\int_{S^{n-1}} \max_{x \notin A_\epsilon} \langle x, y\rangle d\mu(y) &= \int_{(A_\epsilon)^c} + \int_{A_\epsilon} = \mu((A_\epsilon)^c) + \int_0^\alpha \cos\theta d\mu(((A_\epsilon)^c)_\theta) \\ &= \mu((A_\epsilon)^c) + [\cos\theta \mu(((A_\epsilon)^c)_\theta)]_0^\alpha + \int_0^\alpha \sin\theta\mu(((A_\epsilon)^c)_\theta)d\theta \\ &= \cos\alpha + \int_0^\alpha \sin\theta\mu(((A_\epsilon)^c)_\theta)d\theta \\ &\le \cos\alpha + \int_0^\epsilon \sin\theta\mu(((A_\epsilon)^c)_\theta)d\theta C + \int_\epsilon^\alpha \sin\theta d\theta.\end{aligned}$$

But by Lemma 2, $\mu(((A_\epsilon)^c)_\theta) \le \mu((A_{\epsilon-\theta})^c) = 1 - \mu(A_{\epsilon-\theta})$, hence the integral is $\le 1 - \int_0^\epsilon \sin\theta\mu(A_{\epsilon-\theta})d\theta$. Let C be a cap of S^{n-1} with geodesic radius θ_0, i.e., $\mu(C) = \mu$. By Levy's classical inequality [L] (see also [MS]),

$$\mu(A_{\epsilon-\theta}) \ge \mu(C_{\epsilon-\theta}) = \frac{\Gamma(\frac{n}{2})}{\sqrt{\pi}\Gamma(\frac{n-1}{2})}\int_0^{\epsilon+\theta_0-\theta} \sin^{n-2}\varphi d\varphi\ .$$

To prove (ii), let β be the smallest angle such that $\mu(A_{\epsilon+\beta}) = 1$. Since $1 = \mu(A_{\epsilon+\beta}) \ge \mu(C_{\epsilon+\beta})$, we have that $\epsilon + \theta_0 + \beta \le \pi$. It follows that

$$\begin{aligned}&\int_{S^{n-1}} \max_{x \in A_\epsilon} \langle x, y\rangle d\mu(y) = \int_{A_\epsilon} + \int_{(A_\epsilon)^c} \\ &= \mu(A_\epsilon) + \int_0^\beta \cos\theta d\mu(A_{\epsilon+\theta}) = \mu(A_\epsilon) + [\cos\theta\mu(A_{\epsilon+\theta})]_0^\beta + \int_0^\beta \sin\theta\mu(A_{\epsilon+\theta})d\theta \\ &= \cos\beta + \int_0^\beta \sin\theta\mu(A_{\epsilon+\theta})d\theta \ge \cos\beta + \int_0^\beta \sin\theta\mu(C_{\epsilon+\theta})d\theta \equiv f(\beta)\ .\end{aligned}$$

Differentiating $f(\beta)$, we get

$$f'(\beta) = -\sin\beta + \sin\beta\mu(C_{\epsilon+\beta}) \le 0$$

hence $f(\beta)$ is decreasing and attains minimum at $\epsilon + \theta_0 + \beta = \pi$, completing the proof. □

COROLLARY 4. *Fix $0 \le \epsilon \le \pi$. Among all closed subsets $A \subset S^{n-1}$ of equal measure, $s(A_\epsilon)$ is minimal when A is a cap.*

Proof. The proof shows that the right-hand side of Lemma 3 (ii) is $s(C_\epsilon)/a_n$. □

THEOREM 5. *Let A, θ_0, μ, n, k be as in Lemma 3.*

(a) *If $a_k/a_n > 1 - \frac{\Gamma(\frac{n}{2})}{\sqrt{\pi}\Gamma(\frac{n-1}{2})}\int_0^\epsilon \sin\theta \int_0^{\epsilon+\theta_0-\theta} \sin^{n-2}\varphi d\varphi d\theta$, then there is a $k-$codimensional subspace E such that $E \cap S^{n-1} \subseteq A_\epsilon$.*

(b) *If $0 \le \epsilon \le \theta_0 \le \pi - \epsilon$ and*

$$a_k/a_n > 1 - \frac{\Gamma(\frac{n}{2})}{\sqrt{\pi}\Gamma(\frac{n-1}{2})_0}\int_0^\epsilon \sin\ \theta \int_{\theta_0+\theta-\epsilon}^{\epsilon+\theta_0-\theta} \sin^{n-2}\varphi d\varphi d\theta$$

then there is a $k-$codimensional subspace E such that $E \cap S^{n-1} \subset (\partial A)_\epsilon$.

Proof. (a) Apply Theorem 1 (i) and the estimate of Lemma 3 (i) to find E, dimE $= n-k$, with $E \cap (A_\epsilon)^c = \emptyset$. (b) Let $B = A^c \cup \partial A$. Then $A \cup B = S^{n-1}$, $A \cap B = \partial A$. We have

$$1-\mu = \mu(B) = \frac{\Gamma(\frac{n}{2})}{\sqrt{\pi}\Gamma(\frac{n-1}{2})}\int_{\theta_0}^{\pi} \sin^{n-2}\varphi d\varphi$$

Clearly $(\partial A)_\epsilon = (A \cap B_\epsilon) \cup (B \cap A_\epsilon)$ and by Levy's inequality

$$\begin{aligned}\mu((\partial A)_\epsilon) &= \mu(A_\epsilon) + \mu(B_\epsilon) - 1\\ &\geq -1 + \frac{\Gamma(\frac{n}{2})}{\sqrt{\pi}\Gamma(\frac{n-1}{2})}\left\{\int_0^{\theta_0+\epsilon} \sin^{n-2}\varphi d\varphi + \int_{\theta_0-\epsilon}^{\pi} \sin^{n-2}\varphi d\varphi\right\}\\ &= \frac{\Gamma(\frac{n}{2})}{\sqrt{\pi}\Gamma(\frac{n-1}{2})}\int_{\theta_0-\epsilon}^{\theta_0+\epsilon} \sin^{n-2}\varphi d\varphi\ .\end{aligned}$$

As above,

$$\begin{aligned}\int_{S^{n-1}} \max_{x\notin(\partial A)_\epsilon} \langle x,u\rangle d\mu(u) &\leq \mu(((\partial A)_\epsilon)^c) + \int_0^\epsilon \cos\theta d\mu(((\partial A)_\epsilon)^c)_\theta)\\ &= \mu(((\partial A)_\epsilon)^c) + [\cos\theta\mu((((\partial A)_\epsilon)^c)_\theta)]_0^\epsilon + \int_0^\epsilon \sin\theta\mu((((\partial A)_\epsilon)^c)_\theta)d\theta\\ &\leq \mu(((\partial A)_\epsilon)^c) + \cos\epsilon\mu((((\partial A)_\epsilon)^c)_\epsilon) - \mu(((\partial A)_\epsilon)^c)\\ &+ \int_0^\epsilon \sin\theta(1-\mu((\partial A)_{\epsilon-\theta}))d\theta\\ &\leq 1 - \int_0^\epsilon \sin\theta\mu((\partial A)_{\epsilon-\theta})d\theta \quad (\text{since} \mu((((\partial A)_\epsilon)^c)_\epsilon) \leq 1)\\ &\leq 1 - \frac{\Gamma(\frac{n}{2})}{\sqrt{\pi}\Gamma(\frac{n-1}{2})}\int_0^\epsilon \sin\theta \int_{\theta_0-\epsilon}^{\theta_0+\epsilon} \sin^{n-2}\varphi d\varphi d\theta\ ,\end{aligned}$$

the result follows by applying Theorem 1 (i). □

Remark.
(a) If $int(A)$ does not contain antipodal points and $\mu(A) = \mu$, let $B = A \cup (-A)$, then $\mu(B) = 2\mu$, so if $G : \mathbb{R}^n \to \mathbb{R}^k$, $\min_{x\notin A_\epsilon} \|G(x)\| = \min_{x\notin B_\epsilon} \|G(x)\|$, hence

$$\mathbb{E}(\min_{x\notin A_\epsilon} \|G(x)\|) \geq a_k - s((B_\epsilon)^c)$$

thus we can double the measure $\mu(A) = \mu$ to $\mu(B) = 2\mu$, thereby redefining θ_0 to refer to 2μ, and the estimates for ϵ provided by Theorem 5 will be improved.
(b) When $k = n-1$, existence of antipodal points in A_ϵ is a simple case, as was observed by Benyamini, because A must contain two points P_0, Q_0 with $\rho(P_0, Q_0) \geq \theta_0$, otherwise, A is contained in a smaller cap than C, which is impossible. If L is the arc between P_0 and Q_0 and R is the mid-point of L, then $\rho(P_0, R) = \rho(Q_0, R) = \theta_0/2$, so it suffices to take $\epsilon = \pi/2 - \theta_0/2$, and A_ϵ will contain antipodal points.

EXAMPLES

In the following examples we consider various values of small dimensions $1 \le k < n$, and various values of $\mu(A) = \mu$, $0 < \mu < 1$. We compute inequality (1) for these cases, the angles θ_0 are those defined in Theorem 5 and depend on n and μ only (e.g., when $n = 3$, $2\mu = 1 - \cos\theta_0$).

When $n = 3$, and $k = 1$ inequality (1) says

$$2\cos\epsilon + \epsilon\sin(\epsilon + \theta_0) - \sin\epsilon\sin\theta_0 < 0\ .$$

When $n = 4$, and $k = 1, 2$, inequality (1) says

$$\epsilon + \theta_0(1 - \cos\epsilon) - \sin\epsilon + \frac{1}{6}\sin(2\epsilon + 2\theta_0) - \frac{1}{3}\sin\epsilon\cos(2\theta_0) - \frac{1}{6}\cos\epsilon\sin(2\theta_0) > c_k$$

where $c_1 = \pi - \frac{4}{3}$, $c_2 = \frac{\pi}{3}$.

The following table lists various values of μ, n, k and specifies corresponding values of $\epsilon = \epsilon(\mu, n, k)$ such that whenever $A \subset S^{n-1}$, and $\mu(A) = \mu$, there exists a $k-$codimensional subspace E such that $E \cap S^{n-1} \subset A_\epsilon$. The values ϵ are the solutions of the inequalities (1).

One application of Theorem 5 is to continuous functions f defined on the sphere S^{n-1}. Denote by $\omega_f(\epsilon) = \max\{|f(x) - f(y)|;\ x, y \in S^{n-1},\ \rho(x,y) \le \epsilon\}$ the modulus of continuity of $f(x)$. Let $M = \max_x\ f(x)$, $m = \min_x\ f(x)$ and let $m \le t \le M$. For convenience sake only assume $\mu(f^{-1}(t)) = 0$, and let $A^t = f^{-1}([m,t])$, $B^t = f^{-1}([t, M])$.

COROLLARY 6. *Let θ_t be the angle defined in Theorem 5 for the value $\mu_t = \mu(A^t)$. If $0 \le \epsilon \le \theta_t \le \pi - \epsilon$ and if*

$$\frac{a_k}{a_n} > 1 - \frac{\Gamma(\frac{n}{2})}{\sqrt{\pi}\Gamma(\frac{n-1}{2})}\int_0^\epsilon \sin\theta \int_{\theta-\epsilon+\theta_t}^{\epsilon-\theta+\theta_t} \sin^{n-2}\ \varphi\ d\varphi d\theta$$

then there is a $k-$codimensional subspace E such that for all $x \in E \cap S^{n-1}$, $|f(x) - t| \le \omega_f(\epsilon)$.

Proof. E is the space for which $E \cap S^{n-1} \subseteq (\partial A^t)_\epsilon$. Clearly if $x \in E \cap S^{n-1}$, $\exists y \in \partial A^t$ s.t. $\rho(x,y) \le \epsilon$, hence $|f(x) - f(y)| \le \omega_f(\epsilon)$, but $f(y) = t$. □

Remark. When $t = M_f$ is the median of f, $\theta_t = \pi/2$.

The following results relate to Theorem 1 (ii), which gave a condition that guaranteed the existence of a $k-$dimensional subspace $F \subset \mathbb{R}^n$ so that $0 \in \text{int}(P_F(co(A)))$ for any closed subset $A \subset S^{n-1}$. It is obvious that if A has positive measure then any neighborhood U of an interior point of A, can be projected onto an $n-1$ dimensional subspace F so that $0 \in \text{int}P_F(co(A))$, thus the condition is meaningless

μ	ϵ (in degrees)	values of n, k
0.1	84.34	$n = 4,\ k = 2$
0.2	73.77	$n = 4,\ k = 2$
0.3	84.42	$n = 3,\ k = 1$
0.3	86.16	$n = 4,\ k = 1$
0.3	67.01	$n = 4,\ k = 2$
0.4	78.02	$n = 3,\ k = 1$
0.4	80.66	$n = 4,\ k = 1$
0.4	62.10	$n = 4,\ k = 2$
0.5	73.02	$n = 3,\ k = 1$
0.5	76.35	$n = 4,\ k = 1$
0.5	58.31	$n = 4,\ k = 2$
0.6	69.00	$n = 3,\ k = 1$
0.6	72.88	$n = 4,\ k = 1$
0.6	55.28	$n = 4,\ k = 2$
0.7	65.74	$n = 3,\ k = 1$
0.7	70.06	$n = 4,\ k = 1$
0.7	52.82	$n = 4,\ k = 2$
0.8	63.11	$n = 3,\ k = 1$
0.8	50.81	$n = 4,\ k = 2$
0.9	49.22	$n = 4,\ k = 2$

if $\mu(A) > 0$. On the other hand, it is intuitively clear that if $\mu(A)$ is small, there would not be "too many" such subspaces F. To expand this point, let $\mathcal{G}_{n,k}$ denote the Grassman manifold of $k-$dimensional subspaces F of $\mathbb{R}^n$, and $\gamma_{n,k}$ be the Haar measure of $\mathcal{G}_{n,k}$. Denote by $\|T\|$ and $\|T\|_{HS}$ the norm, and the Hilbert-Schmidt norm respectively, of an operator $T = \sum_{i=1}^k \sum_{j=1}^n t_{ij} e_i \otimes e_j$ from $\mathbb{R}^k$ to $\mathbb{R}^n$. Since $\|T - T'\| \leq \|T - T'\|_{HS}$, it follows that the function $\|T\|$ has Lipschitz constant equal to 1. Define for a given closed subset $A \subset S^{n-1}$, $f(T) = \max_{\|y\|=1} \min_{x \in A} \langle Ty, x\rangle$, then clearly also $f(T)$ is a Lipschitz-1 function.

Let G be the Gaussian operator defined in the proof of Theorem 1 (ii). By results of [G], $\mathbb{E}\|G\| \leq a_n + a_k \leq \sqrt{n} + \sqrt{k}$($\frac{k}{\sqrt{k+1}} \leq a_k \leq \sqrt{k}$) and by Theorem 1 (ii), $\mathbb{E}f(G) \geq s - a_k$, where $s = s(A)$.

We now apply the well known inequality proved in [P1]: If $h(T)$ is a Lipschitz-1 function, then for all $\lambda > 0$, $P(h(G) - \mathbb{E}h(G) \geq \lambda) \leq \exp(-\lambda^2/2)$. Combining these facts we obtain:

THEOREM 7. *Let $A \subset S^{n-1}$ be a closed subset, $1 \leq k < n$, $0 < \delta < 1$, and $B(0,r) = \{x \in \mathbb{R}^n;\ \|x\| \leq r\}$. Let*

$$\mathcal{F}_A = \{F \in \mathcal{G}_{n,k}; F \cap B(0, (\frac{1-\delta}{1-\delta})(\frac{s - a_k}{a_k + a_n})) \subseteq P_F(\mathrm{co}(A))\} \ ,$$

then $\gamma_{n,k}(\mathcal{F}_A) \geq 1 - \exp(-\delta^2(s - a_k)^2/2) - \exp(-\delta^2(a_n + a_k)^2/2)$. In particular $\mathcal{F}_A$ has measure close to 1 if $\delta(s - a_k)$ is large.

Proof. Let $h(G) = \min_{y \in S^{k-1}} \max_{x \in A} \langle \frac{Gy}{\|Gy\|}, x \rangle$. Then

$$P(h(G) > (\frac{1-\delta}{1+\delta})(\frac{s - a_k}{a_k + a_n})) \geq P(h(G)\|G\| \geq (1-\delta)\mathbb{E}f(G)$$

and

$$\|G\| \leq (1+\delta)\mathbb{E}\|G\|)$$

$$\begin{aligned}
&\geq P(h(G)\|G\| \geq (1-\delta)\mathbb{E}f(G)) + P(\|G\| \leq (1+\delta)\mathbb{E}\|G\|) - 1 \\
&\geq P(f(G) \geq (1-\delta)\mathbb{E}f(G)) + P(\|G\| \leq (1+\delta)\mathbb{E}\|G\|) - 1 \\
&\geq 1 - \exp(-\delta^2(\mathbb{E}f(G))^2/2) - \exp(-\delta^2(E\|G\|)^2/2) \\
&\geq 1 - \exp(-(\delta(s - a_k))^2/2) - \exp(-(\delta(a_k + a_n))^2/2) .
\end{aligned}$$

Let $F_\omega = G_\omega(\mathbb{R}^n)$, then $\dim F_\omega = k$ a.e., and for every such F_ω which satisfies $h(G_\omega) \geq (\frac{1-\delta}{1+\delta})(\frac{s-a_k}{a_k+a_n})$ we have: Whenever $z \in F_\omega \cap S^{n-1}$, $\exists x \in A$ s.t. $\langle x, z \rangle \geq (\frac{1-\delta}{1+\delta})(\frac{s-a_k}{a_k+a_n})$; this implies that $F_\omega \cap B(0, (\frac{1-\delta}{1+\delta})(\frac{s-a_k}{a_k+a_n})) \subseteq P_{F_\omega}\mathrm{co}(A)$. Finally, the measure $\gamma_{n,k}$ on $\mathcal{G}_{n,k}$ can be identified with the measure P on $\{F_\omega; \omega \in \Omega\}$ (see Remarks (3.1) [G]). □

Remark. If $A = S^{n-1}$ or "close" to being the full sphere, then $s(A)$ is close to a_n, and so, if $n >> k$, $\frac{s-a_k}{a_k+a_n}$ is close to 1. This shows that the estimate $\gamma_{n,k}(\mathcal{F}_A)$ is rather precise in this case.

It may happen that $s(A) - a_k \leq 0$, and yet we are interested in expanding A to A_ϵ so that $s(A_\epsilon) - a_k > 0$. In this case the estimate for $s(A_\epsilon)$ of Lemma 3 (ii) applied to $\mathcal{F}_{A_\epsilon}$ in Theorem 7 becomes useful.

References

[FLM] T. Figiel, J. Lindenstrauss and V.D. Milman, *The dimension of almost spherical sections of convex bodies*, Acta Math. **139** ((1977)), 53–94.

[G] Y. Gordon, *On Milman's inequality and random subspaces which escape through a mesh in* $\mathbb{R}^n$,, GAFA, Israel Seminar 1986–87., Springer Lecture Notes 1317, 1988, pp. 84–106.

[GK] Y. Gordon and N.J. Kalton, *Local structure theory for quasi-normed spaces*, 1992 (preprint).

[GL] Y. Gordon and D.R. Lewis, *Dvoretzky's theorem for quasi-normed spaces*, Illinois J. Math. **35** (1991), 250–259.

[L] P. Lévy, *Probléms concrets d'Analyse fonctionnelle*, Gauthiers, Villars, Paris, 1951.

[M] V.D. Milman, *Almost Euclidean quotient spaces of subspaces of finite dimensional normed spaces*, Proc. Amer. Math. Soc. **94** (1985), 445–449.

[MS] V.D. Milman and G. Schechtman, *Asymptotic theory of finite-dimensional normed spaces*, Springer Lecture Notes 1200, 1986.

[P1] G. Pisier, *Probabilistic methods in the geometry of Banach spaces*, Springer Lecture Notes 1206, 1986, pp. 167–241.

[P2] G. Pisier, *The volume of convex bodies and Banach space geometry*, Cambridge University Press 94, 1989.

[TJ] N. Tomczak-Jaegermann, *Banach Mazur distances and finite dimensional operator ideals*, Pitman, 1988.

Department of Mathematics, Technion, Haifa 32000, Israel
E-mail address: mar2000@technion.technion.ac.il

Contemporary Mathematics
Volume **144**, 1993

THREE-SPACE PROBLEM FOR THE ASYMPTOTIC-NORMING PROPERTY OF BANACH SPACES

ZHIBAO HU AND BOR-LUH LIN

ABSTRACT. Let Y be a subspace of a Banach space X. If X/Y has the asymptotic-norming property I, then for $\kappa = $ I, II or III, X has the asymptotic-norming property κ if and only if Y has the asymptotic-norming property κ.

For a Banach space X, let $S(X)$ [resp. $B(X)$] be the unit sphere [resp. ball] of X. A subset Φ of $B(X^*)$ is called a norming set of X if $\|x\| = \sup_{x^* \in \Phi} x^*(x)$ for all x in X. A sequence $\{x_n\}$ in X is said to be asymptotically-normed by Φ if for every $\mathcal{E} > 0$, there is x^* in Φ and N in $\mathbb{N}$ such that $x^*(x_n) > \|x_n\| - \mathcal{E}$ for all $n \geq N$.

For $\kappa = $ I, II or III, a sequence $\{x_n\}$ in X is said to have the property κ if

I. $\{x_n\}$ is convergent;
II. $\{x_n\}$ has a convergent subsequence;
III. $\cap_{n=1}^{\infty} \overline{co}\{x_k : k \geq n) \neq \phi$.

Let Φ be a norming set of X. For $\kappa = $ I, II or III, X is said to have the asymptotic-norming property κ with respect to $\Phi(\Phi$-ANP-$\kappa)$ if every sequence in $S(X)$ that is asymptotically normed by Φ has the property κ. X is said to have the asymptotic-norming property κ(ANP-κ), $\kappa = $ I, II or III if there is an equivalent norm $\|\cdot\|$ on X such that there is a norming set Φ of $(X, \|\cdot\|)$ and $(X, \|\cdot\|)$ has the Φ-ANP-κ.

In this paper, we show that, for a Banach space X, suppose there is a subspace Y in X such that X/Y has the ANP-I, then for $\kappa = $ I, II or III, X has the ANP-κ if and only if Y has the ANP-κ. For results concerning the asymptotic-norming property, we refer to [JH, GM, HL1, HL2].

We follow the techniques that were used in [GTWZ] for three-space locally uniformly convex renorming of Banach spaces. The basic tool is the following Lemma 1 (see Lemma 1-1 and Remark 1-2 in the chapter on general techniques for renorming some non-separable Banach space of the forthcoming book by R. Deville, G. Godefroy and V. Zizler). The authors wish to thank R. Deville for sending them a preprint of the chapter.

Lemma 1. *Let X be a Banach space and let $\varphi_i : X^* \to \mathbb{R}^+$, $\psi_i : X^* \to \mathbb{R}^+$, $i \in I$, be convex, weak* lower-semicontinuous functions such that $\{\varphi_i\}_{i \in I}$ and $\{\psi_i\}_{i \in I}$ are uniformly bounded on bounded subsets of X^*. Denote, for each $x^* \in X^*$,*

1991 *Mathematics Subject Classification*. Primary 46B20.
Key words and phrases. Asymptotic-norming property, three-space problem.
This paper is in final form and no version of it will be submitted in publication elsewhere.

$$\varphi(x^*) = \sup_{i\in I} \varphi_i(x^*) \ ;$$
$$\theta_{i,k}(x^*) = \varphi_i^2(x^*) + \frac{1}{k}\psi_i^2(x^*) \ ;$$
$$\theta_k(x^*) = \sup_{i\in I} \theta_{i,k}(x^*) \ ;$$
$$\theta(x^*) = \|x^*\|^2 + \sum_{k=1}^{\infty} \frac{1}{2^{k+1}} \{\theta_k(x^*) + \theta_k(-x^*)\} \ .$$

Suppose $\theta(0) < 1$. Let $|||\cdot|||$ denote the Minkowski functional of $B = \{x^* : x^* \epsilon\ X^*, \theta(x^*) \leq 1\}$. Then $|||\cdot|||$ is an equivalent dual norm on X^* with the following properties:

If x_n^*, $x^* \epsilon X^*$, $n \epsilon \mathbb{N}$ satisfy $\lim_n |||\frac{x^*+x_n^*}{2}||| = 1$, then

(i) $$\lim_n \|x_n^*\| = \lim_n \|\frac{x^*+x_n^*}{2}\| = \|x^*\| \ .$$

Furthermore, there is a sequence $\{i_n\}$ in I such that

(ii) $$\lim_n \varphi_{i_n}(x^*) = \lim_n \varphi_{i_n}(x_n^*) = \lim_n \varphi_{i_n}(\frac{x^*+x_n^*}{2}) = \varphi(x^*) \ ;$$

and

(iii) $$\lim_n \{\frac{1}{2}\psi_{i_n}^2(x^*) + \frac{1}{2}\psi_{i_n}^2(x_n^*) - \psi_{i_n}^2(\frac{x^*+x_n^*}{2})\} = 0 \ .$$

Lemma 2. *Let Y be a subspace of a Banach space X. Then every equivalent norm on X/Y is the quotient norm on X/Y induced by some equivalent norm on X.*

Proof. Let $\|\cdot\|_1$ be an equivalent norm on X/Y. Then its dual norm $\|\cdot\|_1$ is an equivalent norm on $(X/Y)^* = Y^\perp \subset (X^*, \|\cdot\|)$. Let $r > 0$ such that $B_r(X^*, \|\cdot\|) \cap Y^\perp \subset B(Y^\perp, \|\cdot\|_1)$ where $B_r(X^*, \|\cdot\|) = \{x^* : x^* \epsilon\ X^*, \|x^*\| \leq r\}$. Let $B = w^*\text{-}\ell co[B_r(X^*, \|\cdot\|) \cup B(Y^\perp, \|\cdot\|_1)]$ and let $|||\cdot|||$ be the Minkowski functional of B. It is clear that $|||\cdot|||$ is an equivalent dual norm on X^*. Let $|||\cdot|||$ be the equivalent norm on X induced by $|||\cdot|||$ on X^*. It is easy to see that $\|\cdot\|_1$ is the quotient norm on X/Y induced by $|||\cdot|||$. □

We, now, prove our main theorem of the paper.

Theorem 3. *Let Y be a subspace of a Banach space X. Suppose X/Y has the asymptotic-norming property I, then for κ = I, II or III, X has the asymptotic-norming property κ if and only if Y has the asymptotic-norming property κ.*

Proof. We only need to show that if Y has the ANP-κ then X has the ANP$-\kappa$.

Step 1. Embedding X into some dual space Z^*.

Since X/Y has the ANP-I, by Corollary 2.8 in [HL1], there is an equivalent locally uniformly convex norm $\|\cdot\|_1$ on X/Y such that $(X/Y, \|\cdot\|_1)$ has the Φ_1-ANP-I for some Φ_1 in $B(Y^{\perp}, \|\cdot\|_1)$. By Lemma 2, we may assume that $\|\cdot\|_1$ is the quotient norm on X/Y induced by an equivalent norm $\|\cdot\|_1$ on X. For $\kappa =$ I, II or III, since Y has the ANP$-\kappa$, let $\|\cdot\|_2$ be an equivalent norm on Y such that Y has the $\Phi_2^{'}$-ANP-κ for some $\Phi_2^{'} \subset B(Y^*, \|\cdot\|_2)$. It is easy to see that every equivalent norm on a subspace can be extended to an equivalent norm of the whole space (see, *e.g.*, [JZ]). Thus, we may assume that $\|\cdot\|_2$ is an equivalent norm on X. Furthermore, by Lemma 3 in [HL2], we may assume that both Φ_1 and Φ_2 are circled subsets of $B(X^*, \|\cdot\|)$ where $\Phi_2 = \{x^* : x^* \epsilon X^*, x^*|_Y \epsilon \Phi_2^{'}$ and $\|x^*\|_2 = \|x^*|Y\|_2\}$. Let $|\cdot|_3$ be the Minkowski functional of $co(\Phi_1 \cup \Phi_2)$ and let Z be the completion of $([\Phi_1 \cup \Phi_2], |\cdot|_3)$ where $[\Phi_1 \cup \Phi_2]$ is the span of $\Phi_1 \cup \Phi_2$. Then Z is a Banach space. Let $T : X \to Z^*$ denotes the evaluation map, *i.e.*, $(Tx)(z) = z(x)$ for all z in Z. It is clear that $|Tx|_3 \leq \|x\|$ for all $x \epsilon X$. On the other hand, suppose $x \epsilon X$ and $\|x\|_2 = 1$. If $\|\bar{x}\|_2 \geq \frac{1}{3}$ in X/Y, then $|Tx|_3 \geq \sup_{x^* \epsilon \Phi_1} x^*(x) = \|\bar{x}\|_1 \geq K\|\bar{x}\|_2 \geq K/3$ for some constant K. If $\|\bar{x}\|_2 < \frac{1}{3}$, choose $y \epsilon Y$ such that $\|x - y\|_2 < \frac{1}{3}$. Then $\|y\|_2 \geq \|x\|_2 - \|x - y\|_2 > \frac{2}{3}$. Thus $|Tx|_3 \geq \sup_{x^* \epsilon \Phi_2} x^*(x) \geq \sup_{x^* \epsilon \Phi_2} x^*(y) - \|x - y\|_2 = \|y\|_2 - \|x - y\|_2 > \frac{1}{3}$. We conclude that T is an isomorphism from X into Z^*. Thus we may regard X, up to isomorphism, a subspace of Z^*.

Step 2. Renorming Z^*.

For $z^* \epsilon Z^*$, define $|z^*|_i = \sup_{z \epsilon \Phi_i} z^*(z)$, $i = 1, 2$. Then $|\cdot|_i$, $i = 1, 2$ are seminorms on Z^*, continuous, weak*-l.s.c, $\|\bar{x}\|_1 = |x|_1$, $x \epsilon X$ and $\|y\|_2 = |y|_2$, $y \epsilon Y$. Define

$$\|z^*\| = (\sum_{i=1}^{3} |z^*|_i^2)^{1/2}, z^* \epsilon Z^* .$$

Then $\|\cdot\|$ is an equivalent dual norm on Z^* with the property that for any sequence $\{z_n^*\}$ and z^* in Z^*,

$$(*) \qquad \lim_n \|z_n^*\| = \lim_n \|\frac{z^* + z_n^*}{2}\| = \|z^*\|$$
$$\Leftrightarrow \lim_n |z_n^*|_i = \lim_n |\frac{z^* + z_n^*}{2}|_i = |z^*|_i \ , i = 1, 2, 3 .$$

Let $B : X/Y \to X$ be the Bartle-Graves continuous selection map with respect to the quotient map from X onto X/Y, *i.e.*, $B(\bar{x}) \epsilon \bar{x}$, $B(\lambda \bar{x}) = \lambda B(\bar{x})$ and $\|B(\bar{x})\|_1 \leq m\|\bar{x}\|_1$ for some constant m, $x \epsilon X$. Let $I = \{\bar{x} : \bar{x} \epsilon X/Y, \|\bar{x}\|_1 = 2\}$. For each $\bar{a} \epsilon I$, since Φ_1 is circled, we may choose a fixed $f_{\bar{a}} \epsilon \Phi_1$ such that $f_{\bar{a}}(\bar{a}) = 1$. Set $P_{\bar{a}}(z^*) = f_{\bar{a}}(z^*)B(\bar{a})$, $z^* \epsilon Z^*$. Then $\|P_{\bar{a}}(z^*)\|_1 \leq 2m\|z^*\|$, $z^* \epsilon Z^*$. Hence $\{P_{\bar{a}}\}_{\bar{a} \epsilon I}$ is uniformly bounded on bounded subsets of Z^*. Define, for each $z^* \epsilon Z^*$,

$$\varphi_{\bar{a}}(z^*) = \frac{1}{2}|z^* + a|_1 \ ,$$

and

$$\psi_{\bar{a}}(z^*) = |z^* - P_{\bar{a}}(z*)|_2 \ .$$

It is easy to check that $\{\varphi_{\bar{a}}\}_{\bar{a}\epsilon I}$ and $\{\psi_{\bar{a}}\}_{\bar{a}\epsilon I}$ satisfy the hypothesis of Lemma 1. Let $|||\cdot|||$ be the equivalent dual norm on Z^* described in Lemma 1 and let $\Phi = \{z : z\epsilon Z, |||z||| \leq 1\}$. Since $|||z^*||| \geq \|z^*\| \geq |z^*|_3$, $z \epsilon Z^*$, it follows that $\Phi_1 \cup \Phi_2 \subset \Phi$.

Step 3. For $\kappa =$ II or III, then $(X, |||\cdot|||)$ has the Φ-ANP-κ.

Suppose $|||x_n||| = 1$, $n \epsilon \mathbb{N}$ in X and $\{x_n\}$ is asymptotically normed by Φ. We only need to show that $\{x_n\}$ has a (resp. weakly) convergent subsequence. Let z^* be a weak*-cluster point of $\{x_n\}$ in Z^*. Then $|||z^*||| = 1$. Since $\Phi \supset \Phi_1 \cup \Phi_2$, there is a subsequence $\{u_n\}$ of $\{x_n\}$ and a countable subset Φ_0 in Φ such that

(1) For all $z \epsilon \Phi_o$, $\lim_n u_n(z) = z^*(z)$;
(2) Φ_o is a norming set of $([u_n, z^*], |||\cdot|||)$;
(3) $\Phi_o \cap \Phi_1$ is a norming set of $([u_n, z^*], |\cdot|_1)$;
(4) $\Phi_o \cap \Phi_2$ is a norming set of $([u_n, z^*, z_1^*], |\cdot|_2)$ where $z_1^* = P_{\frac{2\bar{x}}{|x|_1}}(z^*)$ if there is an element x in X such that $|x|_1 \neq 0$ and $|x - z^*|_1 = 0$; otherwise, $z_1^* = 0$.

From (1) and (2), we have

$$\lim_n |||u_n||| = \lim_n |||\frac{u_n + z^*}{2}||| = |||z^*||| = 1 .$$

By Lemma 1 and (*), there is a sequence $\{\bar{a}_n\}$ in I such that,

(5) $\lim_n |u_n|_i = \lim_n |\frac{u_n+z^*}{2}|_i = |z^*|_i$, $i = 1, 2, 3$;
(6) $\lim_n |z^* + a_n|_1 = \lim_n |u_n + a_n|_1 = \lim_n |\frac{u_n+z^*}{2} + a_n|_1 = \sup_{\bar{a}\epsilon I} |z^* + a|_1$;
(7) $\lim_n \left\{ \frac{1}{2}|z^* - P_{\bar{a}_n}(z^*)|_2^2 + \frac{1}{2}|u_n - P_{\bar{a}_n}(u_n)|_2^2 - |\frac{z^*+u_n}{2} - P_{\bar{a}_n}(\frac{z^*+u_n}{2})|_2^2 \right\} = 0.$

Since $|x|_1 = \|\bar{x}\|_1$, $x\epsilon X$, from (5), we have $\lim_n \|\bar{u}_n\|_1 = |z^*|_1$. Given $\mathcal{E} > 0$, by (3), there is $z \epsilon \Phi_o \cap \Phi_1$ such that $z^*(z) > |z^*|_1 - \mathcal{E}$. Choose N such that $|z^*|_1 > \|\bar{u}_n\|_1 - \mathcal{E}$ and $u_n(z) > |z^*|_1 - \mathcal{E}$ for all $n \geq N$. Then $z(\bar{u}_n) > \|\bar{u}_n\|_1 - 2\mathcal{E}$, $n \geq N$. Hence $\{\bar{u}_n\}$ is asymptotically normed by Φ_1. Since $(X/Y, \|\cdot\|_1)$ has the Φ_1-ANP-I, there is $x\epsilon X$, $\lim_n |u_n - x|_1 = \lim_n \|\bar{u}_n - \bar{x}\|_1 = 0$. By (1) and (3), we have $|x - z^*|_1 = 0$.

Case 1. $\lim_n |u_n|_1 = 0$. Then $\lim_n \|\bar{u}_n\|_1 = 0$. Choose $y_n \epsilon Y$, $\lim_n \|y_n - u_n\|_1 = 0$. By (5), $\lim_n |u_n|_2 = |z^*|_2$. Using (1) and (4) and the argument above when we use (1) and (3) to show that $\{\bar{u}_n\}$ is asymptotically normed by Φ_1, we conclude that $(\{u_n\}, |\cdot|_2)$ is asymptotically normed by Φ_2. Since both $\|\cdot\|_i$, $i = 1, 2$ are equivalent norms on X, we have $\lim_n \|y_n - u_n\|_2 = 0$. Hence $(\{y_n\}, |\cdot|_2 = \|\cdot\|_2)$ is asymptotically normed by Φ_2 in Y. Since Y has the ANP-κ, $\kappa =$ II (resp. III), $\{y_n\}$ has a (resp. weakly) convergent subsequence. Hence $\{u_n\}$ and so $\{x_n\}$ has a (resp. weakly) convergent subsequence.

Case 2. $\lim_n |u_n|_1 \neq 0$. Then $|x|_1 \neq 0$. Let $a_o = \frac{2x}{|x|_1}$. Then $\bar{a}_o \epsilon I$ and $\sup_{\bar{a}\epsilon I} \|\bar{x} + \bar{a}\|_1 = \|\bar{x} + \bar{a}_o\|_1$. By (6),

$$\begin{aligned}\lim_n \|\bar{x} + \bar{a}_n\|_1 &= \lim_n |x + a_n|_1 = \lim_n |z^* + a_n|_1 = \sup_{\bar{a}\epsilon I} |z^* + a|_1 \\ &= \sup_{\bar{a}\epsilon I} |x + a|_1 = \sup_{\bar{a}\epsilon I} \|\bar{x} + \bar{a}\|_1 = \|\bar{x} + \bar{a}_o\|_1 .\end{aligned}$$

Since $(X/Y, \|\cdot\|_1)$ is locally uniformly convex and $\bar{x} \neq o$, hence $\lim_n \|\bar{a}_n - \bar{a}_o\|_1 = 0$. Thus $\lim_n \|B(\bar{a}_n) - B(\bar{a}_0)\|_1 = 0$ and $\lim_n f_{\bar{a}_n}(\bar{a}_o) = \lim_n f_{\bar{a}_n}(\bar{a}_n) = 1$. It follows that $\lim_n f_{\bar{a}_n}(u_n) = \lim_n f_{\bar{a}_n}(x) = \frac{|x|_1}{2} \lim_n f_{\bar{a}_n}(a_o) = \frac{|x|_1}{2} = f_{\bar{a}_o}(x)$. So we have

$$(8) \qquad \lim_n P_{\bar{a}_n}(u_n) = \lim_n P_{\bar{a}_n}(x) = P_{\bar{a}_o}(x) = f_{\bar{a}_o}(x) B(\bar{a}_o) = \frac{|x|_1}{2} B(\bar{a}_o)$$
$$= B(\frac{|x|_1}{2}\bar{a}_o) = B(\bar{x}) \epsilon \bar{x} \ .$$

Hence

$$\overline{\lim_n} |u_n - P_{\bar{a}_n}(\bar{u}_n)|_1 \leq \lim_n \{|u_n - x|_1 + |x - P_{\bar{a}_o}(x)|_1 + |P_{\bar{a}_o}(x) - P_{\bar{a}_n}(u_n)|_1\} = 0 \ .$$

Thus

$$(i) \qquad \lim_n |u_n - P_{\bar{a}_n}(u_n)|_1 = 0 \ .$$

Also, for any $z \in \Phi_o$, by (1) and (8), we have

$$(ii) \qquad \lim_n [u_n - P_{\bar{a}_n}(u_n)](z) = z^*(z) - z[P_{\bar{a}_o}(x)] = [z^* - P_{\bar{a}_o}(z^*)](z).$$

By (7), we have

$$(iii) \qquad \lim_n |u_n - P_{\bar{a}_n}(u_n)|_2 = \lim_n |z^* - P_{\bar{a}_n}(z^*)|_2 = |z^* - P_{\bar{a}_o}(z^*)|_2 \ .$$

Consider the sequence $\{u_n - P_{\bar{a}_n}(u_n)\}$ instead of $\{u_n\}$, as in case 1 and by using (i), (ii), (iii) and (4), we conclude that $\{u_n - P_{\bar{a}_n}(u_n)\}$ has a (resp. weakly) convergent subsequence. By (8), it follows that $\{u_n\}$, hence $\{x_n\}$, has a (resp. weakly) convergent subsequence if Y has the ANP-II (resp. III).

Step 4. X has the ANP-I when both X/Y and Y have the ANP-I.

In this case, both X/Y and Y have equivalent locally uniformly convex norm. Hence X has an equivalent locally uniformly convex norm [GTWZ]. Furthermore, from the above, X has the ANP-II. By Corollary 2.8 of [HL1], we conclude that X has the ANP-I. □

We do not know for $\kappa = $ II or III, if both Y and X/Y have the ANP-κ, does X also have the ANP-κ. However, we have the following result. Before we state the next theorem, let us remark that since every reflexive space admits an equivalent locally uniformly convex norm and it is clear that every reflexive space has the ANP- III, hence every reflexive space has the ANP-I [HL1].

Theorem 4. *Let X be a Banach space. If there is a reflexive subspace Y in X such that X/Y has the ANP-III then X has the ANP-III.*

Proof. Let $|\cdot|$ be an equivalent norm on X/Y and let $\Phi_1 \subset B(Y^\perp, |\cdot|)$ be a norming set of X/Y such that X/Y has the Φ_1-ANP-III. We may assume that $0 \in \Phi_1[HL2]$. Renorm X by defining

$$|||x||| = (|\bar{x}|^2 + \|x\|^2)^{1/2} \ , x \epsilon X \ .$$

Let $\Phi_2 = B(X^*, \|\cdot\|)$ and $\Phi = \{\lambda_1 x_1^* + \lambda_2 x_2^* : x_i^* \epsilon \Phi_i,\ \lambda_i \geq 0,\ i = 1, 2$ and $\lambda_1^2 + \lambda_2^2 = 1\}$. It is clear that Φ is a norming set of $(X, |||\cdot|||)$. We claim that X has the Φ-ANP-III.

Let $x_n \epsilon X$, $|||x_n||| = 1$, $n \ \epsilon \mathbb{N}$ be a sequence asymptotically normed by Φ. Then $\lim_{n,m} |||\frac{x_n + x_m}{2}||| = 1$. Hence $\lim_n \|x_n\| = b$ and $\lim_n |\bar{x}_n| = a$ for some $a, b \geq 0$, $a^2 + b^2 = 1$. We claim that $\{\bar{x}_n\}$ is asymptotically normed by Φ_1 in $(X/Y, |\cdot|)$.

If $a = 0$, since $0 \epsilon \Phi$, it is clear that the claim is true. Suppose the claim is false. Then $a > 0$ and there exists $0 < \delta < a$ such that for every $\varphi \epsilon \Phi_1$, and every $k \epsilon \mathbb{N}$, there is $n_k \geq k$ such that $\varphi(x_{n_k}) < |\bar{x}_{n_k}| - \delta$. Since $\{x_n\}$ is asymptotically normed by Φ in $(X, |||\cdot|||)$, for every $\mathcal{E} > 0$, there exist $\lambda_1\varphi_1 + \lambda_2\varphi_2$ in X and N such that $\lambda_1\varphi_1(x_n) + \lambda_2\varphi_2(x_n) > 1 - \mathcal{E}$, $n \geq N$. For this $\varphi_1 \epsilon \Phi_1$ and for each $k \geq N$, choose $n_k \epsilon \mathbb{N}$ such that $\varphi_1(x_{n_k}) < |\bar{x}_{n_k}| - \delta$. Then

$$\begin{aligned}[(|\bar{x}_{n_k}| - \delta)^2 + \|x_{n_k}\|^2]^{1/2} &\geq \lambda_1(|\bar{x}_{n_k}| - \delta) + \lambda_2\|x_{n_k}\| \\ &\geq \lambda_1\varphi_1(x_{n_k}) + \lambda_2\varphi_2(x_{n_k}) > 1 - \mathcal{E}\ .\end{aligned}$$

Hence $[(a - \delta)^2 + b^2]^{1/2} \geq 1 - \mathcal{E}$ for all $\mathcal{E} > 0$. But then $(a - \delta)^2 + b^2 \geq 1$ which is impossible. This completes the proof that $\{\bar{x}_n\}$ is asymptotically normed by Φ_1. Since X/Y has the Φ_1-ANP-III, hence $\cap_{n=1}^{\infty} \overline{co}\{\bar{x}_k : k \geq n\} \neq \phi$. But Y is reflexive, it follows that $\cap_{n=1}^{\infty} \overline{co}\{x_k : k \geq n\} \neq \phi$. Thus X has the ANP-III. □

References

[B] R. D. Bourgin, *Geometric aspect of convex sets with the Radon-Nikodym property*, Lecture Notes in Math, Springer-Verlag, 1983, pp. 993.

[DGZ] R. Deville, G. Godefroy and V. Zizler, *a forthcoming book on renorming of Banach spaces.*.

[GM] N. Ghoussaub and B. Maurey, *The asymptotic-norming and the Radon-Nikodym properties are equivalent in separable Banach spaces*, Proc. Amer. Math. Soc. **94** (1985), 665-671.

[GTWZ] G. Godefroy, S. Troyanski, J. Whitefield and V. Zizler, *Three-space problem for locally uniformly retund renorming of Banach spaces*, Proc. Amer. Math. Soc. **94** (1985), 647-652.

[HL1] Zhibao Hu and Bor-Luh Lin, *On the asymptotic-norming property of Banach spaces*, Function Spaces, Lecture Notes in Pure and Applied Math, 136, Marcel Dekker, 1991, pp. 195–210.

[HL2] Zhibao Hu and Bor-Luh Lin, *Asymptotic-norming property and Mazur intersection properites in Lebesque-Bochner function spaces* [Submitted].

[JH] R. C. James and A. Ho, *The asymptotic-norming and the Radon-Nikodym properties for Banach spaces*, Ark. Mat. **19** (1981), 53-70.

[JZ] K. John and V. Zizler, *On extension of rotund norms II*, Pacific J. Math. **82** (1979), 451-455.

Zhibao Hu, Department of Mathematics and Statistics, Miami University, Oxford, OH 45056

Bor-Luh Lin, Department of Mathematics, The University of Iowa, Iowa City, IA 52242

Contemporary Mathematics
Volume **144**, 1993

ISOMETRIES OF BLOCH SPACES

K. JAROSZ

ABSTRACT. Let Ω_1 and Ω_2 be domains in $\mathbf{C}^n$ and let $\mathcal{A}$ and $\mathcal{B}$ be associated Bloch spaces. In this note we discuss the question whether any surjective isometry $T : \mathcal{A} \to \mathcal{B}$ is of the form

$$Tf = \lambda(f \circ \phi - f(\phi(0))),$$

where $\phi : \Omega_2 \to \Omega_1$ is a biholomorphic map, and $|\lambda| = 1$. We also introduce a concept of an abstract Bloch space associated with an arbitrary function algebra, and investigate isometries between such spaces.

1. INTRODUCTION

Isometries of various spaces of analytic functions have been studied by a number of authors. At first great deal of work has been devoted to Banach spaces of analytic functions equipped with the sup norm or with an L^p-norm.

The investigation of the sup norm Banach algebras of analytic functions has been facilitated by the general theory of uniform algebras and it is now very well known that in all 'resonable cases' any surjective isometry is given by

$$(*) \qquad Tf = \lambda f \circ \phi.$$

This can be extended to isometries of certain spaces of vector-valued analytic functions with sup norm [2,9]. The isometries of Hardy spaces were studied by de Leeuw-Rudin-Wermer [8], and lately, the vector-valued case has been investigated in [2]. In more recent years the interest switched to Banach spaces of analytic functions equipped with norms more closely related to the analytic structure of the underlying domains in $\mathbf{C}^n$, like the Bloch spaces. For a unit disc $\mathbf{D}$ we define the big Bloch space $\mathcal{B}(\mathbf{D})$ by

$$\mathcal{B}(\mathbf{D}) = \{f \in \mathrm{Hol}(\mathbf{D}) : \sup_{|z|<1} (1 - |z|^2)|f'(z)| = \|f\| < \infty,\ f(0) = 0\},$$

and we define the small Bloch space $\mathcal{B}_0(\mathbf{D})$ by

$$\mathcal{B}_0(\mathbf{D}) = \{f \in \mathcal{B}(\mathbf{D}) : \lim_{|z|\to 1} (1 - |z|^2)|f'(z)| = 0\}.$$

The latter space is equal to the closure, in $\mathcal{B}(\mathbf{D})$, of the algebra of all polynomials. It is also well known that $\mathcal{B}(\mathbf{D})$ can be identified with the bidual of $\mathcal{B}_0(\mathbf{D})$. In 1980 Cima and Wogen [4] proved that any surjective isometry of a small, as well as of a big Bloch space is canonical, that is of the form

$$(**) \qquad Tf = \lambda(f \circ \phi - f(\phi(0))).$$

1991 *Mathematics Subject Classification.* 46E15, 46H25, 46J15.

[1]Partially supported by a grant from the National Science Foundation.

In general the Bloch space $\mathcal{B}(\Omega)$ on a domain Ω in $\mathbf{C}^n$ containing the origin is defined by

$$\mathcal{B}(\Omega) = \{f \in \mathrm{Hol}(\Omega) : \sup \frac{|D_{\xi,z}f|}{H(z,\xi)} = \|f\| < \infty,\ f(0) = 0\},$$

where

$$D_{\xi,z}f = \sum_{j=1}^{n} \xi_j \frac{\partial f(z)}{\partial z_j}, \tag{1}$$

and $H(z,\xi)$ is the length of the vector $\xi \in T_z(\Omega)$ in the Kobayashi metric. The small Bloch space $\mathcal{B}_0(\Omega)$ is defined by

$$\mathcal{B}_0(\Omega) = \{f \in \mathcal{B}(\Omega) : \lim_{z\to\partial\Omega} \sup \frac{|D_{\xi,z}f|}{H(z,\xi)} = 0\}.$$

As in the classical, one dimensional case, $\mathcal{B}_0(\Omega)$ is the closed span of the algebra of all polynomials. In [7] Krantz and Ma proved that any linear surjective isometry of $\mathcal{B}_0(\mathbf{B}^n)$, where $\mathbf{B}^n$ is a unit ball in $\mathbf{C}^n$, is canoni.For general properties of the Bloch spaces see [1,6,10].

The explicit form of $H(z,\xi)$ is known only for some simple domains, and is fairly complicated even for the unit ball [6]. The standard definition of $H(z,\xi)$ is formulated in the language of differential geometry and it is difficult to use it in the context of isometries of Bloch spaces. The following is an equivalent definition which seems to be much better suited for our purpose:

$$H(z,\xi) = \|D_{\xi,z}\|,$$

where the differential map $D_{\xi,z} : H^\infty(\Omega) \to \mathbf{C}$ is defined by (1) and $H^\infty(\Omega)$ is the Banach algebra of all bounded analytic functions on Ω, equipped with the sup norm. It follows directly from the definition that $H(\cdot,\cdot)$ is biholomorphically invariant, that is

$$H(\phi(z), \frac{\phi_*(z)(\xi)}{|\phi_*(z)(\xi)|}) = H(z,\xi) \tag{2}$$

for any automorphism $\phi : \Omega \to \Omega$. Hence for any automorphism $\phi : \Omega \to \Omega$ the canonical map $f \to f\circ\phi - f(\phi(0))$ is an isometry of the Bloch space. The question is whether all the isometries are of this form. We will consider this question in more general setting i.e. for abstract Bloch spaces; this approach apparently simplifies som proofs as we concentrate on the essential ideas rather than on the specific form of the Kobayashi length $H(z,\xi)$. Some ideas used in the proofs are adopted from [4] and [7].

2. Abstract Bloch spaces

Let X be a compact Hausdorff space and let A be a uniform algebra on X, that is an algebra of continuous functions defined on X, such that A contains the constant functions, separates the points of X, and is equipped with the sup norm.

For an x in X, a linear functional D on A is called a derivation at x if

$$D(fg) = f(x)D(g) + D(f)g(x),$$

for $f, g \in A$. Let $\mathcal{D}_x$ be the set of all norm one derivations on A at x, and put $\mathcal{D} = \bigcup \mathcal{D}_x$. We define a seminorm $\|\cdot\|$ on A by $\|f\| = \sup\{|\mathcal{D}(f)| : D \in \mathcal{D}\}$, we define the small Bloch space $\mathcal{B}_0(A)$ as the completion of the quotient space $A/ker\|\cdot\|$. We note that for any $x \in \mathcal{C}hA$, the Choquet boundary of A, we have $\lim D_\alpha(f) = 0$ whenever $D_\alpha \in \mathcal{D}_{x_\alpha}$ and $x_\alpha \to x$. Hence, if we put $A = A(\Omega)$, the above definition reduces to the classical one, given in the previous section. The only difference is that now an element of the Bloch space is an equivalence class of functions wheras in the classical case we imposed the condition $f(0) = 0$ to select a representative of the class. Hence now $(**)$ has even simpler form $(*)$. We define the big Bloch space $\mathcal{B}(A)$ as the bidual of $\mathcal{B}_0(A)$.

We notice that for some function algebras, e.g. the algebra $C(X)$ of all continuous functions on X, there is no non-trivial point derivation so the associated Bloch spaces are trivial; indeed, the existence of a non-trivial point derivation is related to the existence of an analytic structure on the spectrum of the algebra [5]. It is clear from the definition that for any function algebras A and B, any algebra isomorphism $T : A \to B$, and any complex number μ with $|\mu| = 1$, the map μT is an isometry of A onto B as well as of $\mathcal{B}(A)$ onto $\mathcal{B}(B)$, and of $\mathcal{B}_0(A)$ onto $\mathcal{B}_0(B)$. We would like to know for what classes of function algebras any isometry of the corresponding Bloch spaces must be of this form.

We first consider a simple example of an abstract Bloch space. Let $A(\mathbf{D})$ be the disc algebra, that is the algebra of all continuous functions defined on the closed unit disc $\overline{\mathbf{D}}$ and analytic on $\mathbf{D}$. Let K be a compact Hausdorff space, and let A be equal to $C(K)\check{\otimes}A(\mathbf{D})$ the injective tensor product of $A(\mathbf{D})$ and $C(K)$. We have

$$A = \{f \in C(K \times \mathbf{D}) : \forall k \in K, f(k, \cdot) \in A(\mathbf{D})\}.$$

The algebra A has quite simple analytic structure: at any point of $K \times \mathbf{D}$ there is only one point derivation, up to a multiplicative constant, and there is no point derivation at any point of $K \times \partial\mathbf{D}$. On the other hand the structure of $K \times \overline{\mathbf{D}}$, the maximal ideal sp of A, may be more involved as K is an arbitrary compact set. In spite of this we will prove in the next section that an isometry of $\mathcal{B}_0(A)$ onto itself is given by an algebra automorphism of A.

Proposition. *Let K be a compact Hausdorff space and $A = C(K)\check{\otimes}A(\mathbf{D})$. Then any isometry $T : \mathcal{B}_0(A) \to \mathcal{B}_0(A)$ is given by an algebra automorphism of A.*

Conjecture. *Let Ω_1,Ω_2 be bounded domains of holomorphy in $\mathbf{C}^n$. Then:*

(i) *Ω_1, Ω_2 are biholomorphic if and only if $\mathcal{B}_0(A(\Omega_1))$ and $\mathcal{B}_0(A(\Omega_2))$ are isometric;*

(ii) *Any isometry* $T : \mathcal{B}_0(A(\Omega_1)) \to \mathcal{B}_0(A(\Omega_2))$ *is of the form*

$$Tf = \lambda(f \circ \phi - f(\phi(w_0))), \tag{3}$$

for $f \in \mathcal{B}_0(A(\Omega_1))$, *where* $\phi : \Omega_2 \to \Omega_1$ *is an analytic homeomorphism,* $|\lambda| = 1$, *and* w_0 *is a fixed point in* Ω_2.

We were unable to prove the above conjecture in full generality, in the next section we will prove the following special case.

Theorem. *Let* Ω_1, Ω_2 *be totally homogeneous domains in* $\mathbf{C}^n$. *Then any isometry* T *of* $\mathcal{B}_0(A(\Omega_1))$ *onto* $\mathcal{B}_0(A(\Omega_2))$ *is of the form (3).*

We call Ω totally homogeneous if it is a bounded domain of holomorphy, and if for any $z_1, z_2 \in \Omega$, and $D_1 \in \mathcal{D}_{z_1}$, $D_2 \in \mathcal{D}_{z_2}$ there is an automorphism $\phi : \Omega \to \Omega$ such that $\phi(z_1) = z_2$ and $\phi_*(D_2) = D_1$.

3. Proofs

We first make several general observations about isometries of arbitrary Bloch spaces, we will then use them to prove the Example and the Theorem stated in the previous section.

Let E be a Banach space and let K be a subset of the dual space E^*. Assume that for any e in E we have

$$\|e\| = \sup\{|e^*(e)| : e^* \in K\}.$$

Then $\mathrm{ext}(E^*)$, the set of extreme points of the unit ball in E^*, is contained in $\{\mu e^* : |\mu| = 1, e^* \in \overline{K}\}$, where the closure is taken in the weak $*$ topology. Moreover the map from E into $C(\overline{K})$, defined by $\tilde{e}(k) = k(e)$ for $k \in K$, is an isometric embedding.

Let now A and B be function algebras respectively on X and Y, $\mathcal{B}_0(A)$ and $\mathcal{B}_0(B)$ their associated Bloch spaces, $\mathcal{D}_A$ and $\mathcal{D}_B$ the sets of all point derivations with norm one, and let X_0 (Y_0), be the subset of X (Y) such that there exists a non-zero point derivation at any point of X_0 (Y_0). Since a weak $*$ limit of a net of point derivations is a point derivation, the weak $*$ closure of $\mathcal{D}_A$ is contained in the set of all point derivations of norm not greater than one. Hence the map

$$\mathcal{B}_0(A) \ni f \to \tilde{f} \in C(\overline{\mathcal{D}_A}) : \tilde{f}(D) = D(f),$$

is an isometric embedding and $\mathrm{ext}(\mathcal{B}_0(A)^*) \subseteq \mathcal{D}_A$.

Assume that any norm one point derivation on A or B is an extreme point of the unit ball in the dual space of the corresponding Bloch space. It follows that any isometry $T : \mathcal{B}_0(A) \to \mathcal{B}_0(B)$ is of the form,

$$D(Tf) = \phi(D)(f), \tag{4}$$

for $f \in \mathcal{B}_0(A)$ and $D \in \mathcal{D}_B$, where $\phi : \mathcal{D}_B \to \mathcal{D}_A$ is a homeomorphism. Let $D_1, D_2 \in \mathcal{D}_A$ then $\frac{D_1+D_2}{\|D_1+D_2\|} \in \mathcal{D}_A$ if and only if the derivations D_1 and D_2 correspond to the same point of X. Hence ϕ maps point derivations corresponding to the same point of Y onto derivations corresponding to the same point of X. So ϕ produces a homeomorphism ϕ_0 from Y_0 onto X_0.

The strategy to prove that any isometry between two given Bloch spaces is canonical, i.e. is given by an algebra isomorphism, is to show first that any point derivation is an extreme point, and then to use (4). Since both $\mathrm{ext}(\mathcal{B}_0(A)^*)$ and $\mathcal{D}_A$ are preserved by isometries and $\mathrm{ext}(\mathcal{B}_0(A)^*)$ is non-empty, in order to show that these sets are e it is enough to show that A has sufficiently many isometries.

Proof of the Proposition. Fix k in K. Let f be any norm one element of the classical Bloch space $\mathcal{B}_0(\mathbf{D})$. Let g_α be a net of continuous, norm one functions on X convergent uniformly to zero off any neighborhood of k and such that $g_\alpha(k) = 1$. A net f_α of elements of $\mathcal{B}_0(A)$ defined by $f_\alpha(k, z) = f(z)g_\alpha(k)$ is convergent to zero uniformly off any neighborhood of $\{k\} \times \overline{\mathbf{D}}$ so $\{k\} \times \mathbf{D}$ contains at least one element of $\mathrm{ext}(\mathcal{B}_0(\mathbf{D})^*)$. Now fix z and w in $\mathbf{D}$. Let ψ be an analytic automorphism of $\mathbf{D}$ such that $\psi(w) = z$. We define an automorphism φ of $K \times \overline{\mathbf{D}}$ by $\varphi(k, \omega) = \psi(\omega)$. Then $f \to f \circ \varphi$ is an isometry of A onto itself. Hence $\mathrm{ext}(\mathcal{B}_0(A)^*) = \mathcal{D}_A$ (one could prove the last equation more directly by constructing specific functions from $\mathcal{B}_0(A)$ peaking at a given point of $\mathcal{D}_A$, but we chose different arguments to show that the result can be extended to other function algebras with big group of isometries).

It follows that for any isometry $T : \mathcal{B}_0(A) \to \mathcal{B}_0(A)$ we have

$$\frac{\frac{\partial}{\partial w}(Tf)(k, w)}{H(k, w)} = \kappa(k, w)\frac{\frac{\partial}{\partial w'}f(k', w')}{H(k', w')}$$

for $k \in K, w \in \mathbf{D}$, where $|\kappa| \equiv 1$, $H(k, w)$ is the norm of the derivation at $(k, w) \in K \times \mathbf{D}$ and the map $\phi(k, w) = (k', w')$ is an automorphism of $K \times \mathbf{D}$.

Let k_1 and k_2 be distinct elements of K, and w_1, w_2 be elements of $\mathbf{D}$, then the norm distance between the derivation at (k_1, w_1) and at (k_2, w_2) is equal to 2. On the other hand for any fixed $k \in K$ the weak $*$ topology and the norm topology on $\{k\} \times \mathbf{D}$ coincide, therfore the sets of the form $\{k\} \times \mathbf{D}$, $k \in K$ are precisely the connected components of $K \times \mathbf{D}$ with respect to the norm topology, so they are preserved by T^*. Hence the map ϕ is of the following form

$$\phi(k, w) = (\phi_1(k), \phi_2(k, w)),$$

where, for any $k \in K$, $\phi_2(k, \cdot)$ is an automorphism of the unit dis So for any $k \in K$, T produces an isometry from $\mathcal{B}_0(\mathbf{D}) \equiv \mathcal{B}_0(A_{|\{\phi(k)\}\times\mathbf{D}})$ onto $\mathcal{B}_0(\mathbf{D}) \equiv \mathcal{B}_0(A_{|\{k\}\times\mathbf{D}})$. Since any isometry of the classical Bloch space is canonical [4] $\phi_2(k, \cdot)$ is analytic, $\kappa(k, \cdot)$ is constant, and

$$Tf(k, w) = \kappa(k)f(\phi_1(k), \phi_k(w)), \tag{5}$$

for all $f \in \mathcal{B}_0(A)$, where $|\kappa| \equiv 1$, ϕ_1 is an automorphism of K, and ϕ_k is an analytic automorphism of the unit disc for $k \in K$. Note that in (5) we use the notation of abstract Bloch spaces, that is we identify two functions which have the same derivatives at any point.

Proof of the Theorem. Let $A = A(\Omega_1)$, $B = A(\Omega_2)$, $\Gamma = \{z \in \mathbf{C}^n : \|z\| = 1\}$, and let $T : \mathcal{B}_0(A) \to \mathcal{B}_0(B)$ be an isometry. Assume that Ω_1 is totally homogenous, this means that for any $D_1, D_2 \in \mathcal{D}_A$ there is an isometry S of A onto itself such that $D_1 = D_2 \circ S$. Hence $\mathcal{D}_A = \mathrm{ext}(\mathcal{B}_0(A)^*) \cong \Omega_1 \times \Gamma$ (this is the only point where we use the assumption that Ω_1 is totally homogenous). Let us also assume that Ω_2 is totally homogenous. By what we said at the beginning of this section T is of the following form

$$\frac{D_\xi(Tf)(z)}{H(z,\xi)} = \frac{D_{\phi_2(z,\xi)} f(\phi_1(z))}{H(\phi_1(z), \phi_2(z,\xi))}, \tag{6}$$

for $f \in \mathcal{B}_0(A)$, where ϕ_1 is a homeomorphism of Ω_2 onto Ω_1, $D_\xi(f)(z)$ is the derivative of f at the point z in the direction ξ, and $H(z,\xi)$ is the norm of the map $D_\xi(\cdot)(z)$.

Hence for f^2 we get

$$\frac{D_\xi(Tf^2)(z)}{H(z,\xi)} = \frac{D_{\phi_2(z,\xi)} f^2(\phi_1(z))}{H(\phi_1(z), \phi_2(z,\xi))} = 2f(\phi_1(z)) \frac{D_{\phi_2(z,\xi)} f(\phi_1(z))}{H(\phi_1(z), \phi_2(z,\xi))}. \tag{7}$$

Dividing (7) by (6) we get

$$f \circ \phi_1(z) = \frac{D_\xi(Tf^2)(z)}{2D_\xi(Tf)(z)}$$

for all f, z and ξ such that the terms involved in (6) and (7) are not equal to zero. Hence $\phi_1 : \Omega_2 \to \Omega_1$ is an analytic map.

Composing T with the map $f \to f \circ \phi_1^{-1}$ we can assume, without loss of generality, that $\Omega_1 = \Omega_2 = \Omega$, and $\phi_1 = \mathrm{id}_\Omega$, so that T is of the following form

$$\frac{D_\xi(Tf)(z)}{H(z,\xi)} = \frac{D_{\phi_2(z,\xi)} f(z)}{H(z, \phi_2(z,\xi))},$$

for $f \in \mathcal{B}_0(A)$.

We now use an argument adopted from [7]. Let $f \in B_0(A)$ and put $h = T(f^2) - 2fTf$. Since $D_\xi \circ T(\cdot)(z)$ is a derivation at z we get

$$D_\xi(h)(z) = 2f(z) D_\xi \circ T(f) - 2f(z) D_\xi(Tf) - 2D_\xi(f)Tf = -2D_\xi(f)Tf(z).$$

Hence, for $f(z) = z_1$ we get

$$D_{e_1}(h) = -2Tf \text{ and } D_{e_j}(h) = 0, j = 2, ..., n.$$

So h, as well as Tf are functions of z_1 alone. By a similar argument Tf^m is a function of z_1 alone for each positive integer m.

Thus the subspace of $\mathcal{B}_0(A)$ consisting of functions depending on the first variable is mapped onto itself. Hence T^* maps the set of derivations, that anihilate this subspace, onto itself. The same is true for any $j = 1, ..., n$. Hence $\phi_2(z,\xi) = \mu_z(\xi)\xi$, where $|\mu| = 1$ and since $\phi_2(z,\cdot)$ is linear, we get

$$\phi_2(z,\xi) = \mu_z \xi,$$

for $z \in \Omega$, $\xi \in \Gamma$, where $|\mu_z| = 1$. Hence

$$D_\xi(Tf)(z) = \mu_z D_\xi(f),$$

for $f \in \mathcal{B}_0(A)$; therefore $z \to \mu_z$ is an analytic map of absolute value 1, so it is constant. We conclude that $Tf = \mu f$.

REFERENCES

1. J. M. Anderson, J. G. Clunie and Ch. Pommerenkee, *On Bloch functions and normal functions*, J. Reine Angew. Math. **270** (1974), 12–37.
2. M. Cambern and K. Jarosz, *The isometries of $H^1_{\mathcal{H}}$*, Proc. Amer. Math. Soc. **107** (1989), 205–214.
3. M. Cambern and K. Jarosz, *Multipliers and isometries in H^∞_E*, Bull. London Math. Soc. **22** (1990), 463–466.
4. J. A. Cima and W. R. Wogen, *On isometries of Bloch space*, Illinois J. Math. **24** (1980), 313–316.
5. T. W. Gamelin, *Uniform Algebras*, Chelsea Pub. Company, New York, 1984.
6. S. G. Krantz, *Function Theory of Several Complex Variables*, Pure and Applied Mathematics, A Wiley-Interscience Publication, 1982.
7. S. G. Krantz and D. Ma, *On isometric isomorphisms of the Bloch spaces on the unit ball of* $\mathbf{C}^n$, Michigan Math. J. **36** (1989), 173–180.
8. K. de Leeuw, W. Rudin and J. Wermer, *Isometries of some function spaces*, Proc. Amer. Math. Soc. **11** (1960), 694–698.
9. P-K. Lin, *The isometries of $H^\infty(E)$*, Pacific J. Math. **143** (1990), 69-77.
10. R. M. Timoney, *Bloch functions in several complex variables, I*, Bull. London Math. Soc. **12** (1980), 241–267.

DEPARTMENT OF MATHEMATICS AND STATISTICS,
SOUTHERN ILLINOIS UNIVERSITY AT EDWARDSVILLE, EDWARDSVILLE, ILLINOIS 62025

Electronic address: CJ01@SIUEMUS.BITNET

Contemporary Mathematics
Volume **144**, 1993

PROPERTY (β) AND EDELSTEIN'S ALGORITHMS FOR CONSTRUCTING NEAREST AND FARTHEST POINTS

W. A. KIRK

ABSTRACT. Algorithms for constructing points which have nearest and farthest points in closed subsets of uniformly convex Banach spaces are implicit in papers of Edelstein in 1966 and 1968. It is observed here that a condition introduced by Rolewicz in 1987, called property (β), is sufficient for Edelstein's procedure to yield existence of such points. It is also shown that sufficiently smooth Banach spaces always possess property (β).

1. Introduction

In this paper we study a geometric property of Banach spaces introduced by Rolewicz [R2] which lies between uniform convexity and nearly uniform convexity, and in particular how it relates to Edelstein's algorithms for constructing points which have nearest and farthest points in a given closed subset of the space.

Throughout, X will denote a Banach space, B the unit ball of X, and S the unit sphere $(\{x \in X : \|x\| = 1\})$. Recall that X is said to be *nearly uniformly convex* (NUC) (Huff [*H*]) if for each $\epsilon > 0$ there exists $\delta \in (0,1)$ such that for any sequence $\{x_n\} \subset B$ with

$$\text{sep}(x_n) = \inf\{\|x_n - x_n\| : n \neq m\} > \epsilon ,$$

it follows that $\overline{\text{conv}}(\{x_n\}) \cap (1-\delta)B \neq \phi$. In [*H*] Huff characterized (NUC) spaces as those spaces which are reflexive (R) and have uniformly Kadec-Klee norm. (The norm of X is said to be *uniformly Kadec-Klee* (UKK) if for each $\epsilon > 0$ there exist $\delta > 0$ such that if $\{x_n\} \subset B$ converges weakly but not strongly to $x \in B$, then

---

1991 Mathematics Subject Classification: Primary 46B20.
This paper is in final form and no version of it will be submitted for publication elsewhere.

$$\text{sep}(x_n) > \epsilon \Rightarrow \|x\| < 1 - \delta) \ .$$

A property equivalent to uniform convexity was introduced by Rolewicz in [R1]. For $x \in X\backslash B$, set

$$D(x, B) = \text{conv}(\{x\} \cup B) \ ;$$

$$R(x, B) = D(x, B)\backslash B \ .$$

($D(x, B)$ is called the *drop* of x.) Rolewicz [R1] has shown that X is uniformly convex (UC) if and only if for each $\epsilon > 0$ there exists $\delta > 0$ such that if $x \in X\backslash B$ satisfies $1 < \|x\| < \delta$ then $\text{diam}(R(x, B)) < \epsilon$. In connection with this he introduced the following analogous notion in [R2] using the Kuratowski measure of noncompactness. (Recall that the Kuratowski measure of noncompactness $\alpha(A)$ of a set $A \subset X$ is defined as:

$$\alpha[A] = \inf\{\epsilon > 0 : \exists n \in \mathbb{N} \text{ such that } A \subset \cup_{i=1}^{n} A_i \text{with } \text{diam}(A_i) < \epsilon\}.)$$

DEFINITION 1.1. A Banach space X is said to have *property* (β) if for each $\epsilon > 0$ there exists $\delta > 0$ such that if $x \in X\backslash B$ satisfies $1 < \|x\| < \delta$, then $\alpha[R(x, B)] < \epsilon$.

In [R2] Rolewicz shows that (UC) $\Rightarrow$ (β) $\Rightarrow$ (NUC). Our main purpose here is to show that (β) is also sufficient for the use of Edelstein's algorithms in constructing nearest and farthest points in closed subsets of the space.

2. Edelstein's Algorithms

We now state the two theorems which motivated these considerations. The first is due to Edelstein [E1] and the second, independently, to Edelstein [E2] and Steckin [S].

THEOREM 2.1. *If F is a nonempty closed and bounded subset of a uniformly convex Banach space X, then the set*

$$C_1 = \{x \in X : \exists s \in F \text{ such that } \|c - s\| = \sup\{\|c - x\| : x \in F\}\}$$

is dense in X.

THEOREM 2.2. *If F is a nonempty closed subset of a uniformly convex Banach space X, then the set*

$$C_2 = \{x \in X : \exists x \in F \text{ such that } \|c - s\| = \inf\{\|c - x\| : x \in F\}\}$$

is dense in X.

Although each of these theorems has been substantially generalized (*e.g.*, Asplund [A], Borwein [B], Lau [L1; L2], Deville — Zizler [DZ], Zizler [Z]), we are concerned here with Edelstein's original algorithms for generating the sets C_1 and C_2. In particular, Edelstein (and in the latter instance, Steckin) made explicit use of the following fact about uniformly convex spaces.

PROPOSITION 2.1. *Let X be a uniformly convex Banach space and let H be any bounded subset of X. Then for any $d > 0$ the following limits exist uniformly for c, $c' \in H$ satisfying $\|c - c'\| = r > 0$:*

$$\lim_{\xi \to 0} \operatorname{diam}[B(c; d) \cap (X \backslash B(c'; d + r - \xi))] = 0;$$

$$\lim_{\xi \to 0} \operatorname{diam}[B(c; d - r + \xi) \cap (X \backslash B(c'; d))] = 0.$$

Specifically, if ϵ, d, and r are fixed positive numbers, then there exist $\xi = \xi(\epsilon, d, r)$ and $\xi' = \xi'(\epsilon, d, r)$ such that if c, $c' \in H$ satisfy $\|c - c'\| = r$,

(a) $$\operatorname{diam}\,[B(c; d) \cap (X \backslash B(c'; d + r - \xi))] < \epsilon\ ;$$

(b) $$\operatorname{diam}\,[B(c; d - r + \xi) \cap (X \backslash B(c'; d))] < \epsilon\ .$$

We now outline proofs of Theorems 2.1 and 2.2. (Our approach is only implicit in Edelstein's papers; see Kirk [K] for full details.)

Fix $c_0 \in X$ and let $d_0 = \sup\{\|c_0 - x\| : x \in F\}$. Clearly we may suppose $d_0 > 0$. Let $r \in (0, d_0)$ be arbitrary and let $\{r_i\}$ and $\{\epsilon_i\}$ be sequences of positive numbers satisfying $\Sigma r_i < r$ and $\epsilon_i \to 0$. Select $\xi_1 = \xi(\epsilon_1, d_0, r_1)$ as in condition (a) and choose $x_0 \in F$ so that $\|x_0 - c_0\| > d_0 - \xi_1$. Now select c_1 so that

$$c_0 \in \operatorname{seg}[c_1, x_0] \text{ and } \|c_1 - c_0\| = r_1\ ,$$

set $d_1 = \sup\{\|c_1 - x\| : x \in F\}$, and let $\xi_2 = \xi(\epsilon_2, d_1, r_2)$. Proceed by induction to obtain sequences $\{c_n\} \subset X$, $\{x_n\} \subset F$, and $\{d_n\} \subset \mathbb{R}$ such that

$$c_{n-1} \in \operatorname{seg}[c_n, x_{n-1}] \text{ and } \|c_n - c_{n-1}\| = r_n\ ,$$

$d_n = \sup\{\|c_n - x\| : x \in F\}$, and $\|x_n - c_n\| > d_n - \xi_{n+1}$ where $\xi_{n+1} = \xi(\epsilon_{n+1}, d_n, r_{n+1})$.

Now let $d_{n+1} = \sup\{\|c_{n+1} - x\| : x \in F\}$ where $c_{n+1} \in X$ is chosen so that

$$c_n \in \operatorname{seg}[c_{n+1}, x_n] \quad \text{and} \quad \|c_{n+1} - c_n\| = r_{n+1}\ .$$

Since $d_{n+1} \geq \|c_{n+1} - x_n\| = r_{n+1} + \|c_n - x_n\|$ and since the argument terminates if equality holds one may assume there exists $x_{n+1} \in F$ such that both

$$\|c_{n+1} - x_{n+1}\| > d_{n+1} - \xi_{n+2} ;$$

$$\|c_{n+1} - x_{n+1}\| > r_{n+1} + \|c_n - x_n\| ,$$

where $\xi_{n+2} = \xi(\epsilon_{n+2}, d_{n+1}, r_{n+2})$. Thus for $n > k$ the triangle inequality implies

$$\|c_{k+1} - x_{n+1}\| \geq \|c_{n+1} - x_{n+1}\| - \sum_{i=k+2}^{n+1} r_i .$$

But

$$\sum_{i=k+2}^{n+1} r_i < \|c_{n+1} - x_{n+1}\| - \|c_{k+1} - x_{k+1}\| .$$

Therefore

$$\begin{aligned} \|c_{k+1} - x_{n+1}\| &> \|c_{k+1} - x_{k+1}\| \\ &= r_{k+1} + \|c_k - x_k\| \\ &> r_{k+1} + d_k - \xi_{k+1} . \end{aligned}$$

Thus

$$x_{n+1} \in B(c_k; d_k) \cap (X \backslash B(c_{k+1}; d_k + r_{k+1} - \xi_{k+1})) .$$

In the latter instance, if $n, m > k$ then by (a) of Proposition 2.1, $\|x_n - x_m\| \leq \epsilon_{k+1}$ proving that $\{x_n\}$ is a Cauchy sequence. Thus $\{x_n\}$ converges to a point $f \in F$ and, since $\|c_{n+1} - c_n\| = r_n$ with $\Sigma r_i < r$, $\{c_n\}$ is also Cauchy with limit $c \in B(c_0; r)$. Clearly $\|c - f\| = \sup\{\|c - x\| : x \in F\}$, completing the proof.

One may prove Theorem 2.2 using an argument dual to the one just outlined but using (b) instead of (a).

3. Property (β)

The point we wish to emphasize here is that argument outlined above carries over without change if condition (a) in Proposition 2.1 is replaced with

$$(a') \qquad \alpha[B(c; d) \cap (X \backslash B(c'; d + r - \xi))] < \epsilon ,$$

where α is the Kuratowski measure of noncompactness defined in Section 1. In this case one concludes the argument with the fact that lim $\alpha[\{x_n, x_{n+1}, \cdots\}] \to 0$ as $n \to \infty$ and this also leads to the fact that $\{x_n\}$ has a Cauchy subsequence which converges to a farthest point $f \in F$. Indeed, this latter condition would appear to be the canonical one needed to carry out Edelstein's proof to show existence of such a point. We now show that spaces which have property (β) satisfy this condition.

PROPOSITION 3.1. *Let X be a Banach space which has property (β) and let H be a bounded subset of X. Then for any $d > 0$ the following limits exist uniformly for all $c, c' \in H$ satisfying $\|c - c'\| = r > 0$:*

$$\lim_{\xi \to 0} \alpha[B(c; d) \cap (X \backslash B(c'; d + r - \xi))] = 0 \; ; \tag{3.1}$$

$$\lim_{\xi \to 0} \alpha[B(c; d - r + \xi) \cap (X \backslash B(c'; d))] = 0 \; . \tag{3.2}$$

We begin with a non-uniform version of Proposition 3.1. We say that X is *directionally* (β) if given $x \in S$ and $\epsilon > 0$ there exists $\delta > 0$ such that

$$\alpha[R((1 + \delta)x, B)] < \epsilon,$$

and we say that X has *property* (E) if (3.1) and (3.2) hold for each $c, c' \in X$ with $\|c - c'\| = r > 0$.

PROPOSITION 3.2. *If a Banach space X is directionally (β), then X has property (E).*

PROOF. Suppose X does not have property (E). Then there exist $x, c \in X$ with $\|x - c\| = r > 0$ such that $\alpha(B_n) > 2\epsilon$, where

$$B_n = B(x; d) \cap (X \backslash B(c; d + r - \frac{1}{n})) \; .$$

For convenience we assume $r = 1$ and $c = 0$.

Step 1. Let $y \in B_n$, let $\bar{y} = y / \|y\|$, and let $m_y = \frac{1}{2}(x + \bar{y})$. Then

$$\begin{aligned} \|y - m_y\| &\leq \frac{1}{2}(\| y - x \| + \| y - \bar{y} \|) \\ &= \frac{1}{2}(\| y - x \| + \| y \| - 1) \\ &\leq \frac{1}{2} \| y - x \| + \frac{1}{2}(\| x \| + \| y - x \| - 1) \\ &= \| y - x \| = d \; . \end{aligned}$$

Since

$$1 + d - \frac{1}{n} \leq \| y \| \leq \| y - m_y \| + \| m_y \| \leq d + \| m_y \| \; ,$$

we have

$$\| m_y \| \geq 1 - \frac{1}{n} \; .$$

Step 2. Now let $\delta > 0$ be arbitrary, and let $\tilde{m}_y = km_y$ where $k = 2(\frac{1+\delta}{2+\delta})$. Then

$$\|\tilde{m}_y\| = k\|m_y\| \geq k(1 - \frac{1}{n})$$

and since $k > 1$, $\|\tilde{m}_y\| \geq 1$ for n sufficiently large. Also, since

$$\tilde{m}_y = \frac{k}{2}\bar{y} + (1 - \frac{k}{2})(1+\delta)x \ ,$$

$\tilde{m}_y \in R((1+\delta)x, B)$.

Step 3. Now suppose $u, v \in B_n$. Then

$$\| \bar{u} - \bar{v} \| = \| \frac{u}{\| u \|} - \frac{v}{\| v \|} \| \geq \frac{\| u - v \|}{\| u \|} - \| v \| \left| \frac{\| v \| - \| u \|}{\| u \| \| v \|} \right| .$$

Since $\|u\|$, $\|v\| \to 1 + d$ uniformly as $n \to \infty (1 + d - \frac{1}{n} \leq \| u \|, \| v \| \leq 1 + d)$, if $\nu > 0$ is given it is possible to choose n so large that

$$\| \bar{u} - \bar{v} \| \geq \frac{\| u - v \|}{1 + d} - \nu \ .$$

Moreover, if ν is chosen so that

$$\frac{\epsilon}{1 + d + \nu} - \nu \geq \frac{\epsilon}{2(1 + d)}$$

then for $u, v \in B_n$ with $\|u - v\| \geq \epsilon$, then $\| \bar{u} - \bar{v} \| \geq \frac{\epsilon}{2(1+d)}$ and it follows that

$$\| \tilde{m}_u - \tilde{m}_v \| \geq \frac{\epsilon}{4(1 + d)} \ .$$

Now suppose n and ν have been chosen as in Steps 2 and 3. Since $\alpha(B_n) > 2\epsilon$ it is possible to choose distinct points $\{y_1, y_2 \cdots\}$ in B_n in B_n so that $\|y_i - y_j\| \geq \epsilon$ if $i \neq j$. It follows that $\{\tilde{m}_{y_i}\} \subset R((1+\delta)x, B)$; hence

$$\alpha[R((1+\delta)x, B)] \geq \frac{\epsilon}{8(1+d)} \ .$$

Since δ was arbitrary, we conclude that X is not directionally (β) at x.

PROOF OF PROPOSITION 3.1. Suppose the conclusion of Proposition 3.1 fails. Then for some $c \in X$ (assume $c = 0$) there exist positive numbers ϵ, d, r, and a sequence $\{x_n\} \subset X$ with $\|x_n\| \equiv r$ such that $\alpha(B_n) \geq 2\epsilon$, where

$$B_n = B(x_n; d) \cap (X \backslash B(0; d + r - \frac{1}{n})) \ .$$

As above, assume $r = 1$. By repeating the steps in the proof of Proposition 3.2 it is possible to conclude that for n sufficiently large,

$$\alpha[R((1+\delta)x, B)] \geq \frac{\epsilon}{8(1+d)} .$$

Thus for any positive number δ there exists $x \in X \backslash B$ with $\|x\| < 1+\delta$ such that $\alpha[R(x,B)] \geq \frac{\epsilon}{8(1+d)}$; hence X fails to be of type (β).

REMARK 3.1. We note in passing that if a Banach space X has Property (E) then it follows easily that X has (KK) norm (and (UKK) norm if the conclusion of Proposition 3.1 holds). To see this, suppose $\{x_n\}$ is a sequence in B which converges weakly to x, and suppose $\|x\| = 1$. To show that X has (KK) norm it suffices to prove that $\{x_n\}$ converges strongly to x. However this follows from the fact that for arbitrary k, $\{x_n\}$ is eventually in the set

$$B(0;1) \cap (X \backslash B(-x; 2 - \frac{1}{k})) ,$$

and if X has property (E) we have (taking $d = r = 1$) :

$$\alpha[\{x_1, x_2 \cdots\}] = \alpha[\{x_n, x_{n+1} \cdots\}] \to 0 .$$

REMARK 3.2. The converse implications also hold in Propositions 3.1 and 3.2. This fact follows from the following simple estimate. Let $\| x \| = 1$ and let $y \in R((1+\delta)x, B)$ (where $B = B(0;1)$). Then if $\bar{y} = (1-r)y + rx$ and $c = rx$ it follows that

$$\| c - \bar{y} \| = (1-r) \| y \| \leq (1-r)(1+\delta) < 1 + \delta - r ;$$

hence $\bar{y} \in B(c; 1+\delta-r) \cap (X \backslash B)$. Also, for $y_1, y_2 \in R((1+\delta)x, B)$,

$$\| \bar{y}_1 - \bar{y}_2 \| = (1-r) \| y_1 - y_2 \| .$$

Therefore

$$\alpha[B(c; 1+\delta-r) \cap (X \backslash B)] \geq (1-r)\alpha[R((1+\delta)x, B)]$$

4. Smoothness and Property (β).

As noted above, if X has property (β) then X has uniformly Kadec-Klee norm. This fact motivates the observations of this section. For convenience we adopt the following notation: (β) = Property (β); (R) = reflexivity; (F) = Frechet differentiable norm; (UF) = uniformly Frechet differentiable norm; (KK) = Kadec-Klee norm; (UKK) = uniformly Kadec-Klee norm; and (NUC) = nearly uniformly convex norm. Also, let $S(X)$ and $S(X^*)$ denote, respectively, the unit spheres in X and its conjugate space X^*.

THEOREM 4.1. *The following implications hold:*

$$(R) + (F) + (KK) \Rightarrow directionally(\beta) \ ;$$

$$(UF) + (KK) \Rightarrow (\beta) \ .$$

Since it is known that spaces having property (β) are (NUC), a corollary of the above is the (likely known) fact that $(UF) + (KK) \Rightarrow (NUC)$.

PROOF OF THEOREM 4.1. For the first implication we require the following known lemma (cf., $[D]$).

LEMMA 4.1. *Suppose the norm of X is Frechet differentiable at $x_0 \in S(X)$ and let f_0 be the unique member of $S(X)$ such that $\langle x_0, f_0 \rangle = 1$. If $\{f_n\} \subset S(X^*)$ satisfies $\langle x_0, f_n \rangle \to 1$, then $\|f_n - f_0\| \to 0$.*

Now suppose X fails to satisfy (β) in the direction $x \in S$. Then for some $\epsilon > 0$, $\alpha(B_n) \geq 2\epsilon$, $n = 1, 2, \cdots$, where

$$B_n = R((1 + \frac{1}{n})x, B) \ .$$

Choose $x_n \in B_n$ such that $\|x_i - x_j\| \geq \epsilon$ if $i \neq j$, let $z_n = (1 + \frac{1}{n})x$, and let $m_n = \frac{1}{2}(x_n + z_n)$. By passing to a subsequence, we may suppose also that $\{x_n\}$ converges weakly, say to $w \in B$. Now choose $f_n \in X^*$ so that $\|f_n\| = 1$ and so that $\langle m_n, f_n \rangle = \|m_n\|$. Note in particular that $\|m_n\| \to 1$ as $n \to \infty$. Also

$$\|m_n\| = \langle m_n, f_n \rangle = \frac{1}{2}\langle x_n + z_n, f_n \rangle \leq \frac{1}{2}(\|x_n\| + \|z_n\|) \to 1 \quad \text{as} \quad n \to \infty \ ;$$

hence, $\langle x_n, f_n \rangle \to 1$ and $\langle z_n, f_n \rangle \to 1$ as $n \to \infty$. By Lemma 4.1, $\|f_n - f_x\| \to 0$, where $f_x \in X^*$ satisfies $\|f_x\| = \langle x, f_x \rangle = 1$. But

$$|\langle x_n, f_x - f_n \rangle| \leq \|x_n\| \|f_x - f_n\| \to 0, n \to \infty \ .$$

Therefore, $\langle x_n, f_x \rangle \to 1$ as $n \to \infty$; hence $1 = \langle w, f_x \rangle \leq \|w\|$. But since $\{x_n\}$ converges weakly but not strongly to w this contradicts the assumption that X is (KK).

The second implication of Theorem 4.1 is proved in exactly the same way as the first except that the proof requires the following uniform version of Lemma 4.1. Although this fact is likely known as well, we include a proof based upon the proof of Lemma 4.1 given in $[D, p.29]$.

LEMMA 4.2. *Suppose the norm of X is uniformly Frechet differentiable on $S(X)$, let $z_n \in S(X)$, $n = 1, 2, \cdots$, and for each n let $f_n^* \in S(X^*)$ be* *the unique functional for which* $\langle z_n, f_n^* \rangle = 1$. *Suppose* $\{f_n\} \subset S(X^*)$ *satisfies* $\langle z_n, f_n \rangle \to 1$. *Then* $\|f_n^* - f_n\| \to 0$.

PROOF. Suppose the conclusion is false. Then by passing to a subsequence one may assume

$$\langle x_n, f_n^* - f_n \rangle \geq 2r > 0$$

for some sequence $\{x_n\} \subset S(X)$. But then

$$\begin{aligned}\langle z_n, f_n^* - f_n \rangle &\leq (\langle z_n, f_n^* - f_n \rangle)(r^{-1}\langle x_n, f_n^* - f_n \rangle - 1)\\ &= \langle -z_n + r^{-1}\langle z_n, f_n^* - f_n \rangle x_n, f_n^* - f_n \rangle\\ &= \langle z_n + r^{-1}\langle z_n, f_n - f_n^* \rangle x_n, f_n \rangle - \langle z_n + r^{-1}\langle z_n, f_n - f_n^* \rangle x_n, f_n^* \rangle\\ &\leq |z_n + r^{-1}\langle z_n, f_n - f_n^* \rangle x_n, f_n \rangle|\\ &\quad - \| z_n \| - r^{-1}\langle \langle z_n, f_n - f_n^* \rangle x_n, f_n^* \rangle .\end{aligned}$$

Now, setting $y_n = r^{-1}\langle z_n, f_n - f_n^* \rangle x_n$,

$$\| y_n \| = r^{-1}|\langle z_n, f_n - f_n^* \rangle| = r^{-1}\langle z_n, f_n^* - f_n \rangle ;$$

thus,

$$(*) \qquad \begin{aligned} 0 < r = \|y_n\|^{-1}\langle z_n, f_n^* - f_n \rangle &\leq \|y_n\|^{-1}[|\langle z_n + y_n, f_n \rangle| - \|z_n\| - \langle y_n, f_n^* \rangle]\\ &\leq \|y_n\|^{-1}[\|z_n + y_n\| - \|z_n\| - \langle y_n, f_n^* \rangle] .\end{aligned}$$

However the Frechet derivative D_{z_n} of the norm at z_n in the direction y is given by

$$D_{z_n}(y) = \|z_n\|^{-1}\langle y, f_n^* \rangle = \langle y, f_n^* \rangle .$$

Thus

$$\lim_{\|y\| \to 0} \| y \|^{-1} [\| z_n + y_n \| - \| z_n \| - \langle y, f_n^* \rangle] = 0 ,$$

and the convergence is uniform in z_n and $y/\|y\|$. This implies that the right hand side of (*) converges to 0 as $n \to \infty$, and the resulting contradiction establishes the lemma.

ADDENDUM

After this paper was written the author learned of several forthcoming papers on this topic. Our Proposition 3.1 as well as Remark 3.2 follow from results contained in D. N. Kutzarova and P. L. Papini: "A characterization of property (β) and LUR" (to appear in Boll. Un. Mat. Ital.) In their paper, Kutzarova and Papini also show that a Banach space is LUR if and only if it is strictly convex and locally (directionally) (β). Also, Kutzarova has shown that if the norm of a Banach space is both "nearly uniformly smooth" and (NUC), then it possesses property (β) (D. Kutzarova: "An isomorphic characterization of the property (β) of Rolewicz," to appear; also see D.

Kutzarova, E. Maluta and S. Prus: "Property (β) implies normal structure of the dual space", to appear). This latter fact in conjunction with Remark 3.1 provides an alternate proof that $(UF) + (KK) \Rightarrow (\beta)$.

Acknowledgements

The author is grateful to the referee for noticing numerous oversights in the original draft of this manuscript.

References

[A] E. Asplund, Farthest points in reflexive locally uniformly rotund Banach spaces, Israel J. Math. 4(1966), 213-216.

[B] J. M. Borwein, Weak local supportability and applications to approximation, Pacific J. Math. 82(1979), 323-338.

[BG] J. M. Borwein and J. R. Giles, The proximal normal formula in Banach space, Trans. Amer. Math Soc. 302(1987), 371-381.

[DZ] R. Deville and V. Zizler, Farthest points in w^*-compact sets, Bull. Austral. Math. Soc. 38(1988),433-439.

[D] J. Diestel, *Geometry of Banach Spaces — Selected Topics*, Lecture Notes in Mathematic no. 485, Springer-Verlag, Berlin, Heidelberg, New York, Tokyo, 1975.

[E1] M. Edelstein, Farthest points of sets in uniformly convex Banach spaces, Israel, J. Math. 4(1966), 171-176.

[E2] M. Edelstein, On nearest points of sets in uniformly convex spaces, *J.* London Math. Soc. 43(1968), 375-377.

[H] R. Huff, Banach spaces which are nearly uniformly convex, Rocky Mountain *J.* Math. 10(1980), 743-749.

[K] W. A. Kirk, Nearest and farthest points of closed sets in hyperbolic spaces, *Constantin Carathedory: An International Tribute* (Th. M. Rassias, ed.), World Scientific Publ. Co., Singapore, 1991, pp. 581-591.

[L1] K-S. Lau, Farthest points in weakly compact sets, Israel J. Math. 22(1975), 168-174.

[L2] K-S. Lau, Almost Chebyshev subsets in reflexive Banach spaces, Indiana Univ. Math. J. 27(1978), 791-795,

[R1] S. Rolewicz, On drop property, Studia Math. 85(1987), 27-35.

[R2] S. Rolewicz, On Δ-uniform convexity and drop property, Studia Math. 87(1987), 181-191.

[S] S. B. Steckin, Approximation properties of sets in normed linear spaces, Rev. Roum. Math. Pures et Appl. 8(1963), 5-18 (Russian).

[Z] V. Zizler, On some extremal problems in Banach spaces, Math. Scand. 32(1973), 214-224.

Department of Mathematics, University of Iowa, Iowa City, IA 52242

Contemporary Mathematics
Volume **144**, 1993

p-Hilbertian Subsequences in $\ell_1(X)$

HELMUT KNAUST

ABSTRACT. A Banach space X is said to have property (S_p), if every weakly null sequence in X has a p-Hilbertian subsequence. It is shown that $\ell_1(X)$, the Banach space of absolutely summing sequences of elements in X, has property (S_p), whenever X does.

P. Cembranos showed in [**2**] that $\ell_1(X)$ has the hereditary Dunford-Pettis property, if (and only if) the Banach space X is hereditarily Dunford-Pettis. A Banach space X is hereditarily Dunford-Pettis, if every closed subspace Y of X is Dunford-Pettis, i.e. whenever (y_n) is a weakly null sequence in Y and (y_n^*) is a weakly null sequence in its dual Y^*, then $\lim_{n\to\infty} y_n^*(y_n) = 0$. $\ell_1(X)$ is the Banach space of all absolutely summing sequences in X, i.e. $x = (x_i) \in \ell_1(X)$, if $x_i \in X$ for all $i \in \mathbb{N}$ and $||x||_{\ell_1(X)} := \sum_{i=1}^{\infty} ||x_i||_X < \infty$.

The hereditary Dunford-Pettis property is equivalent to property (S), studied in [**3,7**]. Our goal in this note is to extend the result in [**2**] to Banach spaces with property (S_p) [**8**] for $1 < p < \infty$. Moreover we improve on Cembranos's result by showing that the naturally occurring constant connected with property (S) (and property (S_p)) is the same for the Banach spaces X and $\ell_1(X)$.

DEFINITION. *A Banach space X has property $(S_p), 1 < p \le \infty$, if every weakly null sequence (x_n) in X admits of a subsequence (y_n), so that for some constant C the following holds:*

$$(1) \qquad \left\| \sum_{n=1}^{K} a_n y_n \right\| \le C \text{ for all } K \in \mathbb{N}, (a_n) \in \mathbb{R}^K \text{ with } \left(\sum_{n=1}^{K} |a_n|^p \right)^{1/p} \le 1.$$

A sequence (y_n) satisfying (1) is called *p-Hilbertian.* For the case $p = \infty$ we obtain the definition of property (S). (Here we interpret $(\sum_{n=1}^{K} |a_n|^{\infty})^{1/\infty}$ as $\sup_{n=1,\dots,K} |a_n|$.)

1991 *Mathematics Subject Classification.* Primary 46B20, 46E40.

This paper is in final form and no version of it will be submitted for publication elsewhere.

Some examples: ℓ_p has property (S_p) for $1 < p < \infty$, $L_p(0,1)$ has property (S_r), where $r = \min\{2, p\}$; James's spaces J [**5**] and JT [**6**] have property (S_2) [**1**]. c_0 and the tree spaces defined by Hagler [**4**] and Talagrand [**9**] have property (S); Schur spaces, in particular ℓ_1, have property (S_p) for all $1 < p \leq \infty$.

It was shown in [**7,8**] that if X has property (S_p), then the constant C in the definition above can be chosen uniformly for all weakly null normalized sequences in X:

THEOREM 1. *Let X be a Banach space and let $1 < p \leq \infty$. Then the following properties are equivalent:*
(a.) *X has property (S_p).*
(b.) *There is a constant C so that every weakly null sequence (x_n) in X, with $\|x_n\| \leq 1$ for all $n \in \mathbb{N}$, has a subsequence (y_n) satisfying* (1).

We denote by $s_p(X)$ the infimum of all constants C satisfying (b.) of Theorem 1.

REMARK. $s(X) = s_\infty(X)$ is also the infimum of all constants C satisfying: *every normalized sequence in X admits of a subsequence which is C-equivalent to the unit vector basis of c_0.*

We can now state our result:

THEOREM 2. *Let $1 < p \leq \infty$ and let X be a Banach space with property (S_p). Then $\ell_1(X)$ has property (S_p), and moreover, $s_p(X) = s_p(\ell_1(X))$.*

REMARK. If X has property (S_p), then every subspace Y of X satisfies property (S_p), and $s_p(Y) \leq s_p(X)$. Note that X embeds isometrically into $\ell_1(X)$. Thus the converse of Theorem 2 holds as well. Moreover, it follows that it will suffice to prove that $s_p(\ell_1(X)) \leq s_p(X)$, if X has property (S_p).

The proof uses the following standard result about weakly null sequences in $\ell_1(X)$, whose proof we omit. We denote sequences in $\ell_1(X)$ by $(x(n))_{n\in\mathbb{N}}$.

LEMMA 3. *Let $(x(n))$ be a weakly null sequence in $\ell_1(X)$. Then*
(a.) *$(x_i(n))_{n\in\mathbb{N}}$ is weakly null in X for all $i \in \mathbb{N}$.*
(b.) *For every $\epsilon > 0$ there is an $i_0 \in \mathbb{N}$ so that $\sum_{i=i_0}^{\infty} \|x_i(n)\| < \epsilon$ for all $n \in \mathbb{N}$.*

PROOF OF THEOREM 2. We will only consider the case $1 < p < \infty$. The proof for $p = \infty$ is the same (with the usual notational adjustments).

Let $(x(n))$ be a weakly null sequence in $\ell_1(X)$ with $\|x(n)\| \leq 1$ for all $n \in \mathbb{N}$. $x(n) = (x_i(n))$ with $x_i(n) \in X$ for all $i, n \in \mathbb{N}$. Let $\epsilon > 0$ be given. We have to show that $(x(n))$ admits of a subsequence $(y(n))$ so that

$$\left\| \sum_{n=1}^{\infty} a_n y(n) \right\| \leq s_p(X) + \epsilon \text{ for all } (a_n) \in \mathbb{R}^{\mathbb{N}} \text{ with } \sum_{n=1}^{\infty} |a_n|^p \leq 1.$$

We choose a sequence (δ_i) of positive real numbers satisfying

$$\sum_{i=1}^{\infty} \delta_i \leq \epsilon. \tag{2}$$

Next we choose a decreasing sequence (ϵ_i) of positive reals satisfying

$$(1+3\epsilon_1)(s_p(X)+\epsilon_1)+\epsilon_1 \leq s_p(X)+\delta_1 \tag{3}$$

$$\text{and} \quad 2\epsilon_i(s_p(X)+\epsilon_i)+i\epsilon_i \leq \delta_i \qquad \text{for } i \geq 2. \tag{4}$$

We will define an increasing sequence $(N(k))$ of positive integers, so that $(y(k)) = (x(N(k)))$ satisfies the desired upper ℓ_p-estimate.

By Lemma 3(b.) we can find $i_1 \in \mathbb{N}$ so that for all $n \in \mathbb{N}$,

$$\sum_{i=i_1+1}^{\infty} \|x_i(n)\| \leq \epsilon_1. \tag{5}$$

Since $\|x_i(n)\| \leq 1$ for all $i, n \in \mathbb{N}$, we can pass to a subsequence M_1'' of $\mathbb{N}$ so that for some $r_i \geq 0$ $(i = 1, \dots, i_1)$ with $\sum_{i=1}^{i_1} r_i \leq 1+\epsilon_1$ the following holds:

$$\|x_i(n)\| \leq r_i \text{ for all } i = 1, \dots, i_1; n \in M_1''$$

Since X has property (S_p) and $(x_i(n))$ is weakly null for all $i \in \mathbb{N}$ by Lemma 3(a.), we can find a subsequence M_1' of M_1'' so that for all $i = 1, \dots, i_1$:

$$\left\| \sum_{n \in M_1'} a_n x_i(n) \right\| \leq r_i(s_p(X)+\epsilon_1) \text{ for all } (a_n) \text{ with } \sum_{n \in M_1'} |a_n|^p \leq 1.$$

Consequently we obtain the estimate

$$\sum_{i=1}^{i_1} \left\| \sum_{n \in M_1'} a_n x_i(n) \right\| \leq (1+\epsilon_1)(s_p(X)+\epsilon_1) \text{ for all } (a_n) \text{ with } \sum_{n \in M_1'} |a_n|^p \leq 1. \tag{6}$$

We set $N(1) = \min M_1'$ and let $M_1 = M_1' \setminus \{N(1)\}$. This completes the first induction step.

Let us now assume that $N(k-1)$ has been defined as the minimal element of a set $M_{k-1}' \subset \mathbb{N}$. We set $M_{k-1} = M_{k-1}' \setminus \{N(k-1)\}$. Furthermore assume $i_1 < i_2 < \dots < i_{k-1}$ have been defined.

Using Lemma 3(b.) again, we can find $i_k \in \mathbb{N}$ so that for all $n \in M_{k-1}$,

$$\sum_{i=i_k+1}^{\infty} \|x_i(n)\| \leq \epsilon_k. \tag{7}$$

As in the first induction step we can now find a subsequence M_k'' of M_{k-1} and $r_i \geq 0$ $(i = i_{k-1}+1, \dots, i_k)$ with $\sum_{i=i_{k-1}+1}^{i_k} r_i \leq \epsilon_{k-1}+\epsilon_k \leq 2\epsilon_{k-1}$ so that

$$\|x_i(n)\| \leq r_i \text{ for all } i = i_{k-1}+1, \dots, i_k; n \in M_k''.$$

Passing to another subsequence M_k' of M_k'' we obtain

$$(8) \qquad \sum_{i=i_{k-1}+1}^{i_k} \left\| \sum_{n\in M_k'} a_n x_i(n) \right\| \le 2\epsilon_{k-1}(s_p(X)+\epsilon_k)$$

for all (a_n) with $\sum |a_n|^p \le 1$.

We let $N(k) = \min M_k'$. This completes the definition of the integer sequence $(N(k))$.

Let $(a_k) \in \mathbb{R}^{\mathbb{N}}$ be given with $\sum_{k=1}^{\infty} |a_k|^p \le 1$. We obtain the desired estimate as follows:

$$\begin{aligned}
\left\| \sum_{k=1}^{\infty} a_k y(k) \right\| &= \sum_{i=1}^{\infty} \left\| \sum_{k=1}^{\infty} a_k y_i(k) \right\| \\
&= \sum_{i=1}^{i_1} \left\| \sum_{k=1}^{\infty} a_k y_i(k) \right\| + \sum_{i=i_1+1}^{i_2} \left\| \sum_{k=1}^{\infty} a_k y_i(k) \right\| \\
&\quad + \sum_{j=2}^{\infty} \sum_{i=i_j+1}^{i_{j+1}} \left\| \sum_{k=1}^{\infty} a_k y_i(k) \right\| \\
&\le (1+\epsilon_1)(s_p(X)+\epsilon_1) + \left[\sum_{i=i_1+1}^{i_2} \|y_i(1)\| + \sum_{i=i_1+1}^{i_2} \left\| \sum_{k=2}^{\infty} a_k y_i(k) \right\| \right] \\
&\quad + \sum_{j=2}^{\infty} \left[\sum_{i=i_j+1}^{i_{j+1}} \sum_{r=1}^{j} \|y_i(r)\| + \sum_{i=i_j+1}^{i_{j+1}} \left\| \sum_{k=j+1}^{\infty} a_k y_i(k) \right\| \right]
\end{aligned}$$

The last inequality uses (6). By (5), (7) and (8) we get

$$\begin{aligned}
\left\| \sum_{k=1}^{\infty} a_k y(k) \right\| &\le (1+\epsilon_1)(s_p(X)+\epsilon_1) + [\epsilon_1 + 2\epsilon_1(s_p(X)+\epsilon_2)] \\
&\quad + \sum_{j=2}^{\infty} [j\epsilon_j + 2\epsilon_j(s_p(X)+\epsilon_j)].
\end{aligned}$$

Finally we use (3), (4) and (2) to conclude that

$$\left\| \sum_{k=1}^{\infty} a_k y(k) \right\| \le s_p(X) + \delta_1 + \sum_{j=2}^{\infty} \delta_j < s_p(X) + \epsilon.$$

REMARK. It has come to the author's attention that Theorem 2 was recently also obtained by J.M.F. Castillo and F. Sánchez.

REFERENCES

1. I. Amemiya and T. Ito, *Weakly null sequences in James spaces on Trees*, Kodai Math. J. **4** (1981), 418–425.
2. P. Cembranos, *The hereditary Dunford-Pettis property for $\ell_1(E)$*, Proc. A.M.S. **108** (1990), 947–950.

3. P. Cembranos, *The hereditary Dunford-Pettis property on* $C(K, E)$, Illinois J. Math. **31** (1987), 365–373.
4. J. Hagler, *A counterexample to several questions about Banach spaces*, Studia Math. **60** (1977), 289–308.
5. R. C. James, *A non-reflexive Banach space isometric to its second conjugate*, Proc. Nat. Acad. Sci. (USA) **37** (1951), 174–177.
6. R. C. James, *A separable somewhat reflexive Banach space with non-separable dual*, Bull. A.M.S. **80** (1974), 738–743.
7. H. Knaust and E. Odell, *On* c_0*-sequences in Banach spaces*, Israel J. Math. **67** (1989), 152–169.
8. H. Knaust and E. Odell, *Weakly null sequences with upper* ℓ_p*-estimates*, Functional Analysis, Proceedings, The University of Texas at Austin, 1987–89 (E. Odell and H. Rosenthal, eds.), Lecture Notes in Mathematics 1470, Springer-Verlag, Berlin, 1991, pp. 85–107.
9. M. Talagrand, *La propriété de Dunford-Pettis dans* $C(K, E)$ *et* $L^1(E)$, Israel J. Math. **44** (1983), 317–321.

Department of Mathematical Sciences, University of Texas at El Paso, El Paso, TX 79968

E-mail address: helmut@math.ep.utexas.edu

Contemporary Mathematics
Volume **144**, 1993

SOME GEOMETRICAL PROPERTIES OF BANACH SPACES RELATED TO NEARLY UNIFORM CONVEXITY

D. N. Kutzarova, Bor-Luh Lin and Wenyao Zhang

Abstract. A class of Banach spaces between k-nearly uniformly convex spaces and nearly uniformly convex spaces is introduced. This class of spaces also lies strictly between the ω-uniformly rotund spaces and ω-fully convex spaces.

Huff [H] introduced the nearly uniformly convex (NUC) spaces as a generalization of uniformly convex Banach spaces and showed that the NUC spaces are equivalent to reflexive spaces possessing the uniform Kadec-Klee property. Independently, in connection with the notion of measure of noncompactness, Goebel and Sekowski [GSe] introduced the equivalent class of Δ-uniformly convex spaces. Recently, Rolewicz [R] showed that the nearly uniform convexity is, in certain way, a uniformization of the drop perperty for the unit ball of the space. Based on the drop property, Kutzarova [K2] introduced the $k\beta$ and k-NUC spaces where k is a positive integer. She shows that $k\beta$ implies $(k+1)$-NUC and that the $k\beta$ and k-NUC spaces are isomorphically lie strictly between the uniformly convex and the nearly uniformly convex spaces. By extending the nearly uniform convexity from the unit ball to closed convex sets, the nearly uniformly convex sets are studied in [KR].

There are three other generalizations of uniform convexity, namely, k-uniformly rotundity (k-UR) by Sullivan [S], k-fully convexity (kR) by Fan and Glicksberg [FG1 and FG2] where $k \in \mathbb{N}$ and the locally uniform convexity (LUR) by Lovaglia [L]. Recently, Lk-UR and LkR are introduced in [NW]; Lk-NUC are studied in [KLi] and by extending k to infinite, the ωUR and ωR spaces are defined in [LZ].

In this paper, we introduce the corresponding limit cases of k-NUC and Lk-NUC and study the ωNUC and $L\omega$NUC spaces. We show that they possess the similar relationships with ωUR and ωR spaces as the k-NUC spaces are related to k-UR and kR spaces. We show, by example, that the corresponding properties fail for NUC. Furthermore, we observe that ωNUC spaces possess the Banach-Saks (BS) property while there exist NUC [P] and 2R space [LL] that fail to have the BS property. Examples are given to show that ωNUC and $L\omega$NUC spaces are new classes of Banach spaces.

1991 *Mathematics Subject Classification.* Primary 46B20, 46B10.

Key words and phrases. ω-nearly uniformly convex spaces, locally ω-nearly uniformly convex spaces, ω-uniformly rotundity, ω-fully convexity, Banach-Saks property.

This paper is in final form and no version of it will be submitted in publication elsewhere.

I. For a Banach space X, let $B(X)$ (resp. $S(X)$) denotes the unit ball (resp. sphere) of X. For a sequence $\{x_n\}$ in X, let $\text{sep}(x_n) = \inf\{\|x_n - x_m\| : n \neq m \text{ in } \mathbb{N}\}$.

Recall that a Banach space X is called NUC if for each $\epsilon > 0$, there is a δ, $0 < \delta < 1$ such that for any sequence $\{x_n\}$ in $B(X)$ with $\text{sep}(x_n) > \epsilon$ then $(co\{x_n\}) \cap (1-\delta)B(X) \neq \phi$.

For $k \in \mathbb{N}$, X is said to be k-NUC (resp. $Lk-$NUC) if for every $\epsilon > 0$ (resp. and for every $x \in S(X)$), there is a δ, $0 < \delta < 1$ such that for every sequence $\{x_n\}$ in $B(X)$ with $\text{sep}(x_n) > \epsilon$, then there are $n_1, n_2, \ldots, n_k$ satisfying $\frac{1}{k}\|\sum_{i=1}^k x_{n_i}\| < 1-\delta$ (resp. $\frac{1}{k+1}\|x + \sum_{i=1}^k x_{n_i}\| < 1-\delta$).

Definition. A Banach space X is said to be ωNUC (resp. LωNUC) if for every $\epsilon > 0$ (resp. and for every x in $S(X)$), there are a positive integer k and a δ, $0 < \delta < 1$ such that for every sequence $\{x_n\}$ in $B(X)$ with $\text{sep}(x_n) > \epsilon$, there are $n_1, \ldots, n_k$ satisfying $\frac{1}{k}\|\sum_{i=1}^k x_{n_i}\| < 1-\delta$ (resp. $\frac{1}{k+1}\|x + \sum_{i=1}^k x_{n_i}\| < 1-\delta$).

The ωNUC spaces was introduced and was called fNUC spaces in [K2]. We change the name to ωNUC to be consistent with ωUR and ωR spaces.

It is clear that every k-NUC space is ωNUC and that each ωNUC space is NUC.

A Banach space X is called ωUR (resp. LωUR) [LZ] if for (resp. every $x \in S(X)$ and) every lower triangular matrix $\{x_i^{(n)} : i = 1, \cdots, n, n \in \mathbb{N}\}$ in $B(X)$, $\lim_n (n - \|\sum_{i=1}^n x_i^{(n)}\|) = 0$ (resp. $\lim_n[(n+1) - \|x + \sum_{i=1}^n x_i^{(n)}\|] = 0$) implies that $\lim_n d(x_1^{(n)}, \cdots, x_n^{(n)}) = 0$ (resp. $\lim_n d(x, x_1^{(n)}, \cdots, x_n^{(n)}) = 0$) where for any $x_1, \cdots, x_{k+1}$ in X, d_i is defined to be the distance from x_{i+1} to the affine space spanned by $\{x_1, \cdots, x_i\}$, $i = 1, \cdots, k$ and $d(x_1, \cdots, x_{k+1}) = \min\{d_1, \cdots, d_k\}$.

A Banach space X is said to be ωR (resp. LωR) [LZ] if for (resp. every x in $S(X)$ and for) every sequence $\{x_n\}$ in $B(X)$, $\lim_{n_1,\cdots,n_k} \|x_{n_1} + \cdots + x_{n_k}\| = k$ (resp. $\lim_{n_1,\cdots,n_k} \|x + x_{n_1} + \cdots + x_{n_k}\| = k+1$) for all $k \in \mathbb{N}$, then $\{x_n\}$ is convergent (resp. to x) in X.

It is known that every k-UR space is $k\beta$, hence is $(k+1)$-NUC and that every strictly convex k-NUC space is kR [K3]. We show that the similar results also hold for the limit cases. We need the following result which can be proved as in [Lemma 7, LZ].

Proposition. *If $\{x_n\}$ is a sequence in the unit ball of a Banach space X with $sep(x_n) > \epsilon$ for some $\epsilon > 0$, then for any x in X, there is a subsequence $\{x_{n_i}\}$ of $\{x_n\}$ such that for every positive integer k, $d(x, x_{n_1}, \cdots, x_{n_k}) > \epsilon/4$.*

Theorem 1.

(i) *Every ωUR (resp. LωUR) space is ωNUC (resp. LωNUC).*

(ii) *Every strictly convex ωNUC (resp. LωNUC) space is ωR (resp. LωR).*

Proof.

(i) Suppose X is an ωUR space which is not ωNUC. Then there is an $\epsilon > 0$ such that for each $k \in \mathbb{N}$ there is a sequence $\{x_n^{(k)}\}$ in $B(X)$ with $\text{sep}(x_n^{(k)}) > \epsilon$ and $\frac{1}{k}\|x_{n_1}^{(k)} + \cdots + x_{n_k}^{(k)}\| \geq 1 - \frac{1}{k^2}$ for all $n_1, \cdots, n_k$. By Proposition and by taking

subsequence if necessary, we may assume that, in addition, $d(x_1^{(k)}, \cdots, x_k^{(k)}) \geq \frac{\epsilon}{4}$ for all $k \in \mathbb{N}$. But then the existence of the lower triangular matrix $\{x_i^{(k)} : i = 1, \cdots, k, k \in \mathbb{N}\}$ is a contradiction to the hypothesis that X is ωUR.
The proof that LωUR spaces are LωNUC spaces is similar.
(ii) Suppose X is a strictly convex ωNUC space. Let $\{x_n\}$ be a sequence in $B(X)$ with $\lim_{n_1,\cdots,n_k} \|x_{n_1} + \cdots + x_{n_k}\| = k$ for all $k \in \mathbb{N}$. Since X is strictly convex, we only need to show that $\{x_n\}$ has a convergent subsequence. Suppose not. Then, by taking subsequence, we may assume that $\operatorname{sep}(x_n) > \epsilon$ for some $\epsilon > 0$. Since X is ωNUC, there is a $\delta > 0$ and $k \in \mathbb{N}$ such that for each integer N, there is $n_1, \cdots, n_k > N$ with $\frac{1}{k}\|x_{n_1} + \cdots + x_{n_k}\| < 1 - \delta$. But this implies that $\lim_{n_1,\cdots,n_k} \|x_{n_1} + \cdots + x_{n_k}\| \leqq k(1-\delta) < k$. A contradiction.
Similarly, it follows that every strictly convex LωNUC space is LωR. □

The fact that ωNUC spaces possess the BS property was pointed out in [K2]. For a detail proof, see [Th. 9, K3].

II. The two Banach spaces that are critical in our construction of examples are the space E constructed by Schachermayer [S] to show that the BS property is not L_2-hereditary and the reflexive Banach space B that fail to have the BS property constructed by Baernstein [B].

It is known that NUC is preserved by finite sum and infinite sum under certain conditions [see Ba or KL1 and KL2]. For ωNUC the same result is valid for a finite sum as in [KL1]. We state here only the case for ℓ_p-sums.

Theorem 2. *$(\sum_{i=1}^n \oplus X_i)_p$, $1 \leq p < \infty$ is ωNUC if and only if every X_i, $i = 1, \cdots, n$ is ωNUC.*

Proof. We only need to show that if X and Y are ωNUC spaces then $(X \oplus Y)_p$ is ωNUC.

Let $e_n = (x_n, y_n)$, $n \in \mathbb{N}$ be a sequence in $B((X \oplus Y)_p)$ with $\operatorname{sep}(e_n) > \epsilon$ for some $\epsilon > 0$. By taking subsequence and perturbation, we may assume that $\|x_n\| = a$ and $\|y_n\| = b$ for all $n \in \mathbb{N}$. Since $\operatorname{sep}(e_n) > \epsilon$, it follows that either $\operatorname{sep}(x_n) > \frac{\epsilon}{2}$ or $\operatorname{sep}(y_n) > \frac{\epsilon}{2}$.

Suppose $\operatorname{sep}(x_n) > \frac{\epsilon}{2}$. Since X is ωNUC, there are $\delta_1 > 0$, $k_1 \in \mathbb{N}$ and $n_1, \cdots, n_{k_1}$ such that $\frac{1}{k_1}\|\sum_{i=1}^{k_1} x_{n_i}\| < a(1-\delta_1)$. Similarly, if $\operatorname{sep}(y_n) > \frac{\epsilon}{2}$, then there are $\delta_2 > 0$, $k_2 \in \mathbb{N}$ and $m_1, \cdots, m_{k_2}$ such that $\frac{1}{k_2}\|\sum_{j=1}^{k_2} y_{m_j}\| < b(1-\delta_2)$.

We now observe that if Z is any Banach space and $\{z_n\}$ is a sequence in $B(Z)$ with $\frac{1}{k}\|\sum_{i=1}^{k} z_i\| < 1 - \delta$ for some $\delta > 0$ and some $k \in \mathbb{N}$. Then for any $k \in \mathbb{N}$,

$$\frac{1}{k+h}\|\sum_{i=1}^{k+h} z_i\| \leq \frac{k}{k+h}\left(\frac{1}{k}\|\sum_{i=1}^{k} z_i\| + \frac{1}{k}\sum_{i=k+1}^{k+h} \|z_i\|\right)$$
$$\leqq (1-\delta)\frac{k}{k+h} + \frac{h}{k+h} < 1 \, .$$

Hence, by this observation and by letting $k = \max(k_1, k_2)$ and replacing δ_1, δ_2 by suitable $\delta > 0$, we may assume that $k_1 = k_2 = k$, $\delta_1 = \delta_2 = \delta$ and $n_i = m_i$, $i = 1, \cdots, k$. Now we have

$$\frac{1}{k}\|e_{n_1} + \cdots + e_{n_k}\| \le \max([a^p(1-\delta)^p + b^p]^{1/p}, [a^p + b^p(1-\delta)^p]^{1/p}) < 1$$

Hence $(X \oplus Y)_p$ is ωNUC. □

Example 1. There is an ωNUC space that is not LωUR. Hence ωNUC does not imply ωUR.

Let E be the Schachermayer's space. Then E is 8-NUC [Th. 17, K3] and so E is ωNUC. Let $X = (\mathbb{R} \oplus E)_1$. By Theorem 2, X is ωNUC. Since E is not superreflexive, E fails to be ωUR. Let $\{e_i^{(n)}\}_{i=1,\cdots,n; n\in\mathbb{N}}$ be a lower triangular matrix in $B(E)$ such that

$$\lim_n (n - \|e_1^{(n)} + \cdots + e_n^{(n)}\|) = 0$$

and

$$\lim_n d(e_1^{(n)}, \cdots, e_n^{(n)}) \neq 0 .$$

Let $x = (1, 0)$, $x_i^{(n)} = (0, e_i^{(n)})$, $i = 1, \cdots, n$; $n \in \mathbb{N}$. Then

$$\{x, x_i^{(n)}\}_{i=1,\cdots n; n\in\mathbb{N}} \text{ is in } B(X) \text{ with}$$

$$\lim_n ((n+1) - \|x + \sum_{i=1}^{n} x_i^{(n)}\|) = 0$$

and

$$\lim_n d(x, x_1^{(n)}, \cdots, x_n^{(n)}) \neq 0 .$$

Thus X is an ωNUC space that fails to be LωUR.

Example 2. There is a NUC space that fails to have any equivalent ωNUC norm.

Let B be the Baernstein's space. Then B is NUC and B does not have the BS property [P]. Since every ωNUC space has the BS property, hence E does not have any equivalent ωNUC norm.

Example 3. There is a strictly convex NUC space that fails to be LωR. Hence strictly convex NUC space may fail to be either ωR or LωNUC.

Let $(B, |\cdot|)$ be the Baernstein's space. For $x = (a_1, a_2, \cdots)$ in B, let $\tau(x) = [\sum_n (\frac{a_n}{n})^2]^{1/2}$. Define, for $x \in B$,

$$\|x\| = (|x|^2 + \tau(x)^2)^{1/2} .$$

Then $\|\cdot\|$ is an equivalent strictly convex norm on B. As in [P], it can be proved that $(B, \|\cdot\|)$ is NUC. Let $e_n = (0, \cdots, 0, \overset{n^{th}}{1}, 0, \cdots)$, $n \in \mathbb{N}$ and let $x_n = \frac{e_n}{\|e_n\|}$, $n \in \mathbb{N}$. It is easy to show that for all $k \in \mathbb{N}$,

$$\lim_{n_1, \cdots, n_k} \|x_{n_1} + \cdots + x_{n_k}\| = k$$

but $\{x_n\}$ fails to be convergent in B. This shows that $(B, \|\cdot\|)$ is a strictly convex NUC space but is not an ωR space.

To construct a strictly convex NUC space that is not LωR, we combine the above construction and the method in constructing Example 1. Let $(B, |\cdot|)$ be the Baernstein's space. For $(r, x) \in \mathbb{R} \oplus B$, define

$$\|(r,x)\| = |r| + |x| + (r^2 + \tau(x)^2)^{1/2} .$$

Then $(\mathbb{R} \oplus B, \|\cdot\|)$ is a strictly convex NUC space. Choose $r_0 > 0$ and $r_0 + r_0^2 = 1$. Let $x = (r_0, 0)$, $x_n = \frac{(0,e_n)}{\|(0,e_n)\|}$, $n \in \mathbb{N}$ where $\{e_n\}$ is the unit vector basis in B. Then it can be proved that for all $k \in \mathbb{N}$, $\lim_{n_1,\cdots,n_k} \|x + x_{n_1} + \cdots + x_{n_k}\| = k+1$ but $\lim_n \|x_n - x\| \neq 0$. Hence $(\mathbb{R} \oplus B, \|\cdot\|)$ fails to be LωR.

Example 4. There is an ωR space that is not ωNUC.

Let X be the Baernstein's space with the equivalent norm $\|\cdot\|$ such that $(X, \|\cdot\|)$ is 2R [LL]. Then $(X, \|\cdot\|)$ is ωR and since X fails to have the BS property, $(X, \|\cdot\|)$ is not ωNUC.

Example 5. There is an ωR space that is not kR for all $k \in \mathbb{N}$.

For each $k \in \mathbb{N}$, let X_k be a $(k+1)R$ space that is not kR [LY]. Let $X = (\sum_{k=1}^{\infty} \oplus X_k)_p$, $1 < p < \infty$. Then X is not kR for all $k \in \mathbb{N}$. X is an ωR space from the following theorem.

Theorem 3. *Let Y be an ωR space with 1-unconditional normalized basis. If $\{X_n\}_{n\in\mathbb{N}}$ is a sequence of ωR spaces then the space $X = (\sum_n \oplus X_n)_Y$ is ωR.*

Proof. Let $\{x_n\}_{n\in\mathbb{N}}$ be a sequence in $B(X)$ such that for all $k \in \mathbb{N}$, $\lim_{n_1,\cdots,n_k} \|x_{n_1} + \cdots + x_{n_k}\| = k$. We may assume that $\|x_n\| = 1$, $n \in \mathbb{N}$. Let $\{e_n\}_{n\in\mathbb{N}}$ be an 1-unconditional basis of Y such that $\|e_n\| = 1$, $n \in \mathbb{N}$. For each $x_n = (x_n(j))$ in X, let $y_n = \sum_j \|x_n(j)\| e_j \in Y$. Then $\|x_n\| = \|y_n\|$, $n \in \mathbb{N}$ and for every $k \in \mathbb{N}$ and any $n_1, \cdots, n_k$,

$$(*) \qquad \frac{1}{k}\|\sum_{i=1}^{k} x_{n_i}\| \leq \frac{1}{k}\|\sum_{i=1}^{k} y_{n_i}\| \leq \frac{1}{k}\sum_{i=1}^{k} \|y_{n_i}\| = \frac{1}{k}\sum_{i=1}^{k} \|x_{n_i}\| = 1 .$$

Hence $\lim_{n_1,\cdots,n_k} \frac{1}{k}\|\sum_{i=1}^k y_{n_i}\| = 1$ for all $k \in \mathbb{N}$. Since Y is ωR, there exists y in Y such that $\lim_n \|y_n - y\| = 0$. Let $y = \sum_{j=1}^{\infty} a_j e_j$. Then $\lim_n \|x_n(j)\| = a_j$ for all $j \in \mathbb{N}$. Fix $k \in \mathbb{N}$. For each $j \in \mathbb{N}$, let b_j be any cluster point of $\{\frac{1}{k}\|\sum_{i=1}^k x_{n_i}(j)\| : n_1, \cdots, n_k \text{ in } \mathbb{N}\}$. By $(*)$, we have $\|\sum_{j=1}^{\infty} b_j e_j\| = 1 = \|y\|$. Since $b_j \leq a_j$, $j \in \mathbb{N}$ and Y is strictly convex, we conclude that $b_j = a_j$ for all $j \in \mathbb{N}$. Now, $\lim_{n_1,\cdots,n_k} \frac{1}{k}\|\sum_{i=1}^k x_{n_i}(j)\| = b_j = \lim_n \|x_n(j)\|$ for all $k \in \mathbb{N}$ and X_j is ωR, hence there exists $x(j) \in X_j$ such that $\lim_n \|x_n(j) - x(j)\| = 0$, and $\|x(j)\| = a_j$ for all $j \in \mathbb{N}$. Let $x = (x(j))_{j\in\mathbb{N}}$. Then $x \in X$. It remains to show that

$\lim_n \|x_n - x\| = 0$. Given $\epsilon > 0$, choose $N \in \mathbb{N}$ such that for all $n > N$ and for all $j = 1, 2, \cdots, N$,

$$\|\sum_{i \geq n} a_i e_i\| < \frac{\epsilon}{4}, \ \|y_n - y\| < \frac{\epsilon}{4} \ \text{ and } \ \|x_n(j) - x(j)\| < \frac{\epsilon}{4N}.$$

Then for $n \geq N$, we have

$$\begin{aligned}
\|x_n - x\| &\leq \sum_{j=1}^{N} \|x_n(j) - x(j)\| + \|\sum_{j>N} \|x_n(j) - x(j)\| e_j\| \\
&< \frac{\epsilon}{4} + \|\sum_{j>N} (\|x_n(j)\| - \|x(j)\|) e_j\| + 2\|\sum_{j>N} \|x(j)\| e_j\| \\
&\leq \frac{\epsilon}{4} + \|y_n - y\| + 2\|\sum_{j>N} a_j e_j\| < \epsilon .
\end{aligned}$$

This shows that $\lim_n \|x_n - x\| = 0$, which means that X is ωR. □

REFERENCES

[B] A. Baernstein II, *On reflexivity and summability*, Studia Math. (1972), 91–94.

[Ba] J. Banás, *Compactness conditions in the geometric theory of Banach spaces*, Nonlinear Analysis, Theory, Methods and Applications **16** (1991), 669–682.

[FG1] Ky Fan and I. Glicksberg, *Fully convex normed linear spaces*, Proc. Nat. Acad. Sci. U.S.A. **41** (1955), 947–953.

[FG2] Ky Fan and I. Glicksberg, *Some geometric properties of the spheres in a normed linear space*, Duke Math. J. **25** (1958), 553–568.

[GS] R. Geremia and F. Sullivan, *Multi-dimensional columns and moduli of convexity in Banach spaces*, Ann. Mat. Pura. Appl. **127** (1981), 231–251.

[GSe] K. Goebel and T. Sekowski, *The modulus of noncompact convexity*, Ann. Univ. Mariae Curie-Sklodowska, Sect. A **38** (1984), 41–48.

[H] R. Huff, *Banach spaces which are nearly uniform convex*, Rocky Mountain J. Math. **10** (1980), 734–949.

[K1] D. N. Kutzarova, *Every (β)-space has the Banach-Saks property*, C. R. Acad. Bulg. Sci. **42** (1989), 9–12.

[K2] D. N. Kutzarova, *On the k-β and k-NUC Banach spaces*, C. R. Acad. Bulg. Sci. **43** (1990), 13–15.

[K3] D. N. Kutzarova, *k-β and k-nearly uniformly convex Banach spaces*, J. Math. Anal. Appl. **162** (1991), 322–338.

[KL1] D. N. Kutzarova and T. Landes, *NUC and related properties of finite sums*, (preprint).

[KL2] D. N. Kutzarova and T. Landes, *Nearly uniform convexity of infinite direct sums*, (preprint).

[KLi] D. N. Kutzarova and Bor-Luh Lin, *Locally k-nearly uniform convex Banach spaces*, (submitted).

[KR] D. N. Kutzarova and S. Rorlewicz, *On nearly uniformly convex sets*, Archiv. der Math. **57** (1991), 385–394.

[LL] Bor-Luh Lin and Pei-Kee Lin, *A fully convex Banach space which does not have the Banach-Saks property*, J. Math. Anal. Appl. **117** (1986), 273–283.

[LY] Bor-Luh Lin and Xintai Yu, *On the k-uniform rotund and the fully convex Banach spaces*, J. Math. Anal. Appl. **10** (1985), 407–410.

[LZ] Bor-Luh Lin and Wenyao Zhang, *Some geometric properties related to uniform convexity of Banach spaces, Function Spaces*, Lecture Notes in Pure and Applied Math., vol. 136, Marcel Dekker, 1992, pp. 281–293.

[L] A. R. Lovaglia, *Locally uniformly convex Banach spaces*, Trans. Amer. Math. Soc. **78** (1955), 225–238.

[NW] Nan Chao-Xun and Wang Jian-Hua, *On the Lk-UR and L-kR spaces*, Math. Proc. Camb. Phil. Soc. **104** (1988), 521–526.

[P] J. R. Partington, *On nearly uniformly convex Banach spaces*, Math. Proc. Camb. Phil. Soc. **93** (1983), 127–129.

[R] S. Rolewicz, *On Δ-uniform convexity and drop property*, Studia Math. **87** (1987), 181–191.

[S] W. Schachermayer, *The class of Banach spaces which do not have c_0 as spreading model is not L^2-hereditary*, Ann. Inst. Henri. Poincare **19** (1983), 1–8.

[Su] F. Sullivan, *A generalization of uniformly rotund Banach spaces*, Canad. J. Math. **31** (1979), 628–636.

[Y] Xintai Yu, *k-UR spaces are NUC spaces*, Ke Xue Tong Pao **24** (1983), 1473–1475, (Chinese).

D. N. Kutzarova, Institute of Mathematics, Bulgarian Academy of Sciences, 1090 Sofia, Bulgaria

Bor-Luh Lin, Department of Mathematics, The University of Iowa, Iowa City, IA 52242, U.S.A.

Wenyao Zhang, Department of Mathematics, Northern Illinois University, DeKalb, IL 60115, U.S.A.

Contemporary Mathematics
Volume **144**, 1993

ASYMPTOTIC l_p SPACES AND BOUNDED DISTORTIONS

VITALI D. MILMAN AND NICOLE TOMCZAK-JAEGERMANN

ABSTRACT. The new class of Banach spaces, so-called asymptotic l_p spaces, is introduced and it is shown that every Banach space with bounded distortions contains a subspace from this class. The proof is based on an investigation of certain functions, called enveloping functions, which are intimately connected with stabilization properties of the norm.

0. Introduction

During the last year several problems of infinite-dimensional Banach space theory, which remained open for decades, have been finally solved. Some new constructions of Banach spaces have been made which, on one hand, showed limitations of the theory, but on the other hand, also showed how exciting an infinite-dimensional geometry can be. Let us mention a few of them:

(i) a space with no unconditional basic sequence (Gowers–Maurey),
(ii) a space not isomorphic to any of its hyperplanes (Gowers),
(iii) a space such that every bounded operator is a Fredholm operator (Gowers–Maurey).

The problems which were answered by these examples are of a linear-topological nature. Although a thorough study of these properties flourished back in the 1960's, methods developed at that time and later were not sufficient to attack succesfully these problems. The solutions given last year are by-products of a study in a different direction: the infinite-dimensional geometry of convex bodies; that is, the geometry of the unit sphere of a Banach space.

In this introduction we would like to explain the geometry which led to the breakthroughs described above; in the main body of the paper we add some information in this geometric direction.

0.1. In fact, the topic of study which led to the recent development takes its roots, in a large part, in the local theory of Banach spaces; in other words, in the asymptotic theory of finite-dimensional normed spaces. Consider the following question:

Let $f(\cdot)$ be a uniformly continuous real valued function on the unit sphere $S = S(X) = \{x \in X \mid \|x\| = 1\}$ of an infinite-dimensional Banach space

---

1991 *Mathematics Subject Classification.* Primary 46B03, 46B04, 46B20.
The research of the first author was supported in part by the US–Israel Binational Science Foundation.
This paper is in final form and no version of it will be submitted for publication elsewhere.

X. Does the oscillation of f decrease to zero on some sequence E_n of infinite-dimensional subspaces of X?

To state it in a more precise way we need some notation. For a fixed function f as above, and for an arbitrary subspace $E \subset X$, let $I_E(f) = [a(E), b(E)]$, where $a(E) = \inf\{f(x) \mid x \in S \cap E\}$ and $b(E) = \sup\{f(x) \mid x \in S \cap E\}$. Then let

$$\mathcal{O}(f) = \mathcal{O}_X(f) = \inf\{b(E) - a(E) \mid E \subset X, \dim E = \infty\}.$$

The question then becomes: is $\mathcal{O}(f) = 0$?

If the answer is "yes" then

> *there exists a real number s such that for every $\varepsilon > 0$ there is a subspace E with $\dim E = \infty$ such that $|f(x) - s| < \varepsilon$ for all $x \in S \cap E$.*

The collection of all the numbers s is called the *spectrum of f* and denoted by $\gamma_\infty(f)$ or $\gamma_\infty(f, X)$ (see [M.69]).

And so, we are asking whether the spectrum $\gamma_\infty(f)$ is non-empty for all uniformly continuous functions f on the sphere of an arbitrary Banach space X, or of some Banach space X.

Intuition says that the answer is *obviously* negative, at least for $X = l_2$, say, because there is no reason for it to be positive. Uniform continuity is a local geometric condition with no connection to the linear structure of a space, and the existence of $s \in \gamma_\infty(f)$ is a global linear property. One never studies what seems to be obvious and the question is not an exception to this rule.

Note that James [J.64] showed that, in the above terminology, $\gamma_\infty(f, l_1)$ and $\gamma_\infty(f, c_0)$ are non-empty for f being an equivalent norm on these spaces. This result did not contradict the intuition, because the norms in l_1 and c_0 are in a sense extremal, and the proofs deeply depended on this fact. So, at the time, a similar question for, say, l_2, was not even considered.

0.2. However, it was observed in 1967 ([M.67], *cf.* also [M.69], [M.71a]) that a slightly different *finite-dimensional spectrum* $\gamma(f)$ is always non-empty. We say that $s \in \gamma(f)$ whenever

> *for every $\varepsilon > 0$ and for every n there exists an n-dimensional subspace $E_n \subset X$ such that $|f(x) - s| < \varepsilon$ for all $x \in S \cap E_n$.*

We have the following fact valid for every infinite-dimensional Banach space X.

Fact. *For every uniformly continuous real function f on the unit sphere S, $\gamma(f) \neq \emptyset$.*

As we explained above, this somewhat contradicted intuitions of that time. Just to support these intuitions, let us recall the Grinblatt's paper [G.76] where an example was presented of a bounded continuous, but not uniformly continuous, function f on the sphere S in the Hilbert space, which has oscillation at least 1 on every 2-dimensional central section of S. So the fact above indeed fundamentally rests on an interplay of uniform continuity of a function and non-compactness of the sphere.

Thus, since the finite-dimensional spectrum $\gamma(f)$ involves subspaces of arbitrarily high dimensions and it is always non-empty, it eventually became natural to expect that the (infinite-dimensional) spectrum $\gamma_\infty(f)$ is also non-empty.

0.3. Let us now consider the case when the function $f = |||\cdot|||$ is another norm on X, continuous with respect to the original norm. We have the following two mutually exclusive possibilities.

- **(a) Spectrum:** For every norm f we have $\gamma_\infty(f) \neq \emptyset$. This would mean that either on some infinite-dimensional subspace f is arbitrarily small, if $0 \in \gamma_\infty(f)$, or, if $0 \neq s \in \gamma_\infty(f)$, then f is "almost" an isometry on some infinite-dimensional subspace.
- **(b) Distortion:** There is a norm f such that $\gamma_\infty(f) = \emptyset$. This means that the norm f has an oscillation with respect to the original norm which does not tend to zero on *any* sequence of infinite-dimensional subspaces.

In view of the Fact above, the existence of a norm satisfying condition (b) would be clearly connected with some very essential infinite-dimensional effects.

If a uniformly continuous function f satisfies (b) then, obviously, there exists an interval $I = [\beta, \delta]$, with $\beta < \delta$, such that

- **(i)** for every $\varepsilon > 0$ there exists a subspace $Y = Y_\varepsilon$, $\dim Y = \infty$, such that $I_Y(f) \subset [\beta - \varepsilon, \delta + \varepsilon]$;
- **(ii)** For every $E \subset Y$, $\dim E = \infty$ one has $I_E(f) \supset (\beta, \delta)$. (Here Y is a subspace from (i) corresponding to $\varepsilon = 1$, say.)

The collection of all such intervals I is called the *tilda-spectrum* of f and denoted by $\tilde{\gamma}(f)$.

Of course, the case $\delta = \beta$ reduces the interval to one point, $\beta \in \gamma_\infty(f)$, which we also consider as a part of $\tilde{\gamma}(f)$.

Therefore we have (see [M.69])

Fact. *For every uniformly continuous real function f on the unit sphere S, $\tilde{\gamma}(f) \neq \emptyset$.*

0.4. Note that if f is a norm on X as in 0.3, and if $I = [0, \delta] \in \tilde{\gamma}(f)$, then necessarily $\delta = 0$ (see [M.69]).

In the case $\beta > 0$ we introduce the *level of distortion* of an interval $I \in \tilde{\gamma}(f)$ by $d(I) = \delta/\beta$, and the *level of distortion* of an equivalent norm f by

$$d(f) = \sup\{d(I) \mid I \in \tilde{\gamma}(f)\}.$$

We have a similar dichotomy as in 0.3.

- **(a')** Either for any equivalent norm f on X one has $d(f) = 1$, or
- **(b')** there exists an equivalent norm f on X such that $d(f) > 1$.

In terms of the spectrum, condition (a') means that for any equivalent norm f on X and any infinite-dimensional subspace Z of X, the spectrum $\gamma_\infty(f_{|Z})$ of the restriction of f to Z, is non-empty. Similarly, condition (b') means that X contains a distortable infinite-dimensional subspace: there exists an infinite-dimensional subspace Z of X and an equivalent norm f on X such that $\gamma_\infty(f_{|Z}) = \emptyset$; that is, f is a distortion on Z.

It was proved by Milman in 1969 that

Theorem. *Let X be a Banach space. Assume that $d(f) = 1$ for every equivalent norm f on X. Then either for some $1 \leq p < \infty$, X contains a $(1+\varepsilon)$-isomorphic* *copy of l_p (for every $\varepsilon > 0$), or X contains a $(1+\varepsilon)$-isomorphic copy of c_0 (for every $\varepsilon > 0$).*

(The result was stated in [M.69], Section 3.3, with the complete proof in [M.71b].)

And so, alternative (a') would imply an exciting structural theory for Banach spaces. However, in 1974, Tsirelson [Ts.74] constructed a space T which does not contain an isomorphic copy of any l_p $(1 \le p < \infty)$ or of c_0. This means that the space T satisfies the alternative (b'): T contains a distortable infinite-dimensional subspace Z. (In fact, it can be shown by a direct argument that T itself is also distortable.)

An interesting feature of Tsirelson's example is that the norm is not given by an explicit formula but it is defined by an equation. This was the first construction of such a type, and essentially, with only minor modifications, the only one. In the dual form, which has been put forward by Figiel and Johnson [F-J.74], the norm is defined, for a finite sequence of real numbers $x \in \boldsymbol{R}^{(\boldsymbol{N})}$, by

$$\|x\|_T = \max\left\{\|x\|_{c_0}, \frac{1}{2}\sup\sum_{i=1}^{n}\|E_i x\|_T\right\}, \tag{0.1}$$

where the inside supremum is taken over all succesive intervals $\{E_i\}$ of positive integers such that $n < \min E_1 \le \max E_1 < \min E_2 \le \cdots < \max E_{n-1} < \min E_n$ and over all n. For $x = \sum_i t_i e_i \in X$ and an interval E, we set $Ex = \sum_{i\in E} t_i e_i$. Tsirelson's space T is then the completion of $\boldsymbol{R}^{(\boldsymbol{N})}$ under the norm $\|\cdot\|_T$. Most of the important properties of the space T and related spaces can be found in [C-S.89] and references therein.

0.5. Let us return to a distortion situation when $I = [\beta, \delta] \in \tilde{\gamma}(f)$ with $\beta < \delta$ and let us give its geometric interpretation.

Let $\varepsilon < (\delta - \beta)/2$ and let $Y = Y_\varepsilon$ be a corresponding subspace. Define two sets

$$A = \{x \in S \cap Y \mid f(x) < \beta + \varepsilon\} \quad \text{and} \quad B = \{x \in S \cap Y \mid f(x) > \delta - \varepsilon\}.$$

For every infinite-dimensional subspace $E \subset Y$ we have $A \cap E \neq \emptyset$ and $B \cap E \neq \emptyset$. A set satisfying such a property is called an *asymptotic* set (in Y). So in our situation, A and B are two asymptotic sets with positive distance apart, $\operatorname{dist}(A, B) > 0$. The existence of such a pair (A, B) is thus a consequence of distortion. Conversely, this fact also implies some distortion property. The Urysohn function for sets A and B is a uniformly continuous function with an empty spectrum $\gamma_\infty(f)$; to construct an equivalent norm without spectrum some additional convexity assumptions are required.

0.6. Given a distortion situation it is natural to ask a quantitative question, how large can $d(f)$ be. Note that Theorem 0.4 ensures only the existence of distortion but provides no quantitative information on $d(f)$.

It is to Rosenthal's credit that in 1988 he asked the first named author, Odell, and several others, how to find a direct formula for a distortion on Tsirelson's space, and how large such a distortion can be. Odell (unpublished) in 1989/90 constructed two asymptotic sets in T. He also showed that the spaces T_λ, obtained by replacing $1/2$ in the definition (0.1) by $1/\lambda$, have distortions d_λ of order $1/\lambda$, hence $d_\lambda \to \infty$ as $\lambda \to \infty$. So, for every real number d, there is a space with a level of distortion at least d.

Let us mention that an approach to distortions using the theory of Krivine–Maurey types was presented in [H-O-R-S.91]. In particular, this paper contains another proof of Theorem 0.4.

The next step was done by Schlumprecht [S.91], who changed $1/2$ to $1/\ln n$, which allows E_1 to start at any place (not necessarily far out). This has an important effect on the geometry of the space: the unit vector basis became subsymmetric and, as Schlumprecht showed, any distortion level is attained by some equivalent norm. Schlumprecht's space S also has the property of an *infinite distortion*:

there exists a sequence of asymptotic sets $\{A_i\}$ *on the sphere of* S *such that* $\operatorname{dist}(A_i, \operatorname{conv}(\bigcup_{j\neq i} A_j)) \geq 1$.

In fact, S satisfies the still stronger condition that there also exists a sequence of sets $\{A_i^*\}$ on the sphere of the dual space S^* such that the system $\{A_i, A_i^*\}$ is "nearly biorthogonal".

This was the starting point for Gowers' and Maurey's constructions.

Finally, this year, Odell and Schlumprecht [O-S.92] proved that for every $1 < p < \infty$, l_p has an arbitrarily large (and even infinite) distortion, thereby finishing off the problem which originated from [M.69], [M.71b]. Again, they did not construct asymptotic sets far apart in, say, l_2, but transformed them in an ingeneous non-linear way from Tsirelson's space, or, on more advanced level, from Schlumprecht's space. Combining this outstanding result with Theorem 0.4 we see that

Theorem. *Any Banach space X which does not hereditarily contain copies of l_1 and c_0, contains a distortable subspace;* i.e., *there exists an equivalent norm f on X such that $d(f) > 1$.*

Moreover, Odell and Schlumprecht proved that on the sphere $S(l_1)$ there is a Lipschitz function f (not a norm) with an empty spectrum, $\gamma_\infty(f, l_1) = \emptyset$. It was shown earlier by Gowers [G.91] that $\gamma_\infty(f, c_0) \neq \emptyset$, for any uniformly continuous function on $S(c_0)$.

0.7. Let us go back to the quantitative question: does any Banach space X not containing hereditarily copies of l_1 and c_0, have arbitrarily large distortions?

This is not yet clear. To study this problem, we consider in this paper spaces with *bounded distortions.* These are spaces X such that for some constant D we have $d(f) \leq D$, for every infinite-dimensional subspace Z of X and every equivalent norm f on Z. What kind of simple "basic structural blocks" (*i.e.*, subspaces) can such a space X contain? To explain our result let us define the class of *asymptotic l_p* spaces. (The rather standard notation concerning successive blocks of a basis and related concepts will be explained at the beginning of the next section.)

Definition. A Banach space X with a normalized basis $\{x_i\}$ is said to be an *asymptotic l_p* space, for some $1 \leq p < \infty$ (resp., an *asymptotic* c_0 space) if there exists a constant C such that for every n there exists $N = N(n)$ such that any normalized successive blocks $N < z_1 < z_2 < \ldots < z_n$ of $\{x_i\}$ are C-equivalent to the unit vector basis in l_p^n (resp., in l_∞^n). By $\lambda_p(X)$ we denote the infimum of all constants C as above.

Note that Tsirelson's space T is an *asymptotic* l_1 space which does not contain a subspace isomorphic to l_1.

For spaces with bounded distortions, let

$$d(X) = \sup d(f), \tag{0.2}$$

where the supremum is taken over all equivalent norms f on X.

Recall a standard and easy observation that if $Z \subset X$ is an infinite-dimensional subspace and f is an equivalent norm on Z then there exists an equivalent norm $\tilde{f}$ on X such that $\tilde{f}_{|Z} = f$. This immediately implies that $d(Z) \le d(X)$.

Theorem. *Let X be a Banach space with bounded distortions and let $d(X) < D$. There exists a subspace Y of X which is either asymptotic l_p, for some $1 \le p < \infty$, or asymptotic c_0. Moreover, $\lambda_p(Y)$ depends on D only.*

We learned recently that B. Maurey [Ma.92] also proved this theorem and used it to show that every space of type $p > 1$ with an unconditional basis has arbitrarily large distortions.

In contrast with the result for $d(X) = 1$ (Theorem 0.4), the theorem above recognizes, as "basic structural blocks", a *class* of Banach spaces rather than a concrete space, as it was suggested by a "naive" intuition of the 1960's. (It is well-known that by varying λ in the definition of Tsirelson's spaces T_λ we get a sequence of non-isomorphic asymptotic l_1 spaces, and the so-called p-convexified Tsirelson spaces show that the same phenomenon holds for any fixed $1 \le p < \infty$ or c_0.)

Another important point is a difference with the local theory of Banach spaces. The definition of asymptotic l_p spaces is "almost" local, in that it involves finite-dimensional subspaces and parameters not depending on the dimension. However, this isomorphic definition does not imply a $(1+\varepsilon)$-isometric version, in the standard spirit of the local theory. Indeed, we would call a Banach space an *almost isometric asymptotic* l_p space if $\lambda_p(X) = 1$. It is well-known to specialists that every *almost isometric asymptotic* l_p space contains, for every $\varepsilon > 0$, a subspace $(1+\varepsilon)$-isomorphic to l_p (see 6.4 for a short argument).

Our method involves geometry of infinite-dimensional sphere and a suitable geometric language will be introduced in the next section.

1. Preliminaries

1.1. Since in this paper we are concerned with the existence of nice infinite-dimensional subspaces inside Banach spaces from a certain class, we may and will assume, unless stated otherwise, that Banach spaces discussed here have a monotone basis. In such a situation we will use the standard notions of the dual basis, equivalent bases, block bases, block subspaces, basic sequences, etc. They can be found *e.g.*, in [L-T.77]. Let us only mention that we will say that two basic sequences $\{x_i\}$ and $\{e_i\}$ are C-equivalent, for some constant C, if for any (finite) sequence of scalars $\{a_i\}$ we have

$$C^{-1}\|\sum_i a_i x_i\| \le \|\sum_i a_i e_i\| \le C\|\sum_i a_i x_i\|.$$

We will consider only vectors with finite support. For vectors $x, y \in X$ and subspaces E, E_1, E_2, we will freely use the notation $n < x$ to denote that $n < \min\operatorname{supp}(x)$; then $x < y$ if $\max\operatorname{supp}(x) < \min\operatorname{supp}(y)$; then $x < E$ if $x < y$ for every $y \in E$; and $E_1 < E_2$ if $x < y$ for every $x \in E_1$ and $y \in E_2$.

For a Banach space X by B_X and $S(X)$ we denote the unit ball and the unit sphere in X, respectively; for a subspace $E \subset X$ we set $B_E = B_X \cap E$ and $S(E) = S(X) \cap E$.

1.2. Asymptotic sets were defined in 0.5, where their basic connection to distortions was indicated. To get a better understanding of their geometric properties let us make some easy general observations, valid for arbitrary Banach spaces (which may have no basis).

Fact. *Let $(Z, \|\cdot\|)$ be a Banach space and let $|||\cdot|||$ be a seminorm on Z such that $|||z||| \le \|z\|$, for all $z \in Z$. Assume that there exists an asymptotic set $A \subset S(Z, \|\cdot\|)$ such that $\|\cdot\|$ and $|||\cdot|||$ are equivalent on A. Then there exists a subspace E of Z of finite-codimension such that $|||\cdot|||$ is a norm on E equivalent to $\|\cdot\|$.*

Proof. Since A is asymptotic, $|||\cdot|||$ is a norm on some subspace W of finite-codimension. A standard well-known fact (*cf. e.g.*, [K.66], [L-T.77]) implies that if $|||\cdot|||$ and $\|\cdot\|$ were not equivalent on any subspace E of W of finite-codimension then for every $\varepsilon > 0$ there would be an infinite-dimensional subspace F of W such that $|||z||| \le \varepsilon\|z\|$, for all $z \in F$. But for ε sufficiently small this is impossible, since F intersects A. $\square$

Remark. Let Z be a Banach space with bounded distortions, $d(Z) < D$, and let $A \subset S(Z)$ be an asymptotic set symmetric about the origin. Then there exists an infinite-dimensional subspace F of Z such that $(1/D)(B_Z \cap F) \subset \operatorname{conv} A$.

Indeed, let W be the finite-codimensional subspace spanned by A. Applying the fact above to the norm $|\cdot|$ on W whose unit ball is $\operatorname{conv} A \cap W$, we get that $\|\cdot\|$ and $|\cdot|$ are equivalent on a certain subspace E of W of finite codimension. Thus there exists an infinite-dimensional subspace F of E such that

$$\|z\| \le |z| \le D\|z\| \quad \text{for} \quad z \in F. \tag{1.1}$$

This implies the required inclusion of the corresponding unit balls.

1.3. To make the arguments more compact, we introduce several short notations for certain families of subspaces of a given Banach space Z (with a basis). Typically, Z will be a block subspace of the fixed Banach space X.

By $\mathcal{B}_\infty(Z)$ we denote the family of all infinite-dimensional block subspaces $E \subset Z$; next, $\mathcal{B}^t(Z)$ denotes the family of all (block) subspaces $E \in \mathcal{B}_\infty(Z)$ of finite-codimension, *i.e.*, $\dim Z/E < \infty$; finally, if $Y \in \mathcal{B}_\infty(Z)$ and $z \in S(Z)$ (with finite support), then $\mathcal{B}^t(Y, z)$ denotes the family of all subspaces $F \in \mathcal{B}_\infty(Y)$ such that $z < w$ for all $w \in F$.

1.4. Let us recall the geometric notions of asymptotic averages and moduli, which play a major role in our approach. These notions were introduced and studied by Milman in 1967–70. A survey on this subject can be found in [M.71b], *cf.* also the more recent paper [M-P.89].

The moduli are defined relative to a fixed family $\mathcal{B}$ of subspaces of a space X, which satisfies the filtration condition

$$\text{For every } E_1, E_2 \in \mathcal{B} \text{ there exists } \; E_3 \in \mathcal{B} \text{ such that } E_3 \subset E_1 \cap E_2.$$

Typically, the family $\mathcal{B}$ will be $\mathcal{B}^t$, which have been defined in 1.3, although we will make an exception from this rule in Section 2.

For a continuous bounded function $h : S(X) \to \mathbb{R}$ and an infinite-dimensional subspace $E \subset X$ define lower and upper moduli β- and δ-, respectively, by

$$\begin{aligned} \beta[h, \mathcal{B}, E] &= \beta_x[h, \mathcal{B}(E)] = \sup_{F \in \mathcal{B}(E)} \inf_{x \in S(F)} h(x), \\ \delta[h, \mathcal{B}, E] &= \delta_x[h, \mathcal{B}(E)] = \inf_{F \in \mathcal{B}(E)} \sup_{x \in S(F)} h(x). \end{aligned} \tag{1.2}$$

For a continuous bounded function $f : S(X) \times S(X) \to \mathbb{R}$ and $E \in \mathcal{B}_\infty(X)$ we set

$$\begin{aligned} \beta\beta[f, \mathcal{B}, E] &= \beta_x\left[\beta_y[f(x, \cdot), \mathcal{B}(E, x)], \mathcal{B}, E\right] \\ \delta\delta[f, \mathcal{B}, E] &= \delta_x\left[\delta_y[f(x, \cdot), \mathcal{B}(E, x)], \mathcal{B}, E\right]. \end{aligned} \tag{1.3}$$

Observe that if $E \subset F$ then

$$\beta[h, \mathcal{B}, F] \le \beta[h, \mathcal{B}, E] \le \delta[h, \mathcal{B}, E] \le \delta[h, \mathcal{B}, F],$$

hence also

$$\beta\beta[f, \mathcal{B}, F] \le \beta\beta[f, \mathcal{B}, E] \le \delta\delta[f, \mathcal{B}, E] \le \delta\delta[f, \mathcal{B}, F],$$

for all functions h and f as above.

2. Non-distortable spaces

To develop better geometric intuitions and to illustrate the use of the β- and δ- averages we start with the isometric case and we will sketch the proof of Milman's theorem on non-distortable spaces, Theorem 0.4.

2.1. In the isometric situation discussed here there is no real advantage in passing to a subspace with a basis; in fact, this would confuse the geometric picture rather than clarify it. Therefore we present an argument which makes no reference to the existence of a basis and thus it works for an arbitrary Banach space. The averages we will consider here will be taken with respect to the family $\mathcal{B} = \mathcal{B}^0(E)$ of all finite-codimensional subspaces of a given space E. This family clearly satisfies the filtration condition.

We will consider the collection of functions

$$f_\varepsilon(x, y) = \|x + \varepsilon y\| - 1 \quad \text{for} \quad x, y \in S(X). \tag{2.1}$$

The averages of functions f_ε will be called the β- and δ-moduli, as they reflect a geometric behaviour of the sphere in a Banach space.

For a subspace $E \subset X$ and $x \in S(X)$, the notation $\mathcal{B}(E,x)$ used in (1.3) simply means $\mathcal{B}^0(E)$. Also, $\mathcal{B}_\infty(E)$ denotes the family of all infinite-dimensional subspaces of E.

We consider the local modulus $\beta_y[f_\varepsilon(x,\cdot),\mathcal{B}^0(E,x)]$, denoting it by $\beta(\varepsilon,x,E)$; and the global modulus $\beta\beta[f_\varepsilon,\mathcal{B}^0,E]$, denoting it by $\beta\beta(\varepsilon,E)$. Similarly, the local modulus $\delta_y[f_\varepsilon(x,\cdot),\mathcal{B}^0(E,x)]$ is denoted by $\delta(\varepsilon,x,E)$ and the global modulus $\delta\delta[f_\varepsilon,\mathcal{B}^0,E]$, by $\delta\delta(\varepsilon,E)$.

To illustrate the expected behaviour of the moduli, let us observe that for $X = l_p$, $1 \le p < \infty$ we have

$$\beta(\varepsilon,x,l_p) = \delta(\varepsilon,x,l_p) = (1+\varepsilon^p)^{1/p} - 1,$$

for all $x \in S(l_p)$ and all $\varepsilon > 0$. This function has the order ε^p/p as $\varepsilon \to 0$.

Computation of the moduli for some other spaces can be found in [M.71b].

2.2.

Lemma. *Let X be a Banach space such that $d(X) = 1$. There exists an infinite-dimensional subspace F of X such that*

$$\beta\beta(\varepsilon,F) = \delta\delta(\varepsilon,F) \quad \textit{for} \quad \varepsilon > 0. \tag{2.2}$$

Proof. It is not difficult to see that if $d(f) = 1$ for every equivalent norm f on Z then $\mathcal{O}_Z(g) = 0$ for every $Z \in \mathcal{B}_\infty(X)$ and every uniformly continuous convex function $g : Z \to \boldsymbol{R}$. We will show this at the end of the proof.

Observe that for each $\varepsilon > 0$ and $x \in S(X)$, the function $f_\varepsilon(x,\cdot)$ is convex, therefore $\mathcal{O}(f_\varepsilon(x,\cdot)) = 0$. Stabilizing over y with a given $\theta > 0$ we get a subspace $\tilde{E} \in \mathcal{B}_\infty(X)$ such that

$$0 \le \sup_{y \in S(\tilde{E})} f_\varepsilon(x,y) - \inf_{y \in S(\tilde{E})} f_\varepsilon(x,y) < \theta \quad \text{for} \quad x \in S(\tilde{E}). \tag{2.3}$$

Now we take a dense set $\{x_i\}$ in the unit sphere $S(X)$ and a sequence of $\theta_i \downarrow 0$, and we let, for every $i = 1,2,\dots$, ε vary over a finite θ_i-net $\mathcal{N}_i$ in $[\theta_i, 1/\theta_i]$; this way we can construct a sequence $E_1 \supset \ldots \supset E_i \supset \ldots$ of infinite-dimensional subspaces such that on $S(E_i)$ the inequality analogous to (2.3) holds for x_j, with $j = 1,\dots,i$, and θ_i and all $\varepsilon \in \mathcal{N}_i$, $(i = 1,2,\dots)$. Picking $e_i \in S(E_i)$ for $i = 1,2,\dots$ and setting $E = \operatorname{span}[e_i]$, we get, by this diagonal procedure, $E \in \mathcal{B}_\infty(X)$ such that for all $x \in S(E)$ and $\varepsilon > 0$ we have

$$\beta(\varepsilon,x,E) = \delta(\varepsilon,x,E).$$

Now, with a fixed $\varepsilon > 0$, the function $\delta(\varepsilon,\cdot,E)$ is again convex. Stabilizing over x, with a fixed $\varepsilon > 0$, and then passing to a diagonal in a similar way as before, we get an infinite-dimensional subspace F of E on which (2.2) holds.

It remains to show that $\mathcal{O}_Z(g) = 0$ for every $Z \in \mathcal{B}_\infty(X)$ and every uniformly continuous convex function $g : Z \to \boldsymbol{R}$. To this end, fix $Z \in \mathcal{B}_\infty(X)$ and pick $[\beta,\delta] \in \tilde{\gamma}(g)$, and, for an arbitrary (fixed) $\varepsilon > 0$, let $Y \in \mathcal{B}_\infty(Z)$ be a corresponding stabilizing subspace, satisfying conditions (i) and (ii) from 0.3.

Assume first that $g(x) = g(-x)$ for $g \in S(X)$. Consider the symmetric asymptotic sets A and B defined in 0.5. By Remark in 1.2 (with $D = 1+\varepsilon$) we get a subspace $E \in \mathcal{B}_\infty(Y)$ such that $B \cap E \subset B_E = B_X \cap E \subset (1+\varepsilon)\text{conv}\, A$. Thus, by the convexity and uniform continuity of g, we have

$$\delta - \varepsilon - \theta \le \inf_{z \in B \cap E} g(z) \le \sup_{w \in A} g(w) + \theta \le \beta + \varepsilon + \theta,$$

where $\theta = \theta(\varepsilon) \to 0$ as $\varepsilon \to 0$. Letting $\varepsilon \to 0$ we get $\beta = \delta$, hence $\mathcal{O}_Z(g) = 0$.

If g is arbitrary, set $h(x) = (1/2)(g(x) + g(-x))$. Since $\mathcal{O}_Y(h) = 0$, find a subspace $Y_1 \in \mathcal{B}_\infty(Y)$ such that $|h(x) - s| < \varepsilon$ for all $x \in S(Y_1)$. If ε is small enough then $|s - (\beta+\delta)/2| < 2\varepsilon$, hence $\beta + \varepsilon < s < \delta - \varepsilon$. (In fact, we will use the latter inequality only.) Indeed, otherwise $g(x)$ and $g(-x)$ would not compensate each other. Formally, pick $w, v \in S(Y_1)$ such that $g(w) = \beta + \varepsilon$ and $g(v) = \delta - \varepsilon$ and observe that

$$s - (\beta+\delta)/2 \le \big(s - h(w)\big) + \big((g(w) + g(-w))/2 - (\beta+\delta)/2\big) \le 2\varepsilon;$$

and similarly, using $h(v)$, it is easy to establish the lower estimate by -2ε.

Consider the set $A = \{y \in S(Y_1) \mid |g(y) - s| \le \varepsilon\}$, which is asymptotic in Y_1. Observe that for $y \in A$ we have

$$|g(-y) - s| \le |g(y) + g(-y) - 2s| + |g(y) - s| \le 3\varepsilon,$$

so that A is "almost" symmetric. By Remark in 1.2 we get a subspace $E \in \mathcal{B}_\infty(Y_1)$ such that $B_E \subset (1+\varepsilon)\text{conv}\,(A \cup -A)$. Thus for every $z \in S(E)$ we have,

$$g(z) \le \sup_{w \in A \cup -A} g(w) + \theta \le s + 3\varepsilon + \theta,$$

where $\theta = \theta(\varepsilon) \to 0$ as $\varepsilon \to 0$. This, combined with the stabilization property of h, yields that $\mathcal{O}_Z(g)$ is arbitrarily small, hence equal to 0, as required. □

Remark. In the non-distortion situation of the lemma it is not difficult to show that X contains a subspace with a basis $\{u_i\}$ such that for every n and for all blocks $n < w < v$ we have $\max(\|w+v\|, \|w-v\|) \le (1+2^{-n})\|w+v\|$. A standard argument shows that the tails of $\{u_i\}$ are unconditional with the constants as close to 1 as we wish. We could then consider the moduli related to the family $\mathcal{B} = \mathcal{B}^t$ and note that the argument from 2.3 applies for these moduli as well. Moreover, in this situation it would be clearly sufficient to discuss the observation opening the proof of the lemma only for symmetric functions.

2.3. Sketch of the proof of Theorem 0.4. We will show that the equality $\beta\beta(\varepsilon, F) = \delta\delta(\varepsilon, F)$ yields that the space F contains $(1+\varepsilon)$-isomorphic copies of l_p or c_0.

Using the definitions of $\beta\beta(\varepsilon, F)$ and $\delta\delta(\varepsilon, F)$, it is possible, given $\eta > 0$, to construct a basic sequence $\{y_i\}$ in F such that for any finite sequence $\{a_i\} \in \boldsymbol{R}^{(\boldsymbol{N})}$ of real numbers the norm $\|\sum_i a_i y_i\|$ admits an upper estimate by $(1+\eta)\Phi(\{\delta\delta(|a_i|, F)\})$ and a lower estimate by $(1+\eta)^{-1}\Phi(\{\beta\beta(|a_i|, F)\})$, where $\Phi(\{\cdot\})$ is a real function defined on the space of all finite sequences of real numbers (see [M.71b], Theorem 4.5). Moreover, the same estimates are satisfied for every vector of the form $\sum_i a_i u_i$, where $\{u_i\}$ is a block basis of $\{y_i\}$. (This argument is similar to a well-known construction sketched in 6.4.) By Lemma 2.2, this implies that all block bases of $\{y_i\}$ are $(1+\eta)$-equivalent. By Zippin's theorem, the basis $\{y_i\}$ is $(1+\eta)^\alpha$-equivalent to the unit vector basis in l_p, for some $1 \le p < \infty$ or in c_0 (here $\alpha > 0$ is a numerical constant). □

3. Tilda-spectrum in general

3.1. Let X be a Banach space with a basis $\{x_i\}$. Let $f(z_1, z_2, \dots, z_l)$ be a uniformly continuous real function defined on sequences of l normalized block vectors $z_1 < z_2 < \dots < z_l$. First let us describe a rough intuition of an interval $[\beta, \delta]$ of tilda-spectrum of f on a subspace $\tilde{Y} \in \mathcal{B}_\infty(X)$, leaving the precise definition for later parts of this section.

Let $\tilde{I}$ be the closure of the interval of values of f on $\tilde{Y}$. By restricting the domain of the variable z_l to any subspace of $\tilde{Y}$, with other variables fixed, we do not increase $\tilde{I}$. Therefore, for any $z_1 < z_2 < \dots < z_{l-1}$ fixed, let Y_1 be a subspace of $\tilde{Y}$ such that passing with z_l to Y_1 corresponds to the "maximal decrease" of $\tilde{I}$. Let $I^{(1)}$ be the closure of the interval of values of this restricted f. Continue the procedure of restricting z_{l-1}, with $z_1 < z_2 < \dots < z_{l-2}$ fixed. The closed interval $I^{(l)} = [\beta, \delta]$ obtained after the l-th step is called an interval of the tilda-spectrum of f in $\tilde{Y}$.

3.2. Let $\tilde{Y} \in \mathcal{B}_\infty(X)$. The precise definition of the tilda-spectrum of f on $\tilde{Y}$ involves the notions of the β- and δ- averages, introduced in 1.4. These averages will be applied to functions of the form $h(z_1, \dots, z_k)$, considered as functions of z_k with $z_1 < \dots < z_{k-1}$ fixed, and with respect to the family $\mathcal{B}^t(Y, z_{k-1})$ of all finite-codimensional block subspaces of Y with the support after z_{k-1} ($Y \in \mathcal{B}_\infty(X)$ is a subspace). To make the formulas more compact, we will indicate the variable z_k and the subspace Y in the subscripts, leaving z_{k-1} to be understood from the context. Thus we will write, *e.g.* $\beta_{z_k,Y}(h(z_1, \dots, z_{k-1}, z_k))$ for $\beta[h(z_1, \dots, z_{k-1}, \cdot), \mathcal{B}^t(Y, z_{k-1})]$, and so on.

We say that an interval $[\beta, \delta]$ is in the tilda spectrum of f on $\tilde{Y}$ if there is a subspace $Y \in \mathcal{B}_\infty(\tilde{Y})$ such that the following two conditions are satisfied:

(i) $\beta = \beta_{z_1,Y}\,(\beta_{z_2,Y}(\dots(\beta_{z_l,Y}(f(z_1, z_2, \dots, z_l)))\dots))$
and
$\delta = \delta_{z_1,Y}\,(\delta_{z_2,Y}(\dots(\delta_{z_l,Y}(f(z_1, z_2, \dots, z_l)))\dots))$;

(ii) for all $H_1, H_2, \dots, H_l \in \mathcal{B}_\infty(Y)$, each of the averages $\beta_{z_i,Y}$ and $\delta_{z_i,Y}$ in (i) can be replaced by β_{z_i,H_i} and δ_{z_i,H_i}, respectively; that is, we have
$\beta = \beta_{z_1,H_1}\,(\beta_{z_2,H_2}(\dots(\beta_{z_l,H_l}(f(z_1, z_2, \dots, z_l)))\dots))$
and
$\delta = \delta_{z_1,H_1}\,(\delta_{z_2,H_2}(\dots(\delta_{z_l,H_l}(f(z_1, z_2, \dots, z_l)))\dots))$.

A subspace Y for which the above conditions hold is called a spectrum subspace corresponding to $[\beta, \delta]$. Then any further subspace Y' of Y is a spectrum subspace as well.

3.3. To prove the existence of the tilda-spectrum defined in 3.2, it is convenient to introduce modified averages β^{st} and δ^{st}. The definition requires several steps.

3.3.1. Let $h(z_1, \dots, z_k)$ be a uniformly continuous function and let $E \in \mathcal{B}_\infty(X)$. Fix normalized blocks $z_1 < \dots < z_{k-1}$ and set

$$\gamma(E) = \inf\big(\delta_{z_k,G}(h(z_1, \dots, z_{k-1}, z_k)) - \beta_{z_k,G}(h(z_1, \dots, z_{k-1}, z_k))\big),$$

where the infimum is taken over all subspaces $G \in \mathcal{B}_\infty(E)$. Pick $\varepsilon_i \downarrow 0$ and construct a sequence $G_0 = E \supset G_1 \supset \dots$, such that $G_i \in \mathcal{B}_\infty(G_{i-1})$ and

(3.1)
$$\delta_{z_k,G_i}(h(z_1,\dots,z_{k-1},z_k)) - \beta_{z_k,G_i}(h(z_1,\dots,z_{k-1},z_k)) \le \gamma(G_{i-1}) + \varepsilon_i,$$

for $i = 1, 2, \dots$. Set

$$\begin{aligned} \beta^{st}_{z_k,E}(h) &= \beta^{st}_{z_k,E}(h; z_1,\dots,z_{k-1})) \\ &\qquad = \lim_{i\to\infty} \beta_{z_k,G_i}(h(z_1,\dots,z_{k-1},z_k)), \\ \delta^{st}_{z_k,E}(h) &= \delta^{st}_{z_k,E}(h; z_1,\dots,z_{k-1})) \\ &\qquad = \lim_{i\to\infty} \delta_{z_k,G_i}(h(z_1,\dots,z_{k-1},z_k)). \end{aligned} \tag{3.2}$$

Let $G = \operatorname{span}[u_i]$ be a diagonal subspace for $\{G_i\}$; that is, $u_i \in G_i$ for $i = 1, 2, \dots$. It is easy to check from (3.1) that for every subspace $H \in \mathcal{B}_\infty(G)$ we have

$$\beta^{st}_{z_k,E}(h) = \beta_{z_k,H}(h) \quad \text{and} \quad \delta^{st}_{z_k,E}(h) = \delta_{z_k,H}(h). \tag{3.3}$$

3.3.2. Now let $\{(z_1^{(i)},\dots,z_{k-1}^{(i)})\}$ be a dense countable subset of $(k-1)$-tuples of normalized blocks $z_1 < \dots < z_{k-1}$. Let $G^{(1)} = G$ be the subspace constructed at the end of 3.3.1 for $(z_1^{(1)},\dots,z_{k-1}^{(1)})$. Starting from $G^{(1)}$, construct $G^{(2)} \in \mathcal{B}_\infty(G^{(1)})$ for the tuple $(z_1^{(2)},\dots,z_{k-1}^{(2)})$. Proceeding by induction and using (3.3) we get a sequence of subspaces $E \supset G^{(1)} \supset G^{(2)} \supset \dots$ such that for every $i = 1, 2, \dots$ and all $H, H' \in \mathcal{B}_\infty(G^{(i)})$ we have

$$\begin{aligned} \beta_{z_k,H}(h(z_1^{(i)},\dots,z_{k-1}^{(i)},z_k)) &= \beta^{st}_{z_k,E}(h; z_1^{(i)},\dots,z_{k-1}^{(i)}) \\ &\qquad = \beta_{z_k,H'}(h(z_1^{(i)},\dots,z_{k-1}^{(i)},z_k)), \\ \delta_{z_k,H}(h(z_1^{(i)},\dots,z_{k-1}^{(i)},z_k)) &= \delta^{st}_{z_k,E}(h; z_1^{(i)},\dots,z_{k-1}^{(i)}) \\ &\qquad = \delta_{z_k,H'}(h(z_1^{(i)},\dots,z_{k-1}^{(i)},z_k)). \end{aligned}$$

Taking once more a diagonal subspace we get $F = \operatorname{span}[v_i]$, with $v_i \in G^{(i)}$ for $i = 1, 2, \dots$ such that for all $(k-1)$-tuples $(z_1,\dots,z_{k-1})$ and for all subspaces $H \in \mathcal{B}_\infty(F)$ we have

$$\begin{aligned} \beta_{z_k,H}(h(z_1,\dots,z_{k-1},z_k)) &= \beta^{st}_{z_k,E}(h; z_1,\dots,z_{k-1}) \\ \delta_{z_k,H}(h(z_1,\dots,z_{k-1},z_k)) &= \delta^{st}_{z_k,E}(h; z_1,\dots,z_{k-1}). \end{aligned} \tag{3.4}$$

3.3.3. Coming back to the definition 3.2 of the tilda-spectrum, fix a function $f = f(z_1,\dots,z_l)$ and a subspace $\tilde{Y} \in \mathcal{B}_\infty(X)$.

The existence of a tilda-spectrum interval $[\beta,\delta]$ will be proved by providing explicit formulae for β and δ in terms of the stabilized averages β^{st} and δ^{st}. This is done by backward induction.

Let

$$\begin{aligned} b_1(z_1,\dots,z_{l-1}) &= \beta^{st}_{z_l,\tilde{Y}}(f; z_1,\dots,z_{l-1}) \\ d_1(z_1,\dots,z_{l-1}) &= \delta^{st}_{z_l,\tilde{Y}}(f; z_1,\dots,z_{l-1}), \end{aligned}$$

and let $F_1 \in \mathcal{B}_\infty(\tilde{Y})$ be the subspace constructed at the end of 3.3.2 for which (3.4) is satisfied.

Repeat the procedure inside F_1 by setting

$$\begin{aligned} b_2(z_1,\dots,z_{l-2}) &= \beta^{st}_{z_{l-1},F_1}(b_1; z_1,\dots,z_{l-2}) \\ d_2(z_1,\dots,z_{l-2}) &= \delta^{st}_{z_{l-1},F_1}(d_1; z_1,\dots,z_{l-2}), \end{aligned}$$

and let $F_2 \in \mathcal{B}_\infty(F_1)$ be the corresponding subspace.

Proceed by an obvious induction to get functions b_i and d_i for $i = 1,\dots,l$ and subspaces $\tilde{Y} \supset F_1 \supset \dots \supset F_l$. Set

$$\begin{aligned} \beta = b_l &= \beta^{st}_{z_1,F_{l-1}}(b_{l-1}) \\ &= \beta^{st}_{z_1,F_{l-1}}\left(\beta^{st}_{z_2,F_{l-2}}\left(\dots\left(\beta^{st}_{z_l,\tilde{Y}}(f; z_1,\dots,z_{l-1})\right)\dots\right)\right), \\ \delta = d_l &= \delta^{st}_{z_1,F_{l-1}}(d_{l-1}) \\ &= \delta^{st}_{z_1,F_{l-1}}\left(\delta^{st}_{z_2,F_{l-2}}\left(\dots\left(\delta^{st}_{z_l,\tilde{Y}}(f; z_1,\dots,z_{l-1})\right)\dots\right)\right). \end{aligned} \tag{3.5}$$

It is easy to see, using (3.4), that with these definitions of β and δ, the interval $[\beta,\delta]$ satisfies conditions (i) and (ii) of 3.2 for the subspace $Y = F_l$.

3.4. Using the definition of the β- and δ- averages it is easy to see that condition (i) of the definition of the tilda-spectrum in 3.2 is equivalent to the following:

(i') $\forall \theta > 0\ \exists Y_1 \in \mathcal{B}^t(Y)\ \forall z_1 \in S(Y_1)\ \exists Y_2 \in \mathcal{B}^t(Y_1, z_1)\ \forall z_2 \in S(Y_2)$ $\exists Y_3 \in \mathcal{B}^t(Y_2, z_2)\ \forall z_3 \in S(Y_3)\ \dots\ \exists Y_l \in \mathcal{B}^t(Y_{l-1}, z_{l-1})\ \forall z_l \in S(Y_l)$ $f(z_1, z_2,\dots,z_l) \in [\beta - \theta, \delta + \theta]$.

We will show below that this condition implies in fact a stronger property; that is, the existence of a stabilizing subspace for f. Given an interval $[\beta,\delta]$ satisfying condition (i) on a subspace Y we can construct, for any $\theta > 0$, a subspace $G \in \mathcal{B}_\infty(Y)$ such that for all normalized blocks $z_1 < z_2 < \dots < z_l$ in G we have

$$f(z_1, z_2,\dots,z_l) \in [\beta - \theta, \delta + \theta]. \tag{3.6}$$

Fix $\theta' > 0$ and $\eta > 0$ to be defined later. Let $E_1 = Y_1 \in \mathcal{B}^t(Y)$ be the subspace satisfying condition (i') for θ'. Pick an arbitrary vector $u_1 \in S(E_1)$, and let $E_2 = Y_2 \in \mathcal{B}^t(Y_1, u_1)$ be the subspace from condition (i') (again for θ'). Pick an arbitrary vector $u_2 \in S(E_2)$.

In the next step we would like to find a subspace $E_3 \in \mathcal{B}^t(E_2, u_2)$ which would satisfy condition (i') in several ways: it could be taken as Y_3, for vectors $z_1 = u_1$ and $z_2 = u_2$, and it could be taken as Y_2, for an arbitrary vector z_1 running over some finite η-net $\mathcal{N}$ (in the original norm) on the sphere $S(\operatorname{span}[u_1, u_2])$. Since subspaces appearing in (i') are always of finite codimension, it is clear that a required subspace E_3 exists. Then pick an arbitrary $u_3 \in S(E_3)$.

Continuing in an obvious manner we construct a subspace $G = \operatorname{span}[u_i]$ such that $f(z_1, z_2,\dots,z_l) \in [\beta - \theta', \delta + \theta']$ for all $z_1 < \dots < z_l$ with z_is running over all finite η-nets on the spheres $S(\operatorname{span}[z_1,\dots,z_k])$ (with $k = 2, 3,\dots$). Choosing suitable $\eta > 0$ and $\theta' > 0$ depending on θ, we complete the proof of (3.6).

Remark. Given a sequence $\theta_n \downarrow 0$ we can repeat the above construction for every n and then pass to a diagonal subspace. We then obtain a subspace

$Z \in \mathcal{B}_\infty(Y)$ with a (block) basis $\{v_i\}$ such that for every n and for arbitrary normalized blocks $n < z_1 < \ldots < z_l$ of $\{v_i\}$ we have

$$f(z_1, \ldots, z_l) \in [\beta - \theta_n, \delta + \theta_n].$$

3.5. The role of condition (ii) of the definition of the tilda-spectrum is to ensure the existence of large sets of vectors on which the value of the function f is close to extremal. In fact, these sets turn out to be asymptotic in some stabilizing subspace for f.

Let us start by observing that condition (ii) from 3.2 is equivalent to the following:

(ii') for all $\eta > 0$ and all subspaces $H_1, H_2, \ldots, H_l \in \mathcal{B}_\infty(Y)$ we have: $\exists w_1 \in S(H_1)\ \exists w_2 \in S(H_2), w_2 > w_1\ \ldots\ \exists w_l \in S(H_l), w_l > w_{l-1}$ $f(w_1, w_2, \ldots, w_l) \le \beta + \eta$
and
$\exists v_1 \in S(H_1)\ \exists v_2 \in S(H_2), v_2 > v_1\ \ldots\ \exists v_l \in S(H_l), v_l > v_{l-1}$ $f(v_1, v_2, \ldots, v_l) \ge \delta - \eta$.

Let $[\beta, \delta]$ be in the tilda-spectrum of f and let Z be the corresponding stabilizing subspace constructed in Remark 3.4. Condition (ii') leads to the natural definition of sets asymptotic in Z.

With a fixed $\eta > 0$ define $A_1 \subset S(Z)$ by

$$\begin{aligned} A_1 = \big\{ w_1 \in S(Z) \bigm| & \forall H_2, \ldots, H_l \in \mathcal{B}_\infty(Z) \\ & \exists w_2 \in S(H_2), w_2 > w_1\ \exists w_3 \in S(H_3), w_3 > w_2\ \ldots \\ & \exists w_l \in S(H_l), w_l > w_{l-1}\ \ f(w_1, w_2, \ldots, w_l) \le \beta + \eta \big\}. \end{aligned}$$

By (ii'), the set A_1 has a non-empty intersection with every subspace $H_1 \in \mathcal{B}_\infty(Z)$, hence A_1 is asymptotic in Z.

By induction, let $1 \le k < l$, and assume that for any fixed $w_1 < \ldots < w_{k-1}$, with $w_i \in A_i$ for $i = 1, \ldots, k-1$, the set $A_k = A_k(w_1, \ldots, w_{k-1})$ has been defined by the formula

$$\begin{aligned} A_k = \big\{ w_k \in S(Z) \bigm| & \forall H_{k+1}, \ldots, H_l \in \mathcal{B}_\infty(Z) \\ & \exists w_{k+1} \in S(H_{k+1}), w_{k+1} > w_k\ \exists w_{k+2} \in S(H_{k+2}), w_{k+2} > w_{k+1}\ \ldots \\ & \exists w_l \in S(H_l), w_l > w_{l-1}\ \ f(w_1, w_2, \ldots, w_l) \le \beta + \eta \big\}. \end{aligned} \tag{3.7}$$

Moreover, assume that A_k is asymptotic in Z. Then for any fixed $w_1 < \ldots < w_k$, with $w_i \in A_i$ for $i = 1, \ldots, k$, define A_{k+1} by the formula analogous to (3.7). It clearly follows from the form of A_k that A_{k+1} is asymptotic in Z.

Similarily, we can define sets $U_k \subset S(Z)$ for $j = k, \ldots, l$, which are also asymptotic in Z, and such that if $v_1 < \ldots < v_l$ and $v_i \in U_i$ for $i = 1, \ldots, l$ then $f(v_1, \ldots, v_l) \ge \delta - \eta$.

3.6. The notion of tilda-spectrum has the following unconditionality property. For a given function f and a finite sequence $\varepsilon = (\varepsilon_1, \varepsilon_2, \ldots)$ with $\varepsilon_1 = \pm 1$, $\varepsilon_2 = \pm 1$, …, define the function f_ε by

$$f_\varepsilon(z_1, z_2, \ldots, z_l) = f(\varepsilon_1 z_1, \varepsilon_2 z_2, \ldots, \varepsilon_l z_l),$$

for normalized blocks $z_1 < \ldots, z_l$. Then if $[\beta, \delta]$ is in the tilda-spectrum of f and Y is a corresponding spectrum subspace, then for all the f_εs, $[\beta, \delta]$ is again a spectrum interval with the same spectrum subspace Y. Moreover, the stabilizing subspace $Z \subset Y$ of 3.4 is also preserved for all the f_εs. Note however, that the asymptotic sets A_i and U_i described in 3.5 are not the same.

3.7. The final important step in our discussion of tilda-spectrum is an observation that a construction of stabilizing subspaces in 3.4 can be done "almost" simultaneously for any countable family of uniformly continuous real functions, $f_k = f_k(z_1, \dots, z_{l_k})$. For a subspace $\tilde{Y} \in \mathcal{B}_\infty(X)$ and a sequence $\theta_k \downarrow 0$, there exists $Z \in \mathcal{B}_\infty(\tilde{Y})$ such that for every $n = 1, 2, \dots$, if $L_n = \max_{k \le n} l_k$, then for arbitrary normalized blocks $n < z_1 < \dots < z_{L_n}$ in Z we have

$$f_k(z_1, \dots, z_{l_k}) \in [\beta_k - \theta_n, \delta_k + \theta_n] \quad \text{for} \quad 1 \le k \le n.$$

Here $[\beta_k, \delta_k]$ is an interval in the tilda-spectrum of f_k in $\tilde{Y}$. Moreover, all the sets $A_i^{(k)}$ and $U_i^{(k)}$ for $i = 1, 2, \dots$, constructed in 3.5 for the function f_k, are asymptotic in Z.

This follows from 3.4 and 3.5 by the standard diagonal procedure. The details are left for the reader.

4. Spaces with bounded distortions

We now pass to the main theorem on spaces with bounded distortions, Theorem 0.7.

4.1. By passing to an infinite-dimensional subspace of X and considering a suitable renorming of X we may assume, without loss of generality, that X has a monotone basis. The proof of the theorem relies on stabilization properties of a family of real functions which we introduce now and fix throughout the rest of the argument. This family is indexed by the set $\mathbb{Q}_+^{(\mathbb{N})}$ of all finite sequences with positive rational coordinates; for $a = (a_1, \dots, a_l) \in \mathbb{Q}_+^{(\mathbb{N})}$ define the function f_a on a sequence $z_1 < \dots < z_l$ of normalized blocks by

$$f_a(z_1, \dots, z_l) = \Big\| \sum_{i=1}^{l} a_i z_i \Big\|.$$

Let $\{a^{(k)}\}$ be an enumeration of $\mathbb{Q}_+^{(\mathbb{N})}$. We will write f_k for $f_{a^{(k)}}$. Fix $\theta_k \downarrow 0$ satisfying $\theta_k < (1/2D)\|a^{(k)}\|_\infty$ for $k = 1, 2, \dots$. Let $Z \in \mathcal{B}_\infty(X)$ be the stabilizing subspace for all the f_ks, constructed in 3.7. Let $[\beta_k, \delta_k]$ denote the corresponding spectrum intervals for f_k $(k = 1, 2, \dots)$.

The major role in our approach is played by two positive real valued functions on $\mathbb{Q}^{(\mathbb{N})}$ defined via tilda-spectrum of the f_ks as follows. With fixed θ_k, Z, and $[\beta_k, \delta_k]$, as above, we let, for $a = a^{(k)} \in \mathbb{Q}_+^{(\mathbb{N})}$,

$$g(a) = \beta_k \qquad \text{and} \qquad r(a) = \delta_k. \tag{4.1}$$

These definitions can be naturally extended to all $\mathbb{Q}^{(\mathbb{N})}$, by setting, for $\pm a = (\pm a_1, \dots, \pm a_l)$, $g(\pm a) = g(a)$ and $r(\pm a) = r(a)$. We call the functions g and r the *enveloping functions* of X.

4.2. The main part of the proof of the theorem is contained in the following proposition concerning the behaviour of functions g and r for spaces with bounded distortions.

Proposition. *Assume that a Banach space X has bounded distortions and let $d(X) < D$. With the notation from 4.1, either there exists $1 \le p < \infty$ such that*

$$(1/D')\left(\sum |a_i|^p\right)^{1/p} \le g(a) \le r(a) \le D'\left(\sum |a_i|^p\right)^{1/p} \quad \textit{for } a \in \mathbf{Q}^{(\mathbf{N})},$$

or

$$(1/D')\max|a_i| \le g(a) \le r(a) \le D' \max|a_i| \quad \textit{for } a \in \mathbf{Q}^{(\mathbf{N})}.$$

where $D' = 4D$.

4.3. Assuming the truth of Proposition 4.2 let us complete the proof of the theorem.

Proof of Therem 0.7. Let $\{x_i\}$ denote the block basis for the stabilizing subspace Z, which has been fixed in 4.1. Assume that the conclusion of Proposition 4.2 is satisfied for some $1 \le p \le \infty$ (with the obvious convention for $p = \infty$). We will then show that Z is asymptotic-l_p (or asymptotic-c_0, if $p = \infty$).

Fix n and fix $\varepsilon = \varepsilon(n) > 0$ to be defined later. Let $b^{(1)}, \dots, b^{(M)}$ be an ε-net in the unit sphere $S(l_p^n)$ of l_p^n in the l_∞-norm, and assume without loss of generality that $b^{(i)} \in \mathbf{Q}^{(\mathbf{N})}$ for $1 \le i \le M$. Thus for every i there is $k = k_i$ such that $b^{(i)} = a^{(k)}$. Let $N = \max_{1 \le i \le M} k_i$.

Let $N < y_1 < \dots < y_n$ be arbitrary normalized blocks of $\{x_k\}$. By 3.7 and Proposition 4.2 we have, for every $1 \le i \le M$,

$$\begin{aligned}(1/D)\|b^{(i)}\|_p - \theta_k &\le g(b^{(i)}) - \theta_k \le \|b_1^{(i)} y_1 + \dots + b_n^{(i)} y_n\| \\ &\le r(b^{(i)}) + \theta_k \le D\|b^{(i)}\|_p + \theta_k.\end{aligned}$$

By the choice of θ_k this implies

$$(1/2D) \le \|b_1^{(i)} y_1 + \dots + b_n^{(i)} y_n\| \le 2D.$$

An easy approximation argument shows that if ε is sufficiently small (it is enough to take $\varepsilon = (4Dn)^{-1}$) then the latter estimates imply

$$(1/4D) \le \|c_1 y_1 + \dots + c_n y_n\| \le 4D,$$

for any $(c_1, \dots, c_n) \in S(l_p^n)$. Thus $y_1 < \dots < y_n$ are D'-equivalent to the unit vector basis in l_p^n, as required. □

5. Inequalities for enveloping functions

The proof of Proposition 4.2 is based on specific properties of functions g and r in spaces with bounded distortions, which will be established in this section. In what follows we keep the notation from 4.1, and in particular, $Z \in \mathcal{B}_\infty(X)$ is the stabilizing subspace for the functions $\{f_k\}$, constructed in 3.7.

5.1. A space Y with a basis $\{y_i\}$ is said to be *asymptotically unconditional* if there exists a constant D' such that for every n there exists $N = N(n)$ such that for any normalized blocks $N < z_1 < \ldots < z_n$ of $\{y_i\}$ and any sequence of reals $(c_1, \ldots, c_n)$ we have

$$\sup_{\varepsilon_i=\pm 1} \|\varepsilon_1 c_1 z_1 + \ldots + \varepsilon_n c_n z_n\| \le D' \inf_{\varepsilon_i=\pm 1} \|\varepsilon_1 c_1 z_1 + \ldots + \varepsilon_n c_n z_n\|. \tag{5.1}$$

Lemma. *Let X be a Banach space with bounded distortions. Then it contains a subspace $Y \in \mathcal{B}_\infty(X)$ which is asymptotically unconditional.*

Proof. Assume that X has a basis $\{x_i\}$. Given $\alpha > 0$, we will construct a block basis $\{y_i\}$ for which (5.1) holds with the constant $D' = (1+\alpha)\, d(X)$ and $N(n) = n$.

We will use a common and convenient notation that if I and J are intervals of positive integers then $I < J$ means $\max_{i\in I} i < \min_{j\in J} j$. Moreover, for $x = \sum_i t_i x_i \in X$, we set $Ix = \sum_{i\in I} t_i x_i$.

We may assume that X does not contain c_0, otherwise the proof would be finished. For a positive integer n define the norm $|||\cdot|||_n$ on X by

$$|||x|||_n = \sup\Big\|\sum_{i=1}^{n} \varepsilon_i I_i x\Big\|,$$

where the supremum is taken over all intervals $I_1 < \ldots < I_n$ and all $\varepsilon_i = \pm 1$, $i = 1, \ldots, n$. Clearly, $\|x\| \le |||x|||_n \le n\|x\|$ for $x \in X$, so $|||\cdot|||_n$ is an equivalent norm on X. We will show that the set

$$A_n = \Big\{x \in S(X) \mid |||x|||_n \le (1+\alpha)\Big\} \tag{5.2}$$

is asymptotic in X. Thus, by the Remark in 1.2 and (1.1), for every $X' \in \mathcal{B}_\infty(X)$ there is $F_n \in \mathcal{B}_\infty(X')$ such that $|||x|||_n \le (1+\alpha)D\|x\|$ for $x \in F_n$. This leads to the inductive construction of subspaces $X = F_0 \supset F_1 \supset \ldots \supset F_n \supset \ldots$ with $F_n \in \mathcal{B}_\infty(F_{n-1})$ and $\|x\| \le |||x|||_n \le (1+\alpha)D\|x\|$ for $x \in F_n$ $(n = 1, 2, \ldots)$. It is easy to check that any block basis $\{y_n\}$ such that $y_n \in S(F_n)$ satisfies (5.1) with $D' = (1+\alpha)D$.

To show that the set A_n given by (5.2) is asymptotic in X, let $W = \operatorname{span}[w_i] \in \mathcal{B}_\infty(X)$. Fix $N > n$ to be defined later. Let

$$a_N = \sup\Big\{\Big\|\sum_{j=1}^{N} \eta_j w_j\Big\| \mid \eta_j = \pm 1\Big\} = \Big\|\sum_{j=1}^{N} \eta_j^0 w_j\Big\|.$$

Since X does not contain c_0, we have $a_N \to \infty$, as $N \to \infty$.

Set $w = \sum_{j=1}^N \eta_j^0 w_j$. Given intervals $I_1 < \ldots < I_n$, let L_i be the set of all j such that $\operatorname{supp} w_j \subset I_i$ and let K_i be the set of all j such that $I_i w_j \ne 0$, for $i = 1, \ldots, n$. For every $\varepsilon_i = \pm 1$, with $i = 1, \ldots, n$, we have

$$\Big\|\sum_{i=1}^{n} \varepsilon_i I_i w\Big\| \le \Big\|\sum_{i=1}^{n} \varepsilon_i\Big(\sum_{j\in L_i} \eta_j^0 w_j\Big)\Big\| + 2n \le a_N + 2n.$$

Therefore $|||w/a_N|||_n \le 1 + 2n/a_N$. Thus, if $a_N > 2n/\alpha$, then $w/a_N \in A_n \cap S(W)$ hence A_n is asymptotic in X. $\square$

5.2. Let X be a Banach space with bounded distortions, let $Y \in \mathcal{B}_\infty(X)$ be an asymptotically unconditional subspace of X, for some constant D' arbitrarily close to D which can be chosen later, and let $Z \in \mathcal{B}_\infty(Y)$ be the stabilizing subspace for $\{f_k\}$ constructed in 3.7.

The following lemma investigates the behaviour of enveloping functions g and $r : \boldsymbol{Q}^{(\boldsymbol{N})} \to \boldsymbol{R}$.

Lemma. *Assume that a Banach space X has bounded distortions and let $d(X) < D$. Then*

$$g(a) \le r(a) \le 3Dg(a) \qquad \text{for } a \in \boldsymbol{Q}^{(\boldsymbol{N})}.$$

Proof. The left hand side inequality is obvious. To prove the right hand side inequality, for $i = 1, \dots, l_k$ and $k = 1, 2, \dots$, let $A_i^{(k)}$ and $U_i^{(k)}$ denote the ith asymptotic sets, constructed for the function f_k and the subspace Z, as in 3.5.

We will prove that, with fixed $a = a^{(k)} \in \boldsymbol{Q}_+^{(\boldsymbol{N})}$, we have

$$\exists G_1 \in \mathcal{B}_\infty(Z) \ \ \forall z_1 \in S(G_1) \ \exists G_2 \in \mathcal{B}_\infty(G_1, z_1) \ \forall z_2 \in S(G_2) \dots$$

$$f_k(z_1, z_2, \dots, z_{k_l}) \le D(\beta_k + \theta_k). \tag{5.3}$$

Applying (5.3) to vectors from the appropriate sets $U_i^{(k)}$, for $i = 1, \dots, l_k$, we get

$$\delta_k - \theta_k \le f_k(z_1, z_2, \dots, z_{l_k}) \le D(\beta_k + \theta_k).$$

By the choice of θ_k from 4.1 we have $\theta_k \le (1/2D)\|a^{(k)}\|_\infty \le \beta_k/2$, the latter inequality yields

$$r(a) = \delta_k \le 3D\beta_k \le 3Dg(a). \tag{5.4}$$

To prove that (5.3) holds, first note that since Z is asymptotically unconditional, we can assume, without loss of generality that all the sets $A_i^{(k)}$s are symmetric about the origin. Consider $A_1^{(k)} \subset S(Z)$. By the Remark in 1.2 there exists $G_1 \in \mathcal{B}_\infty(Z)$ such that

$$(1/D)(B_X \cap G_1) \subset \operatorname{conv} A_1^{(k)}.$$

Thus for every $z_1 \in S(G_1)$ there exist $w^1, \dots, w^m \in A_1^{(k)}$ and $t_1, \dots, t_m$, with $t_j > 0$ and $\sum_j t_j = 1$, such that $(1/D)z_1 = \sum_j t_j w^j$.

Now set $G_{2,0} = G_1$ and proceed by induction in $j = 1, \dots, m$. For $j \ge 1$ consider the set $A_2^{(k)}(w^j) \cap G_{2,j-1}$ constructed for the vector w^j.Arguing as before we get a subspace $G_{2,j} \in \mathcal{B}_\infty(G_{2,j-1}, w^j)$ such that

$$(1/D)(B_X \cap G_{2,j}) \subset \operatorname{conv}(A_2^{(k)}(w^j)).$$

Let $G_2 = G_{2,m} \in \mathcal{B}_\infty(G_1, z_1)$. For a fixed $j = 1, \dots, m$, we have, for an arbitrary vector $z_2 \in S(G_2)$,

$$(1/D)z_2 = \sum_{n=1}^{m'} s_{n,j} u^{n,j},$$

with $u^{n,j} \in A_2^{(k)}(w^j)$ and $s_{n,j} > 0$ for $n = 1, \dots, m'$ and $\sum_n s_{n,j} = 1$.

We repeat the process l_k times, for all subsets $A_i^{(k)}$ with $i = 1, 2, \ldots, l_k$. We then have, by the definitions of f_k and of the sets $A_i^{(k)}$,

$$\begin{aligned} f_k(z_1, z_2, \ldots, z_{l_k}) &= D\left\|(1/D)a_1 z_1 + (1/D)a_2 z_2 + \cdots\right\| \\ &\le D\sum_j t_j \left\|a_1 w^j + (1/D)a_2 z_2 + \cdots\right\| \\ &\le D\sum_j t_j \sum_n s_{n,j}\left\|a_1 w^j + a_2 u^{n,j} + \cdots\right\| \\ &\le \cdots \le D(\beta_k + \theta_k), \end{aligned}$$

which is the required estimate (5.3). □

5.3. Next lemma establishes general properties of the function $r : \boldsymbol{Q}^{(\boldsymbol{N})} \to \boldsymbol{R}$. It says that r can be extended in a natural way to the 1-unconditional and 1-subsymmetric norm on the space $\boldsymbol{R}^{(\boldsymbol{N})}$ of all real finite sequences. Moreover, denoting by $\{e_i\}$ the standard unit vector basis in $\boldsymbol{R}^{(\boldsymbol{N})}$, the extended norm has certain blocking property.

Recall that a norm $|\cdot|$ on $\boldsymbol{R}^{(\boldsymbol{N})}$ is 1-subsymmetric if for every sequence $a = (a_1, a_2, a_3, \ldots) \in \boldsymbol{R}^{(\boldsymbol{N})}$ we have $|(a_1, 0, \ldots, 0, a_2, 0, \ldots, 0, a_3, \ldots)| = |a|$.

Lemma. *The function $r(\cdot)$ can be extended to the 1-unconditional and 1-subsymmetric norm on $\boldsymbol{R}^{(\boldsymbol{N})}$. If $\{u_i\}$ is a block basis of the standard unit vector basis with $r(u_i) = 1$ for $i = 1, 2, \ldots$, then for every $a \in \boldsymbol{R}^{(\boldsymbol{N})}$ we have*

$$g(a) \le r(d) \le r(a). \tag{5.5}$$

where for $a = (a_1, \ldots, a_l) = \sum_i a_i e_i$, by d we denote the sequence $d = \sum_i a_i u_i$.

Proof. Consider $r(\cdot)$ as a function on $\boldsymbol{Q}^{(\boldsymbol{N})}$. It is clearly positively homogeneous and we will prove the triangle inequality by showing that if $a, b \in \boldsymbol{Q}^{(\boldsymbol{N})}$ then $r(a + b) \le r(a) + r(b)$.

Indeed, let $a = a^{(i)} = (a_1, a_2, \ldots)$, $b = a^{(j)} = (b_1, b_2, \ldots)$, and $a + b = a^{(m)}$. Fix an arbitrary $\eta > 0$. Pick arbitrary vectors $w_j \in U_j^{(m)}$, for $j = 1, 2, \ldots$. Then

$$r(a + b) - \eta \le \left\|(a_1 + b_1)w_1 + (a_2 + b_2)w_2 + \cdots\right\|,$$

moreover,

$$\|a_1 w_1 + a_2 w_2 + \cdots\| \le r(a) + \eta \quad \text{and} \quad \|b_1 w_1 + b_2 w_2 + \cdots\| \le r(b) + \eta.$$

Since $\eta > 0$ is arbitrary, this shows the triangle inequality.

Recall that for a sequence $a = (a_1, a_2, \ldots)$, we set $\pm a = (\pm a_1, \pm a_2, \ldots)$. Then we have $r(a) = r(\pm a)$, hence the norm $r(\cdot)$ is 1-unconditional. It is also clearly 1-subsymmetric.

To prove (5.5), let $u_i = \sum_{j=k_i+1}^{k_{i+1}} b_j e_j$, for some $0 \le k_1 < k_2 < \ldots$, be a block basis with rational coefficients of the standard unit vector basis, with $r(u_i) = 1$ for $i = 1, 2, \ldots$. Let $a = (a_1, \ldots, a_l) = \sum_i a_i e_i$, then $d = \sum_i a_i u_i = (d_1, \ldots, d_{k_{l+1}})$, where $d_j = a_i b_j$ for $k_i < j \le k_{i+1}$ and $i = 1, 2, \ldots$.

Let $\eta > 0$. There exist vectors $w_1 < w_2 < \dots < w_{k_{l+1}}$ in appropriate asymptotic sets such that

$$r(d) - \eta \leq \Big\| \sum_{j=1}^{k_{l+1}} d_j w_j \Big\|.$$

Since all the vectors belong to Z, we also have, from the form of d,

$$c_i = \Big\| \sum_{j=k_i+1}^{k_{i+1}} d_j w_j \Big\| \leq |a_i|(r(u_i) + \eta) = |a_i|(1+\eta),$$

for $i = 1, \dots, l$. By the triangle inequality and the unconditionality of the norm $r(\cdot)$ this implies $r(c_1, \dots, c_l) \leq r(a)(1+\eta)$.

Setting $w^i = (1/c_i) \sum_{j=k_i+1}^{k_{i+1}} d_j w_j$ we get

$$r(d) - \eta \leq \Big\| \sum_{i=1}^{l} c_i w^i \Big\| \leq r(c_1, \dots, c_l) + \eta$$

which combined with the previous inequality shows the right hand side of (5.5).

The proof of the left hand side inequality is similar. For an arbitrary $\eta > 0$, pick vectors $w'_1 < w'_2 < \dots < w'_{k_{l+1}}$ in appropriate asymptotic sets (for appropriate functions f_i) such that for every $i = 1, \dots, l$ one has

$$1 - \eta = r(u_i) - \eta \leq \Big\| \sum_{j=k_i+1}^{k_{i+1}} b_j w'_j \Big\| \leq r(u_i) + \eta = 1 + \eta.$$

Then

$$\Big\| \sum_i a_i \sum_{j=k_i+1}^{k_{i+1}} b_j w'_j \Big\| = \Big\| \sum_{j=1}^{k_{l+1}} d_j w'_j \Big\| \leq r(d) + \eta.$$

On the other hand, setting $v_i = \sum_{j=k_i+1}^{k_{i+1}} b_j w'_j$ for $i = 1, \dots, l$ we have $1 - \eta \leq \|v_i\| \leq 1 + \eta$. Then

$$\Big\| \sum_i a_i \sum_{j=k_i+1}^{k_{i+1}} b_j w'_j \Big\| = \Big\| \sum_{i=1}^{l} a_i v_i \Big\| \geq (g(a) - \eta)(1 - \eta).$$

Combining the last two estimates we complete the proof of the left hand side of (5.5). □

5.4. Now we can easily complete the proof of Proposition 4.2.

Proof of Proposition 4.2. Let L denotes the completion of $(\boldsymbol{R}^{(\boldsymbol{N})}, r(\cdot))$, and let $\{e_i\}$ be the standard unit vector basis in L. Lemmas 5.2 and 5.3 yield that all normalized block bases of $\{e_i\}$ are (3D)-equivalent to $\{e_i\}$. By Zippin's theorem, this implies that the $\{e_i\}$ is equivalent to the standard unit vector basis in l_p, for some $1 \leq p < \infty$ or in c_0. Moreover, the equivalence constant depends on D only. □

6. Asymptotic l_p spaces, general properties

We conclude this paper with few simple remarks on general asymptotic l_p spaces. To avoid tiresome repetitions, when talking about spaces l_p or asymptotic l_p, respectively, we adopt the convention that the case of $p = \infty$ corresponds to the space c_0 or asymptotic c_0, respectively.

Let Y with a basis $\{y_i\}$ be an asymptotic l_p space for $1 \le p \le \infty$, and let $\lambda_p(Y)$ be the asymptotic l_p constant, as defined in 0.7.

6.1. It is well-known and easy to see that any block subspace of l_p is complemented. The same is true in Tsirelson space and in its convexifications (*cf. e.g.*, [C-S]), although in this case the argument is much more complicated. An analogous fact for arbitrary asymptotic l_p space says that finite-dimensional block subspaces far out are uniformly complemented.

More precisely, for $C > \lambda_p(Y)$, if $N(n) = N < z_1 < \ldots < z_n$ are normalized blocks C-equivalent to the standard unit vector basis in l_p^n, then there exists a projection P from Y onto $\operatorname{span}[z_i]_{i=1}^n$ with $\|P\| \le 2C^2$.

Indeed, pick $z_i^* \in Y^*$ such that $\|z_i^*\| = z_i^*(z_i) = 1$ for $i = 1, \ldots, n$. Let $N < E_1 < \ldots < E_n$ be intervals of positive integers such that $\operatorname{supp} z_i \subset E_i$, for $i = 1, \ldots, n$, and that the union of all the E_is is an interval. For $x \in Y$ set

$$Px = \sum_{i=1}^n z_i^*(E_i x) z_i .$$

Since $E_j z_i = 0$ if $i \ne j$, then P is a projection. Moreover, we have

$$\begin{aligned} \|Px\| &\le C\Big(\sum_{i=1}^n |z_i^*(E_i x)|^p\Big)^{1/p} \le C\Big(\sum_{i=1}^n \|E_i x\|^p\Big)^{1/p} \\ &\le C^2 \Big\| \sum_{i=1}^n \|E_i x\| \, \Big(\frac{E_i x}{\|E_i x\|}\Big) \Big\| \le C^2 \Big\| \sum_{i=1}^n E_i x \Big\| \le 2C^2 \|x\|, \end{aligned}$$

as required.

6.2. If Y is an asymptotic l_p space (for $1 < p \le \infty$) then the dual Y^* is an asymptotic $l_{p'}$ space. This follows from 6.1 by a general duality argument. A direct calculation is just as standard and simple and we leave it to the reader.

6.3. It is easy to observe that if $p > 1$, the basis in Y is shrinking and if $p < \infty$, the basis is boundedly complete. Hence for $1 < p < \infty$, an asymptotic l_p space is reflexive.

Assume to the contrary that the basis $\{y_i\}$ is not shrinking. There exists $x^* \in Y^*$ with $\|x^*\| = 1$, and $\delta > 0$, and a normalized block basis $\{u_i\}$ of $\{y_i\}$ such that $|x^*(u_i)| > \delta$. Fix n to be defined later. Then for every k we have

$$\Big\| \sum_{i=k}^{k+n-1} u_i \Big\| \ge \Big| x^*\Big(\sum_{i=k}^{k+n-1} u_i \Big) \Big| \ge n\delta .$$

On the other hand, if k is large enough, the left hand side is smaller than or equal to $C\, n^{1/p}$. Chosing appropriate n we get a contradition, if $p > 1$.

Assume the basis is not boundedly complete. Then there exists normalized block basis $\{u_i\}$ of $\{y_i\}$ such that $\sup_n \|\sum_{i=1}^n u_i\| = M < \infty$. On the other hand if $u_k < \ldots < u_{k+n-1}$ is far enough, then

$$\Big\|\sum_{i=k}^{k+n-1} u_i\Big\| \geq (1/C)\, n^{1/p}.$$

If $p < \infty$, we again come to a contradition by an appropriate choice of n.

6.4. The notion of asymptotic l_p spaces is fundamentally an isomorphic concept and it cannot be reduced to a $(1+\varepsilon)$- isometric one. As mentioned in 0.7, to the contrary to the local concept of finite representability of l_p, asymptotic l_p does not imply any related almost isometric property of a block subspace. To be more precise let us discuss the quantity $\lambda_p(Y)$ in more detail.

Note that the isometric condition $\lambda_p(Y) = 1$ is equivalent to the fact that for every $\varepsilon > 0$ and for every n there exists $N = N(\varepsilon, n)$ such that any normalized blocks $N < z_1 < z_2 < \ldots < z_n$ are $(1+\varepsilon)$-equivalent to the standard unit vector basis in l_p^n. (In 0.7 we called such a space an *almost isometric asymptotic* l_p space.)

Recall that if a space Y merely satisfies a weaker condition: for every $\varepsilon > 0$ there exists $N = N(\varepsilon)$ such that any two normalized blocks $N < z_1 < z_2$ are $(1+\varepsilon)$-equivalent to the standard unit vector basis in l_p^2, then, for every $\varepsilon > 0$, Y contains an $(1+\varepsilon)$-isomorphic copy of the l_p-space. Let us sketch this well-known and standard argument.

Given $\varepsilon > 0$, fix $\varepsilon_i \downarrow 0$ such that $\prod_i (1+\varepsilon_i) \leq (1+\varepsilon)$. By an easy induction pick a sequence of normalized blocks $u_1 < u_2 < \ldots < u_i < \ldots$ such that $N(\varepsilon_i) < u_i$ for $i = 1, 2, \ldots$. Then the block basis $\{u_i\}$ is $(1+\varepsilon)$-equivalent to the standard unit vector basis in l_p. Indeed, for any finite sequence of scalars $\{a_i\}$ we have

$$\begin{aligned}\Big\|\sum_{i=1}^{\infty} a_i u_i\Big\| &\leq (1+\varepsilon_1)\left(|a_1|^p + \Big\|\sum_{i=2}^{\infty} a_i u_i\Big\|^p\right)^{1/p}\\ &\leq (1+\varepsilon_1)(1+\varepsilon_2)\left(|a_1|^p + |a_2|^p + \Big\|\sum_{i=3}^{\infty} a_i u_i\Big\|^p\right)^{1/p}\\ &\leq \ldots \leq \prod_i (1+\varepsilon_i)\Big(\sum_{i=1}^{\infty} |a_i|^p\Big)^{1/p}.\end{aligned}$$

In a similar way we get the lower estimate, hence

$$(1+\varepsilon)^{-1}\Big(\sum_{i=1}^{\infty} |a_i|^p\Big)^{1/p} \leq \Big\|\sum_{i=1}^{\infty} a_i u_i\Big\| \leq (1+\varepsilon)\Big(\sum_{i=1}^{\infty} |a_i|^p\Big)^{1/p},$$

as required.

Clearly, $\lambda_p(Y)$ is an isomorphic invariant. However, there exist spaces Y such that $\lambda_p(Y) < \infty$ but there is no equivalent norm $|||\cdot|||$ on Y such that for some block subspace $Z \in \mathcal{B}_\infty(Y)$ the equality $\lambda_p(Y, |||\cdot|||) = 1$ would hold. The construction above obviously yields that this is true for every asymptotic l_p space which does not contain subspaces isomorphic to l_p. In particular, Tsirelson space and its convexifications have this property.

References

[C-S.89] CASAZZA, P. G. & SHURA, T., "Tsirelson's space". Lecture Notes in Math., 1363, Springer Verlag, 1989.

[F-J.74] FIGIEL, T. & JOHNSON, W. B., A uniformly convex Banach space which contains no l_p. *Compositio Math.*, 29 (1974), 179–190.

[G.92] GOWERS, W. T., Lipschitz functions on classical spaces. *European J. Comb.*, to appear

[Gr.76] GRINBLAT, L. S., Circles in the unit sphere of a Hilbert space. *J. of Funct. Anal.*, 31 (1979), 272–274.

[G-M.92] GOWERS, W. T. & MAUREY, B., The unconditional basic sequence problem. Preprint.

[H-O-R-S.91] HAYDON, R. & ODELL, E. & ROSENTHAL, H. P. & SCHLUMPRECHT, TH., On distorted norms in Banach spaces and the existence of l_p types. Preprint.

[J.64] JAMES, R. C., Uniformly non-square Banach spaces. *Ann.of Math.*, 80 (1964), 542–550.

[K.66] KATO, T., "Perturbation theory for linear operators". Springer Verlag, 1966.

[L-T.77] LINDENSTRAUSS, J. & Tzafriri, L., " Classical Banach spaces, Vol. I". Springer Verlag, 1977.

[M.67] MILMAN, V. D., Infinite-dimensional geometry of the unit sphere in a Banach space. *Soviet Math. Dokl.*, 8 (1967), 1440–1444.

[M.69] MILMAN, V. D., Spectrum of continuous bounded functions on the unit sphere of a Banach space. *Funct. Anal. Appl.*, 3 (1969), 67–79.

[M.71a] MILMAN, V. D., New proof of the theorem of Dvoretzky. *Funct. Anal. Appl.*, 5 (1971), 28–37.

[M.71b] MILMAN, V. D., The geometric theory of Banach spaces, Part II. *Usp. Math. Nauk* 26 (1971), 73–149.(in Russian) (English translation: Russian Math Surveys 26 (1971), 79–163.)

[M-P.89] MILMAN, V. D.& PERELSON, A., Infinite dimensional geometric moduli and type–cotype theory. in "Geometric aspects of Banach spaces", (ed. E. Martinez-Peinador and Rodes), London Math. Soc. Lecture Notes Series, 140 (1989), 11–38.

[Ma.92] MAUREY, B., Private communication.

[O-S.92] ODEL, E. & SCHLUMPRECHT, TH., The distortion problem. Preprint.

[S.91] SCHLUMPRECHT, TH., An arbitrarily distortable Banach space. To appear

[Ts.74] TSIRELSON, B. S. Not every Banach space contains l_p or c_0. *Funct. Anal. Appl.*, 8 (1974), 138–141.

DEPARTMENT OF MATHEMATICS, RAYMOND AND BEVERLY SACKLER FACULTY OF EXACT SCIENCES, TEL AVIV UNIVERSITY, TEL AVIV, ISRAEL, *and* DEPARTMENT OF MATHEMATICS, OHIO STATE UNIVERSITY, COLUMBUS, OHIO, 43210, USA

E-mail address: vitali@taurus.bitnet

DEPARTMENT OF MATHEMATICS, UNIVERSITY OF ALBERTA, EDMONTON, ALBERTA, CANADA T6G 2G1

E-mail address: ntomczak@vega.math.ualberta.ca; ntomczak@vm.ucs.ualberta.ca

Contemporary Mathematics
Volume **144**, 1993

ON SCHREIER UNCONDITIONAL SEQUENCES

E. ODELL

ABSTRACT. Let (x_n) be a normalized weakly null sequence in a Banach space and let $\varepsilon > 0$. We show that there exists a subsequence (y_n) with the following property:

$$\text{if } (a_i) \subseteq \mathbb{R} \text{ and } F \subseteq \mathbb{N}$$

satisfies $\min F \geq |F|$ then

$$\Big\|\sum_{i \in F} a_i y_i\Big\| \leq (2+\varepsilon)\Big\|\sum a_i y_i\Big\| .$$

Let (x_n) be a basic sequence in a Banach space and let $K < \infty$. We say that (x_n) is *K-Schreier unconditional* if whenever F is an admissible subset of $\mathbb{N}$ (*i.e.*, $\min F \geq |F|$) and (a_i) is a finitely nonzero sequence of scalars, then

$$\Big\|\sum_{i \in F} a_i x_i\Big\| \leq K\Big\|\sum_{i=1}^{\infty} a_i x_i\Big\| .$$

We prove the

Theorem. *Let (x_n) be a normalized weakly null sequence in a Banach space. Then for any $\varepsilon > 0$, (x_n) contains a $(2+\varepsilon)$-Schreier unconditional subsequence.*

Remarks.

1. It follows easily from our theorem, Rosenthal's ℓ_1 theorem [R] and James' theorem that ℓ_1 cannot be distorted [J] that every infinite dimensional Banach space contains a $(2+\varepsilon)$-Schreier unconditional sequence. By the recent remarkable example of Gowers and Maurey [GM], fully unconditional sequences need not exist.

2. Even before [GM], the example of Maurey and Rosenthal [MR] showed that (x_n) need not have an unconditional subsequence. However some positive results are already known. First by the theory of spreading models (see e.g., [BL]) it is known that one can extract a subsequence (y_n) satisfying

$$\Big\|\sum_{i \in F} a_i y_i\Big\| \leq (1+\varepsilon)\Big\|\sum_{i \in G} a_i y_i\Big\|$$

whenever $(a_i) \subseteq \mathbb{R}$, $F \subseteq G$ and G is admissible. Our proof will yield a different proof of this fact. A second result is that (x_n) contains a "nearly unconditional"

1991 *Mathematics Subject Classification.* 46B15.
Research partially supported by the National Science Foundation Grant DMS-8903197

subsequence. This is due to Elton [E]. A rough sketch of the proof is contained in the expository paper [O].

Both of these results require Ramsey theory as does our proof.

3. It is not possible, in general, to get the Schreier unconditional constant less than 2. Indeed, the example (x_n) of [MR] (see [LT], p.28) has the property that in any subsequence, for any $\varepsilon > 0$, there exists a two element block basis (z_1, z_2) with support (z_2) admissible and such that (z_1, z_2) is $(1+\varepsilon)$-equivalent to (s_1, s_2) where $\|a_1 s_1 + a_2 s_2\| = \max(|a_1|, |a_1 + a_2|)$.

4. The theorem yields the following fact. Let (x_n) be a normalized weakly null basic sequence in a Banach space X. Then for $\varepsilon > 0$ there exists a subsequence (y_n) of (x_n) having biorthogonal functionals (f_n) satisfying $\sup_n \|f_n\| < 1 + \varepsilon$.

5. The theorem solves a problem posed to us by P.G. Casazza. Cassaza [C] has used the theorem to prove the following result.

Let X be a Banach space of type p (respectively, cotype q). then for all $\varepsilon > 0$ and $1 < r < p$ (respectively, $q < r < \infty$) there exists a normalized basic sequence (x_i) in X satisfying $\|\sum a_i x_i\| \leq (1+\varepsilon)(\sum |a_i|^r)^{1/r}$ for all scalars (a_i) (respectively, $\|\sum a_i x_i\| \geq (1+\varepsilon)^{-1}(\sum |a_i|^r)^{1/r}$).

6. The term "Schreier unconditional" derives from the use of admissible sets in the definition of Schreier's space (see [CS], p.1), the historical forerunner of Tsirelson's space.

Proof of the theorem. Let $\varepsilon > 0$. By passing to a subsequence we may assume that (x_n) is basic with basis constant less than $1 + \varepsilon$. Thus the tail projections have norm less than $2 + \varepsilon$. Hence it suffices to prove the

Proposition. *Let (x_n) be a normalized weakly null basic sequence. Let $k \in \mathbb{N}$ and $\varepsilon > 0$. There exists a subsequence (y_n) of (x_n) such that for $F \subseteq \mathbb{N}$ with $|F| \leq k$ and $(a_i) \subseteq \mathbb{R}$, a finitely nonzero sequence,*

$$\Big\| \sum_{i \in F} a_i y_i \Big\| \leq (1+\varepsilon) \Big\| \sum a_i y_i \Big\| .$$

Indeed the theorem follows easily from the proposition and a standard diagonal argument. We introduce some terminology before proceeding.

Let $p \in \mathbb{N}$. $\vec{a} = (a_i)_{i=1}^p \subseteq \mathbb{R}$ is called a *p-pattern.* If $(z_i) \subseteq X$, our Banach space, $f \in X^*$ and $n_1 < n_2 < \cdots < n_p$ are integers then we say *f has pattern $\vec{a}$ on $F = (n_i)_{i=1}^p$ with respect to (z_n)* if $f(z_{n_i}) = a_i$ for $i \leq p$.

$\vec{a} = \vec{\emptyset}$ is called a *0-pattern* and every $f \in X^*$ has pattern $\vec{\emptyset}$ on $F = \emptyset$.

Lemma 1. *Let (z_n) be a normalized weakly null basic sequence in X and let p be a nonnegative integer. Let $\vec{a}$ be a p-pattern, $\mathcal{F} \subseteq 2Ba\, X^*$ and let $\delta > 0$. Then there exists $M = (m_i)_{i=0}^\infty$, a subsequence of $\mathbb{N}$, with the following property.*

If $F \subseteq (m_i)_{i=1}^\infty$, $|F| = p$ and there exits $f \in \mathcal{F}$ having pattern $\vec{a}$ on F with respect to (z_n), then there exists $g \in \mathcal{F}$ having pattern $\vec{a}$ on F with respect to (z_n) and satisfying $|g(z_{m_0})| < \delta$.

Proof. We first observe that there exists $m \in \mathbb{N}$ such that for all $f \in 2Ba\, X^*$ there exists $i \leq m$ with $|f(z_i)| < \delta$. Indeed, if not then for all $m \in \mathbb{N}$ there exists

$f_m \in 2Ba\,X^*$ with $|f_m(z_i)| \geq \delta$ for $i \leq m$. But then if f is any w^*-cluster point of (f_m), $|f(z_i)| \geq \delta$ for all i and hence (z_i) is not weakly null.

Let $N = (m+1, m+2, \dots)$. Divide the p-element subsets of N as follows: $[N]^p = \bigcup_{i=0}^m \mathcal{A}_i$ where for $1 \leq i \leq m$,

$$\mathcal{A}_i = \{F \in [N]^p : \text{ there exists } f \in \mathcal{F} \text{ having pattern } \vec{a} \text{ on } F \text{ and } |f(z_i)| < \delta\}$$

and

$$\mathcal{A}_0 = \{F \in [N]^p : \text{ no } f \in \mathcal{F} \text{ has pattern } \vec{a} \text{ on } F\} .$$

By Ramsey's theorem there exist a subsequence $\overline{M}$ of N and $0 \leq i \leq m$ such that $[\overline{M}]^p \subseteq \mathcal{A}_i$. Let $\overline{M} = (m_i)_{i=1}^\infty$ and set $m_0 = i$ if $1 \leq i \leq m$ and $m_0 = 1$ otherwise. □

Remark. Actually our proof shows that if there exists $f \in \mathcal{F}$ having pattern $\vec{a}$ on some $F \subseteq (m_i)_1^\infty$ then for every $G \subseteq (m_i)_1^\infty$, $|G| = p$, there exists $g \in \mathcal{F}$ having pattern $\vec{a}$ on G (and also $|g(z_{m_0})| < \delta$).

Lemma 2. *Let (z_n) be a normalized weakly null sequence in X. Let $\vec{A}$ be a finite set of patterns, $\mathcal{F} \subseteq 2Ba\,X^*$ and let $\delta > 0$. There exists a subsequence (y_n) of (z_n) with the following property.*

Let $\vec{a} \in \vec{A}$ and let $f \in \mathcal{F}$ have pattern $\vec{a}$ on $F \subseteq \mathbb{N}$ with respect to (y_n). Let i be arbitrary if $F = \emptyset$ and let $i < \min F$ if $F \neq \emptyset$. Then there exists $g \in \mathcal{F}$ having pattern $\vec{a}$ on F with respect to (y_n) and $|g(y_i)| < \delta$.

Proof. Fix $\vec{a} \in \vec{A}$ and apply Lemma 1 to (z_n) to obtain $M^1 = (m_i^1)_{i=0}^\infty$. Apply Lemma 1 again to $(z_{m_i^1})_{i=1}^\infty$ to obtain $M^2 = (m_i^2)_{i=0}^\infty$. Continue in this fashion and let $M_{\vec{a}} = (m_0^i)_{i=1}^\infty$.

Note that if M is any subsequence of $M_{\vec{a}}$ then the subsequence $(y_n) \equiv (z_n)_{n \in M}$ satisfies the conclusion of the lemma for the fixed pattern $\vec{a}$. Thus by repeating the argument a finite number of times, now starting with $(z_n)_{n \in M_{\vec{a}}}$, for each element of $\vec{A}$ we obtain the desired result. □

Proof of the proposition. Let $k \in \mathbb{N}$ and $\varepsilon > 0$. We may assume that if (f_n) is the sequence of associated biorthogonal functionals of (x_n), $\sup_n \|f_n\| < 3$. Choose $0 < \eta < 1$ such that $(1 - 2\eta)(1 + \eta)^{-1} > (1 + \varepsilon)^{-1}$. Choose $(\delta_i)_{i=0}^\infty$ decreasing to 0 with

$$3k\delta_0 + \sum_{i=1}^\infty 3\delta_i < \eta .$$

Let A_0 be a finite δ_0-net in $[-1, 1]$ containing 0 and let

$$\vec{A} = \{\vec{a} = (a_i)_{i=1}^p : 1 \leq p \leq k \text{ and } a_i \in A_0 \text{ for } i \leq p\} \cup \{\vec{\emptyset}\} .$$

For $i \in \mathbb{N}$ choose a δ_i-net B_i for $[-1, 1]$ with $A_0 \subseteq B_1 \subseteq B_2 \subseteq \cdots \subseteq [-1, 1]$.

From our set-up we note that if $\overline{f} \in Ba\,X^*$ and $F \in [\mathbb{N}]^k$ then there exists $f \in X^*$ with $\|f - \overline{f}\| < \eta$, $f(x_i) \in A_0$ if $i \in F$ and $f(x_i) \in B_i$ if $i \notin F$. Set $\mathcal{F} = \{f \in (1 + \eta)Ba\,X^* : f(x_i) \in B_i \text{ for all } i\}$.

Apply Lemma 2 to $\vec{A}$, $\mathcal{F}$, δ_1 and (x_n) to obtain a subsequence (x_n^1) which satisfies: if $f \in \mathcal{F}$ has pattern $\vec{a} \in \vec{A}$ on F with respect to (x_n^1) and $1 < \min F$ or $F = \emptyset$ then there exists $g \in \mathcal{F}$ having pattern $\vec{a}$ on F with respect to (x_n^1) and $|g(x_1^1)| < \delta_1$.

Set $y_1 = x_1^1 = x_{p_1}$ (this defines p_1) and for $b \in B_{p_1}$ let

$$\mathcal{F}_b = \{f \in \mathcal{F} : f(y_1) = b\} .$$

Apply Lemma 2 successively to $\vec{A}$, δ_2, $\mathcal{F}_b$ ($b \in B_{p_1}$) starting with the sequence $(x_i^1)_{i\geq 2}$ to obtain $(x_i^2)_{i=1}^\infty \subseteq (x_i^1)_{i\geq 2}$ satisfying:

if $b \in B_{p_1}$ and $f \in \mathcal{F}_b$ has pattern $\vec{a} \in \vec{A}$ on F with respect to (x_i^2) and $x_1^2 < \min F$ or $F = \emptyset$, then there exists $g \in \mathcal{F}_b$ having pattern $\vec{a}$ on F with respect to (x_i^2) and $|g(x_1^2)| < \delta_2$.

Let $y_2 = x_1^2 = x_{p_2}$ (note $p_1 < p_2$) and for each $\vec{b} = (b_1, b_2) \in B_{p_1} \times B_{p_2}$ set $\mathcal{F}_{\vec{b}} = \{f \in \mathcal{F} : f(y_i) = b_i \text{ for } i = 1, 2\}$. Apply Lemma 2 again successively to $\vec{A}$, δ_3, $\mathcal{F}_{\vec{b}}$ ($\vec{b} \in B_{p_1} \times B_{p_2}$) starting with $(x_i^2)_{i\geq 2}$ to obtain $(x_i^3)_{i=1}^\infty$. Set $y_3 = x_1^3 = x_{p_3}$ and continue inductively in this manner.

Let $(a_i) \subseteq \mathbb{R}$ be a finitely supported sequence, $F \in [\mathbb{N}]^k$ and $\|\sum_{i\in F} a_i y_i\| = 1$. We shall prove that $\|\sum a_i y_i\| > (1+\varepsilon)^{-1}$ which will prove the proposition. Without loss of generality, $\|\sum a_i y_i\| \leq 1$. Choose $\overline{f} \in Ba\, X^*$ with $\overline{f}(\sum_{i\in F} a_i y_i) = 1$. As noted above there exists $f \in \mathcal{F}$ with $\|f - \overline{f}\| < \eta$, $f(\sum_{i\in F} a_i y_i) > 1 - \eta$, $f(y_i) \in A_0$ if $i \in F$ and $f(y_i) \in B_{p_i}$ if $i \notin F$. Let f have pattern $\vec{a} \in \vec{A}$ on F with respect to (y_n).

If $1 \in F$ set $g_1 = f$ and if $1 \notin F$ choose $g_1 \in \mathcal{F}$ having pattern $\vec{a}$ on F with respect to (y_n) and $|g_1(y_1)| < \delta_1$. Let $F_1 = F \cap \{2, 3, 4, \dots\}$ and let $\vec{a}_1$ be the pattern of g_1 on F_1. Of course $g_1(y_1) \in B_{p_1}$ and $\vec{a}_1 \in \vec{A}$.

If $2 \in F_1$ set $g_2 = g_1$ and if $2 \notin F_1$, choose $g_2 \in \mathcal{F}_{g_1(y_1)}$ having pattern $\vec{a}_1$ on F_1 with respect to (y_n) and satisfying $|g_2(y_2)| < \delta_2$. Let $F_2 = F \cap \{3, 4, \dots\}$ and let g_2 have pattern $\vec{a}_2$ on F_2. If $3 \in F_2$ let $g_3 = g_2$ and otherwise choose $g_3 \in \mathcal{F}_{(g_2(y_1), g_2(y_2))}$ having pattern $\vec{a}_2$ on F_2 with respect to (y_n) and $|g_3(y_3)| < \delta_3$.

Let $\ell = \max\{i : a_i \neq 0\}$ and continue as above until obtaining $g_\ell \equiv g$. g has pattern $\vec{a}$ on F and thus $g(\sum_{i\in F} a_i y_i) > 1 - \eta$. Also $\|g\| < 1 + \eta$ and $|g(y_i)| < \delta_i$ if $i \leq \ell$ and $i \notin F$. Thus

$$\begin{aligned}\Big\|\sum a_i y_i\Big\| &\geq (1+\eta)^{-1} g\Big(\sum a_i y_i\Big) \\ &\geq (1+\eta)^{-1}\left(g\Big(\sum_{i\in F} a_i y_i\Big) - \sum_{\substack{i\notin F\\ i\leq \ell}} |a_i|\,|g(y_i)|\right) .\end{aligned}$$

Since $\|\sum a_i y_i\| \leq 1$ and $\sup_n \|f_n\| < 3$, $|a_i| < 3$ for all i. Thus we continue

$$\begin{aligned}&> (1+\eta)^{-1}\left[(1-\eta) - 3\sum_{i=1}^\infty \delta_i\right] \\ &> (1+\eta)^{-1}(1-2\eta) > (1+\varepsilon)^{-1} \quad \text{by our choice of } \eta . \qquad \square\end{aligned}$$

Remark. Using the results of [ORS] and the proof above the following theorem can be proved.

Let (F_n) be a weakly null FDD with $\dim F_n \to \infty$. Then for $\varepsilon > 0$ there exists $n_k \uparrow \infty$ and subspaces $G_k \subseteq F_{n_k}$ with $\dim G_k \to \infty$ such that (G_k) is $2+\varepsilon$-Schreier unconditional, i.e., $\|\sum_{i\in F} x_i\| \leq (2+\varepsilon)\|\sum x_i\|$ whenever $x_i \in G_i$ and F is admissible.

References

[BL] B. Beauzamy and J.-T. Lapresté, *Modèles étalés des espaces de Banach*, Travaux en Cours, Hermann, Paris, 1984.

[C] P.G. Casazza, private communication.

[CS] P.G. Casazza and T.J. Shura, *Tsirelson's Space*, LNM 1363, Springer-Verlag, 1989.

[E] J. Elton, *Weakly null normalized sequences in Banach spaces*, Doctoral Thesis, Yale University, 1978.

[GM] T. Gowers and B. Maurey, *The unconditional basic sequence problem*, preprint.

[J] R.C. James, *Uniformly non-square Banach spaces*, Ann. of Math. **80** (1964), 542–550.

[LT] J. Lindenstrauss and L. Tzafriri, *Classical Banach Spaces I*, Springer-Verlag, New York, 1977.

[MR] B. Maurey and H. Rosenthal, *Normalized weakly null sequences with no unconditional subsequence*, Studia Math. **61** (1977), 77–98.

[O] E. Odell, *Applications of Ramsey theory to Banach space theory*, Notes in Banach Spaces (H.E. Lacey, ed.), Univ. of Texas Press, 1980, pp. 379–404.

[R] H. Rosenthal, *A characterization of Banach spaces containing ℓ_1*, Proc. Nat. Acad. Sci. (U.S.A.) **71** (1974), 2411–2413.

Department of Mathematics, The University of Texas at Austin, Austin, TX 78712-1082

E-mail address: odell@math.utexas.edu

Recent Titles in This Series

(Continued from the front of this publication)

114 **Jeffrey C. Lagarias and Michael J. Todd, Editors,** Mathematical developments arising from linear programming, 1990
113 **Eric Grinberg and Eric Todd Quinto, Editors,** Integral geometry and tomography, 1990
112 **Philip J. Brown and Wayne A. Fuller, Editors,** Statistical analysis of measurement error models and applications, 1990
111 **Earl S. Kramer and Spyros S. Magliveras, Editors,** Finite geometries and combinatorial designs, 1990
110 **Georgia Benkart and J. Marshall Osborn, Editors,** Lie algebras and related topics, 1990
109 **Benjamin Fine, Anthony Gaglione, and Francis C. Y. Tang, Editors,** Combinatorial group theory, 1990
108 **Melvyn S. Berger, Editor,** Mathematics of nonlinear science, 1990
107 **Mario Milman and Tomas Schonbek, Editors,** Harmonic analysis and partial differential equations, 1990
106 **Wilfried Sieg, Editor,** Logic and computation, 1990
105 **Jerome Kaminker, Editor,** Geometric and topological invariants of elliptic operators, 1990
104 **Michael Makkai and Robert Paré,** Accessible categories: The foundations of categorical model theory, 1989
103 **Steve Fisk,** Coloring theories, 1989
102 **Stephen McAdam,** Primes associated to an ideal, 1989
101 **S.-Y. Cheng, H. Choi, and Robert E. Greene, Editors,** Recent developments in geometry, 1989
100 **W. Brent Lindquist, Editor,** Current progress in hyperbolic systems: Riemann problems and computations, 1989
99 **Basil Nicolaenko, Ciprian Foias, and Roger Temam, Editors,** The connection between infinite dimensional and finite dimensional dynamical systems, 1989
98 **Kenneth Appel and Wolfgang Haken,** Every planar map is four colorable, 1989
97 **J. E. Marsden, P. S. Krishnaprasad, and J. C. Simo, Editors,** Dynamics and control of multibody systems, 1989
96 **Mark Mahowald and Stewart Priddy, Editors,** Algebraic topology, 1989
95 **Joel V. Brawley and George E. Schnibben,** Infinite algebraic extensions of finite fields, 1989
94 **R. Daniel Mauldin, R. M. Shortt, and Cesar E. Silva, Editors,** Measure and measurable dynamics, 1989
93 **M. Isaacs, A. Lichtman, D. Passman, S. Sehgal, N. J. A. Sloane, and H. Zassenhaus, Editors,** Representation theory, group rings, and coding theory, 1989
92 **John W. Gray and Andre Scedrov, Editors,** Categories in computer science and logic, 1989
91 **David Colella, Editor,** Commutative harmonic analysis, 1989
90 **Richard Randell, Editor,** Singularities, 1989
89 **R. Bruce Richter, Editor,** Graphs and algorithms, 1989
88 **R. Fossum, W. Haboush, M. Hochster, and V. Lakshmibai, Editors,** Invariant theory, 1989
87 **Laszlo Fuchs, Rüdiger Göbel, and Phillip Schultz, Editors,** Abelian group theory, 1989
86 **J. Ritter, Editor,** Representation theory and number theory in connection with the local Langlands conjecture, 1989
85 **Bor-Luh Lin, Editor,** Banach space theory, 1989

(See the AMS catalog for earlier titles)